MACROALGAL BIOREFINERIES
FOR THE BLUE ECONOMY

MACROALGAL BIOREFINERIES FOR THE BLUE ECONOMY

Alexander Golberg
Tel Aviv University, Israel

Arthur Nils Robin
Tel Aviv University, Israel

Meiron Zollmann
Tel Aviv University, Israel

Hadar Traugott
Tel Aviv University, Israel

Ruslana Rachel Palatnik
University of Haifa, Israel

Alvaro Israel
Israel Oceanographic & Limnological Research Institute, Israel

World Scientific

EW JERSEY · LONDON · SINGAPORE · BEIJING · SHANGHAI · HONG KONG · TAIPEI · CHENNAI · TOKYO

Published by

World Scientific Publishing Co. Pte. Ltd.

5 Toh Tuck Link, Singapore 596224

USA office: 27 Warren Street, Suite 401-402, Hackensack, NJ 07601

UK office: 57 Shelton Street, Covent Garden, London WC2H 9HE

Library of Congress Cataloging-in-Publication Data
Names: Golberg, Alexander, author. | Robin, Arthur Nils, author. | Zollmann, Meiron, author. |
    Traugott, Hadar, author. | Palatnik, Ruslana Rachel, author. | Israel, Alvaro, author.
Title: Macroalgal biorefineries for the blue economy / Alexander Golberg, Tel Aviv University, Israel;
    Arthur Nils Robin,Tel Aviv University, Israel; Meiron Zollmann,Tel Aviv University, Israel;
    Hadar Traugott, Tel Aviv University, Israel; Ruslana Rachel Palatnik, University of Haifa, Israel;
    Alvaro Israel, Israel Oceanographic & Limnological Research Institute, Israel.
Description: New Jersey : World Scientific, [2020] | Includes bibliographical references and index.
Identifiers: LCCN 2020034551 | ISBN 9789811224287 (hardcover) |
    ISBN 9789811224294 (ebook for institutions) | ISBN 9789811225406 (ebook for individuals)
Subjects: LCSH: Algal biofuels. | Marine algae--Refining. | Marine algae--Biotechnology. |
    Marine algae industry. | Marine algae culture. | Sustainable aquaculture. | Biochemical engineering.
Classification: LCC TP339 .G64 2020 | DDC 660.6/3--dc23
LC record available at https://lccn.loc.gov/2020034551

British Library Cataloguing-in-Publication Data
A catalogue record for this book is available from the British Library.

For any available supplementary material, please visit
https://www.worldscientific.com/worldscibooks/10.1142/11937#t=suppl

Desk Editors: George Vasu/ Steven Patt

Typeset by Stallion Press
Email: enquiries@stallionpress.com

Preface

For thousands of years, humanity has prospered mostly due to mastering terrestrial agriculture. Food, clothing, housing, energy, materials, and drugs are mostly derived from land grown plants and crops. Moreover, for years, plant-derived paper and animal skins were the major tools for accumulated knowledge transmission through generations. However, in recent years we see the unpreceded growth in the human population. Such growth, in turn, generated the new pressures on all supply chains that are required for human well-being. These supply chains require providing more and more plant-derived products and biomass. Such an increase of the raw biomass requires more land and freshwater for its cultivation on land. The question whether we have sufficient water and land to feed the growing population is still open, and the current answer by most experts is that regional disparities in access to such resources in quantity and quality will create immense tensions between human societies in the foreseen future. However, the price we will pay in environmental damages for the conversion of new lands to cropland and freshwater for irrigation is enormous. Further expansion and intensification of agriculture present direct damages to biodiversity and water and soil quality.

At the same time, for the past millennia, multiple coastal communities have been using marine biomass — seaweeds or macroalgae — as biomass that supplies their needs in food, clothing, housing, materials, and drugs. The global production of this marine biomass, which does not require any arable land or freshwater for cultivation, is still five orders of magnitude smaller than the terrestrial crops. However, only a tiny amount of seas' and oceans' surfaces are used for cultivation. The most substantial potential of this offshore biomass production is still to be revealed.

In addition, besides being more than a sustainable replacement or alternative to land biomass, seaweeds can provide unique benefits. Indeed, from their uncommon polysaccharides to pigments including their mineral content and other peculiar bioactive molecules and structures, macroalgae are an untapped source of innovative solutions for human societies.

How can seas and oceans provide the biomass that will power the further sustainable development of humanity? How will it power the bio- and blue economies?

In this book, we attempt to concisely show what has already been achieved in the field of offshore macroalgae cultivation, and which building blocks already exist that will enable seaweeds to supplement and replace terrestrial crops for future economic development.

The fundamental production unit of the emerging bioeconomies is the BIOREFINERY — a system that results in multiple essential-for-humans products from the biomass. In this book, we describe the offshore biorefineries and explain the motivation to go offshore in Chapter 1. Why should we even think about trying to develop cultivation and processing systems that will be installed in this challenging, high-energy environment? In Chapter 2, we explain what is the blue economy and how marine biorefineries can contribute to it. Chapter 3 is about marine crops — macroalgae. We discuss briefly species diversity, their spatial and chemical diversity, that is vital for their use in economics. Chapter 4 introduces the concept of the marine biorefinery and provides a theory behind its design and efficiencies optimization. In Chapter 5, we discuss the environmental impact of offshore seaweed production and provide some recently developed tools for sustainable management of seaweed cultivation. In Chapter 6, (Upstream) we discuss various options for getting a sufficient amount of the seaweed feedstock for their processing. In Chapter 7, (Downstream) we discuss very common but very important issues related to biomass handling and processing. We do not attempt to cover all existing technology to process seaweeds — this was done in several recent books. Rather, we attempt to show the broad range of applicable and environmentally friendly technologies. Chapter 8 is all about applications. From seaweed aquaculture to climate engineering — these and many other uses is what makes seaweed biomass an attractive and important resource for future economic development. In Chapter 9, we show an approach for the potential valorization of the macroalgae products and show economic models behind the development of large-scale seaweed industry. The major focus is on the incorporating of uncertainty in the economic models.

This book is for the readers at all levels, and we hope it will provide guidelines for those who decide to contribute to this exciting and emerging field of the blue economy which is an essential part of sustainable development.

About the Authors

Alexander Golberg is an Associate Professor at the Porter School of Environmental Studies, Faculty of Exact Sciences Tel Aviv University, Israel and Head of the Environmental Bioengineering. He major research interests are in the field of applied thermodynamics with applications in environment and health. His major interest is in developing technologies to enable blue economy by the transformation of agriculture to offshore seagriculture, thus reducing the environmental impacts of biomass production on arable lands and drinking water. In 2012 he introduced the concept of distributed seaweed biorefineries to enable energy transition in low-income coastal communities. Currently he works on the development of novel offshore seaweed cultivation methods and energy efficient processing technologies such as non-thermal pulsed electric fields and subcritical water extractions.

Arthur Nils Robin is a PhD candidate at Tel Aviv University. He is part of the Environmental Bioengineering Laboratory and works on the bioprocessing of seaweed and on seaweed biorefinery. His main expertise is on the application of Pulsed Electric Field on seaweed biomass. His research also involves designing green seaweed cascading biorefinery and the development of innovative technologies from seaweed for food, advanced biomaterial and biomedical applications. He graduated in 2015 with a master of food and bioprocess engineering degree from the Ecole Nationale Supérieure d'Agronomie et des Industries Alimentaires in Nancy, France.

Meiron Zollmann is a PhD candidate at Porter School of the Environmental and Earth sciences, Tel Aviv University, Israel. He is part of the environmental bioengineering laboratory and is currently working on modeling of seaweed cultivation for biorefinery and bioremediation. His models are based on lab, outdoor and offshore experiments and are used to simulate growth rates, chemical composition and nitrogen sequestration in various scales. He graduated in 2016 with a Bachelor of Science in Environmental Engineering degree from the Technion Israel Institute of Technology, Haifa, Israel.

Hadar Traugott is a Ph.D. student in TelAviv University, Israel. She has a BSc in chemical engineering from Ben-Gurion University and MSc in mechanical engineering from TelAviv University. Her research area is intensified offshore seaweed cultivation.

Ruslana Rachel Palatnik, PhD, is Senior Lecturer at the Department in Economics and Management at the Yezreel Valley College (YVC); the Head of SEED- Sustainable Economic and Environmental lab at YVC; Israel, and the Senior Research Fellow at the Natural Resources and Environmental Research Center (NRERC), University of Haifa, Israel. Her research interests are climate change economics, environmental policy evaluation, energy and natural resource economics, and water management. She developed a Computable General Equilibrium model of the Israeli economy (IGEM) and has been conducting climate change adaptation and mitigation policy analyses. In 2015–16 she was the visiting researcher at the Department of Agricultural and Resource Economics, University of California at Berkeley where she was analyzing the economic aspects of biofuels. In 2019 Dr Palatnik was granted IIASA-Israel Program Associate Visiting Professor position where she joined the Energy group to adopt the global energy model to Israel for energy related policy analysis.

Alvaro Israel is a senior scientist at Israel Oceanographic & Limnological Research, Ltd. The National Institute of Oceanography, Haifa, Israel. During his career he engaged in studying photosynthesis, carbon fixation and ecology of marine macroalgae, and has a vast experience in seaweed aquaculture and strain selection intended for the production of food and valuable molecules. Seaweed taxonomy using molecular tools, seaweed invasions and issues related to global change are under current investigation in his lab.

Contents

Chapter 1

Why Go for Seagriculture? The Need for New Sources of Biomass for Bioeconomy and What the Seas Can Offer Humanity?

1.1 The Use of Land to Feed the World Population

Life is the geological force that shapes the Earth. Humans, a tiny percent of the total biomass on Earth, have a continuous enormous impact on the Earth's landscapes. Control of fire, the transition from foraging to pastoralism, and the shifting from traditional to modern intensified farming, subsidized by fossil fuels, completely changed the planet over and underground (Blitzer and Redman, 2001; Ramankutty *et al.*, 2018; Smil, 2008). Only in the last 300 years, between 1700 and 2007, croplands expanded from ~3 to ~15 million km^2 and pasturelands from ~5 to ~27 million km^2 — almost 5-fold expansion! (Ramankutty *et al.*, 2018). The expansion came on the account of forests, grasslands, savannas, and shrublands (Ramankutty *et al.*, 2018). A clear example of Earth's reshaping by human activity is the raising of grasslands as the world's most extensive terrestrial biome, at the expense of forests (Ellis and Ramankutty, 2008; Smil, 2008).

Agricultural lands occupy a third of the ice-free land area (Ramankutty *et al.*, 2008), not only to feed the androgenic food chains but also to provide income to 40% of the world's population, contributing to ~30% of gross domestic product (GDP) in low-income countries providing also construction materials, energy, and additional high-value products for the human species (Ramankutty *et al.*, 2018). Models of future land requirements for agriculture for developing countries project a net increase in the arable area of some 120 million ha (from 966 million ha in the base year to 1,086 million ha in 2050) or 12.4%. The bulk of this projected expansion is expected to occur in sub-Saharan Africa (64 million ha) and Latin America (52 million ha), with almost no land expansion in East and South Asia, and even a small decline in the Near East and North Africa.

Forty percent increase in the world population and increased average food consumption to 130 kcal per person per day by 2050 are predicted. Although the growth in agricultural production will slow due to the projected slowdown in population growth and world population reaching medium to high levels of food consumption, a 70% increase in agricultural production would still be needed by 2050. This demand translates into additional production of 1 billion tones of cereals and 200 million tones of meat a year by 2050. About 90% of the growth in crop production would be a result of higher yields and increased cropping intensity, with the remainder coming from land expansion. Arable land is expected to expand by 70 million ha (less than 5%), an expansion of about 120 million ha (12%) in developing countries being offset by a decline of 50 million ha (8%) in developed countries.

1.2 Do We Have Enough Land?

But do we have enough land to supply the needs of current and future generations? And by how much does the arable land need to increase, for example, in the next 30 years? In 2009, arable land and land under permanent crops required ~1.5 billion ha from the total 13.4 billion ha available. An additional 2.7 billion ha are thought to be possible to convert for crop productions. Thus, this picture shows that the land resource for agriculture is still far from the required amount and that land availability is not a limiting factor for food supply for the human population. Important to mention here is that instead of crop yields increases, the land expansion will continue to be a critical factor for the increase of agricultural produce in areas where environmental, demographic, and socio-economic conditions are favorable.

However, this projected affluence of land for conversion raised several questions and was challenged. This calculation ignored non-agriculture land uses such as forest cover, built area, protected areas, and military areas. Studies estimated that only 40% of the net land balance is available in reality, while 45% is covered by forests, 12% by protected areas and 3% by build areas (Ruttan and Alexandratos, 1996). An additional study at the global level showed that build areas occupy up to 60 million ha of the gross land balance, protected areas 200 million ha, and forests 800 million ha, so the net land balance would be 1.5 billion ha and not 2.7 billion ha as suggested before (Bruinsma, 2009).

Also, the criteria for inclusion of the novel, undeveloped land as potential agricultural land was the land's suitability to support a single, often marginal, crop to the minimal yield level. Thus, large areas of land in North Africa that allow for the cultivation of only minor crops (and other minor products such as olive trees) are counted as suitable, even though their real use to produce agricultural products to feed the populations is negligible.

The geographic distribution of the land with crop production potential is uneven. Globally, 60% of the remaining undeveloped land suitable for agriculture belongs only to 13 countries (Smil, 2008). On the one hand, some 90% of the remaining 1.8 billion ha are in Latin America and sub-Saharan Africa, and half is concentrated in seven countries: Brazil, the Democratic Republic of the Congo, Angola, Sudan, Argentina, Colombia and the Plurinational State of Bolivia. However, and on the other hand, there is no spare land available for agricultural expansion in South Asia and the Near East and North Africa.

The remaining unused land suffers from reduced yields because of ecological fragility, low fertility, toxicity, high incidence of disease, or lack of infrastructure. Its conversion to suitable food production land is possible but at a high economic cost. Both natural events and human intervention lead to deterioration of the land's productive potential, for example through soil erosion or salinization of irrigated areas. For example, more than 70% of the land with rainfed crop production potential in sub-Saharan Africa and Latin America suffers from one or more soil and terrain constraints (International Institute for Applied Systems Analysis, 2012).

Therefore, evaluation of suitability may be overestimated, and much of the remaining used land cannot be considered as a resource that is readily usable for agricultural production on demand. Although 820 million people remain undernourished (*State Food Security and Nutrition World 2019*, 2019) and 2 billion suffer from micronutrient deficiencies, overall Agriculture provides more than enough calories for all people on the planet (FAO, IFAD, UNICEF, 2019). In addition, the predicted population growth to 10 billion by 2050 (United Nations, 2019), coupled with the increasing quality of life and wealth, which leads to the increased consumption of agricultural products (Tilman and Clark, 2014), poses new pressures on agriculture and land as its major resource. One of the studies conducted by the Food and Agriculture Organization (FAO) (Alexandratos and Bruinsma, 2012) projected the 60% increase in agricultural production by 2050 compared with a 2005–2007 baseline. This study estimated that between 2005–2007 and 2050, global demand for meat production and sugarcane and sugarbeet production will increase by 76%, oil crop production will increase by 90%, and cereal production will increase by 50% (Alexandratos and Bruinsma, 2012). An additional study, which used the coupled future projections of population growth and GDP with income-dependent estimates of per capita crop demand, projected a 100% increase in global demand for calories and a 110% increase in protein by 2050 (Tilman *et al.*, 2011; Tilman and Clark, 2014). Although these studies led to the general projection of doubling food demand by 2050 (Foley *et al.*, 2011; Ray *et al.*, 2013), several recent works challenge this number with new/old projections of 25–70% increase (Hunter *et al.*, 2017; Tomlinson, 2013).

However, even these new numbers are very high. Does this mean that similar to other resources we will reach the "peak arable land" or "peak farmland"? And if yes, will this happen soon? Or will climate change allow to convert new, currently unsuitable to cultivation areas to croplands, expanding the land resource (Hannah *et al.*, 2020), similar to what horizontal drilling did to fossil fuels? A 2013 study suggests that we are very close or already within the peak farmland time, a term used to describe a time when humanity might reach the maximum usable Earth's land surface area for agriculture (Ausubel *et al.*, 2013). In this study, the authors showed a reduction in rates of cropland expansion over 1961–2010, with expansions of 0.24% per year over the whole of 1961–2010 but only 0.04% per year from 1995–2010. They showed that the slow down was a result of rising yields and relatively slower than expected growth in consumption countering increased pressure on croplands from growth in population and affluence. Projecting forward under different assumptions, they showed possible scenarios whereby cropland areas would peak and then decline. This was also supported by the FAO study (Alexandratos and Bruinsma, 2012) which analyzed historical trends that suggest that 77% of increased production over the 1961–2005 period came from increased yields, 14% from the expansion of croplands, and 9% from increases in cropping intensity. The study projected that by 2050, 80% of future production growth will come from yield growth and 10% will come from cropland expansion and increases in cropping intensity (Alexandratos and Bruinsma, 2012). Therefore, the contribution of cropland expansion to production growth is expected to decrease by ~4% in the future (Alexandratos and Bruinsma, 2012), which is still in the limit of 25–70% predicted by Hunter *et al.* (2017) and Tomlinson (2013).

Nevertheless, similarly to other natural resources, we will never exhaust all the recoverable land reserves at a known cost (Smil, 2008). Recoverable reserves exhaustion is not a matter of actual physical depletion but rather a burden of persistent and insupportable real costs increases resulting in the declining availability of a resource (Smil, 2008). Thus, as in the case with fossil fuels and with strategic metals (Smil, 2008), we have no early onset of unbearable costs on land resources. Thus, there are no sudden ends, but gradual shifts into new supply directions (Smil, 2008), such as changing diets to lower red meat levels, increasing local agriculture in non-arable areas (for example on commercial roofs), developing low-footprint protein sources, and also moving the production offshore. Further growth of global arable farmlands pushes us to consider an unprecedented possibility: future limits on farmlands arise not from resource shortage but from the necessity to keep these cycles compatible with long-term habitability of the biosphere. Indeed, agriculture and land conversion are among the leading causes of environmental degradation and pollution, which is in many cases not recoverable in the foreseen future, leading to changes in the biosphere and thus related changes in the lithosphere, atmosphere, and hydrosphere. We must slow cropland expansion, as most of the new lands available for clearing are in the tropics and of high carbon and biodiversity value (Ramankutty *et al.*, 2018). The threat of climate change is an especially important reason to avoid deforestation for agriculture, a trend that has continued intensively for the last 300 years (Smil, 2008).

1.3 The Environmental Costs of Land Conversion

Removing forests and other natural vegetation (LEPERS *et al.*, 2005) results in climate change and biodiversity loss (Ramankutty *et al.*, 2018). Thirty percent of forest removal worldwide was for agriculture (Ramankutty *et al.*, 2008). In the last twenty years of the 20th century, more than 55% of new agricultural land came at the expense of intact forests, and additional 28% came from disturbed forests (Gibbs *et al.*, 2010). Globally, between 2000 and 2010, it is thought that 80% of deforestation resulted from conversion to agriculture and grazing lands (Kissinger *et al.*, 2012). This ongoing deforestation and land conversion contribute about 9% of the current greenhouse gas emissions (GHG), out of a total of 22% contributed by all agricultural sectors (Intergovernmental Panel on Climate Change and Intergovernmental Panel on Climate Change, 2015). For instance, the conversion of tropical forests to cropland releases approximately three times more carbon into the atmosphere compared with temperate forests (West *et al.*, 2010). Agriculture also creates massively fragmented habitats, including forests (Haddad *et al.*, 2015) with large stretches of natural habitats, such as the Brazilian Atlantic Forest, now existing in degraded fragments of <1,000 ha in size, all within 1 km of the forest edge (Ramankutty *et al.*, 2018). This leads to habitat replacement and management choices on converted lands. Across biomes and taxonomic groups, conversion to pasture and cropland results in losses of ~20–30% of local species richness (Newbold *et al.*, 2015). Furthermore, this biodiversity loss is nonrandom, with marked declines in species that are functionally important to ecosystems, including large-bodied pollinators (Larsen *et al.*, 2005).

In addition to habitat loss and fragmentation, agricultural management impacts biodiversity through management choices, such as the use of pesticides, fertilizers, and crop choice. Fertilization, from nitrogen-fixing legumes and application of manure and synthetic fertilizers, has contributed to a global increase in nitrogen (N) flow (Vitousek *et al.*, 1997). This results in species loss in terrestrial and freshwater environments (Moreno-Mateos *et al.*, 2017; Stevens *et al.*, 2004). Pesticide application has also been linked to declines in populations of non-target plants and insects (Brittain and Potts, 2011).

The impact of agriculture on soils is tightly linked to land-use change and agricultural management. The alteration of vegetative cover, through the replacement of forests or grasslands with annual crops, influences infiltration, erosion, and organic matter inputs. It is estimated that already by 1990 ~15% of the world's soils were in some way degraded (Oldeman, 1994). Current rates of erosion of agricultural land are estimated to be in an order of magnitude higher than that of natural erosion or soil formation processes. Land clearing for agriculture has also led to soil degradation through other means. For example, vegetation removal in semiarid

Western Australia resulted in recharging of groundwater at a rate two orders of magnitude above the background rate, causing water tables to rise, and salinization of ~10% of agricultural lands in the region (George *et al.*, 1997). In addition, the loss of soil organic matter, which results from fortifying/supplementing soil nutrients with synthetic mineral fertilizers (N-P-K) without replenishing organic material, has pushed agricultural systems into a state of rapid nutrient cycling with high rates of nutrient loss. This, in combination with shorter rotations and loss of cover crops, has led to increases in soil-borne pathogens (Veresoglou *et al.*, 2013), increases in crop susceptibility to droughts (de Vries *et al.*, 2012) and crop yield declines (Bennett *et al.*, 2012).

Furthermore, agriculture is the biggest user of freshwater on this planet and is the major cause of freshwater eutrophication (Smil, 2008). Agricultural production accounts for 92% of the human water footprint, ~77% of which can be attributed to rain-fed agricultural systems (Hoekstra and Mekonnen, 2012). Of the agricultural water use, 12% is in freshwater, with irrigation accounting for ~64% of withdrawals worldwide (Döll *et al.*, 2014). Agricultural water use has had catastrophic impacts on freshwater resources, for example, the complete loss of the 68,000 km^2 of the Aral Sea at the end of the last century (Micklin, 1988, 2007) and groundwater depletion crises in northwest India (Chen *et al.*, 2014; Dalin *et al.*, 2017; Rodell *et al.*, 2009). In addition to the effects of used quantities, loading of nutrients (Carpenter *et al.*, 1998), pesticides (Alavanja *et al.*, 2013), and livestock antibiotics (Kemper, 2008) all have negative effects on water quality and pose public health problems for humans.

To summarize land production of agricultural products: Cultivated land fed over 4 people/ha on average globally, in China 8.5 people/ha (Smil, 2008). This translates into the carrying capacity of 200 kg/ha/year of antropomass (given 45 kg/person) on average and 400 kg/ha in China (with some provinces 500 kg/ha). In comparison, African primates, chimpanzees and gorillas, occupy <1 kg/ha of their limited and disappearing habitats. This trend cannot continue even if the energy flux is available. To replicate the growth of yields during the 20th century, there is a need for a sevenfold increase of crop yields, which is close to near or above photosynthetic maxima even with the maximum explanation of cultivated land with intensive multi-cropping. Genetic manipulation of photosynthesis is not yet available and thus, feeding the growing population requires the development of novel approaches that will not cause a revolution, but rather provide soil for a slower shift to new practices (Smil, 2008).

1.4 Seagriculture as an Alternative

So, what is the alternative? How and where additional biomass can be produced to minimize the new land conversion and erosion? May be at sea? Can marine biomass — seaweed — provide new major crops for future generations? Human and seaweed interactions seem to date back to the Neolithic period (Ainis *et al.*, 2014; Dillehay *et al.*, 2008; Erlandson *et al.*, 2015). The earliest known written record of seaweed usage by humans originates from China, about 1700 years ago (Yang *et al.*, 2017). For centuries, coastal populations harvested a wide variety of seaweeds for food and feed. Next, industrial uses (gels, fertilizers) emerged. This industrial use of seaweed biomass has shifted over the years, from exploiting beach-cast seaweeds as fertilizers and a source of potash, via iodine production, to hydrocolloid extraction (Polikovsky and Golberg, 2019; Synytsya *et al.*, 2015). At all stages, the "potential" of the future industry has been viewed as being larger than its actual scale and this is as relevant today as it was 100 years ago (Hafting *et al.*, 2015; Ramachandra and Hebbale, 2020). To date, macroalgae still present only a tiny percent of the global biomass supply ($\sim 17 \cdot 10^6$ wet weight macroalgae in comparison to $16 \cdot 10^{11}$ tons of terrestrial crops, grasses, and forests) (Pimentel, 2012; Pimentel and Pimentel, 2008; Roesijadi *et al.*, 2010b). The current, biorefinery (Fig. 1.1), suggests that seaweed biomass can be used for both high and low-value products. These include, for example, higher value uses as raw materials for specialty polysaccharides and agricultural biostimulants (Buschmann *et al.*, 2017). In addition, further down in the value pyramid one can find functional products such as valuable ingredients for food and feed, cosmeceuticals, nutraceuticals, pharmaceuticals, and bioenergy (Buschmann *et al.*, 2017).

1.5 Global-Scale Analysis of Seagriculture Potential

Global projection of the metabolic model of a green model seaweed *Ulva*, coupled to the mapping of large scale environmental conditions, shows that theoretically, without taking into consideration any technological or ecological limitations, *Ulva* biomass can be produced in approximately 10% of the World Ocean (Fig. 1.2(a)), largely in regions that are relatively rich in nitrate and phosphate as the North Pacific and North Atlantic subpolar gyres, Southern Ocean and Eastern Equatorial Pacific (Moore *et al.*, 2013).

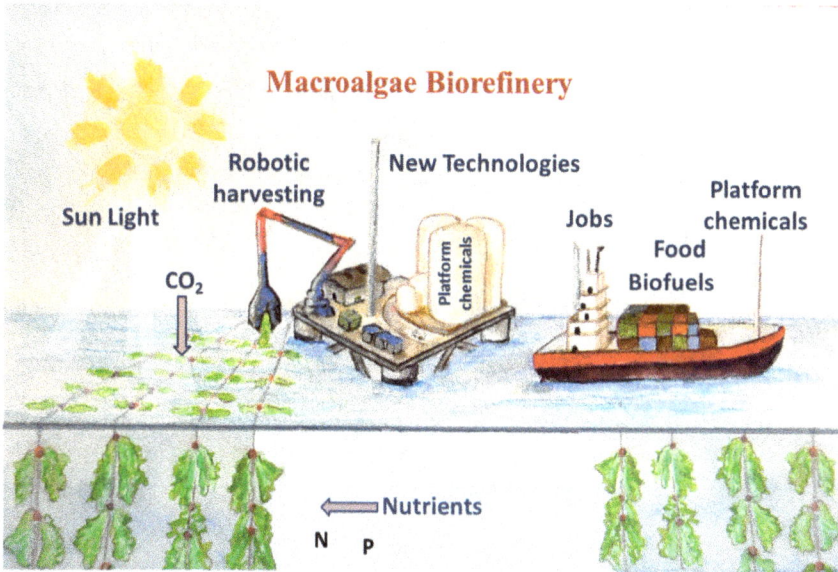

Fig. 1.1 The concept of off-shore biorefineries for the production of food, platform chemicals, and biofuels in the ocean (Graphical design courtesy Mark Polikovsky).

Source: Adapted with permission from Lehahn *et al.* (2016a).

The potential of the offshore biorefinery to produce biomass and various products critically depends on the cultivation stocking density with the variation of the yields in the order of magnitude (Table 1.1). At 4 kg m^{-2} (Nikolaisen *et al.*, 2011c) stocking density, the global annual biomass production potential is ~10^{11} ton DW year^{-1}, and the total theoretical primary energy of *Ulva* biomass from the offshore cultivation is 2052 EJ year^{-1} (Fig. 1.2(a) and Tables 1.1, 1.2, based on the low heating value (LHV) for *Ulva* of 19 MJ per kg DW (Bruhn *et al.*, 2011c)). The total theoretical primary energy of *Ulva* biomass from the shallow, near-shore waters cultivation is 18 EJ year^{-1} (Table 1.2). In comparison, the global potential of energy crops for biofuels is estimated at 125–760 EJ per year (Doornbosch and Steenblik, 2007).

The deployment of biomass cultivation systems in the ocean is a highly complex problem whose feasibility depends on technological readiness and environmental factors, like water depth and distance from the shore (Fao *et al.*, 2013). Limiting the area used for oceanic biomass cultivation according to these two environmental parameters may lead to

Offshore bioenergy production potential

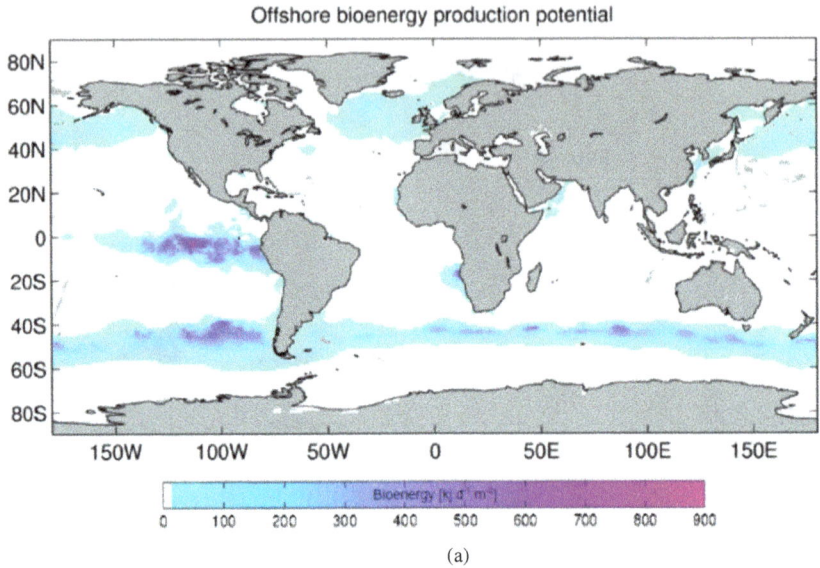

(a)

Biomass production potential

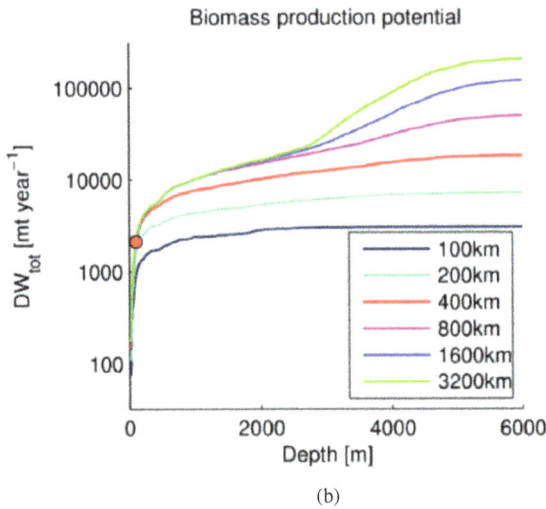

(b)

Fig. 1.2   The global potential for offshore biorefinery. (a) Potential for the daily production of bioenergy over the World Ocean, taking an optimum biomass stocking density of 4 kg m^{-2}. Values in the map will change with changes in stocking density in a nonlinear way (Bruhn *et al.*, 2011c). (b) Impact of water depth and distance from the shore on total biomass production potential, taking biomass stocking density of 1 kg m^{-2} and 4 kg m^{-2} (dashed and solid lines, respectively). The red circle marks the 100 m depth and 400 km distance that define the limits for near-future offshore cultivation. The red triangle marks global potential regardless of any depth or distance from coast limitation.

Source: Adapted with permission from Lehahn *et al.* (2016a).

Table 1.1 Potential for offshore production of biomass and derived products for various cultivation stocking densities. The notion "All waters" refers to all locations regardless of water depth and distance from the coast, while "Shallow nearshore waters" refers to areas associated with water depths smaller than 100 m and located less than 400 km from the coast. The experimentally found optimum stocking density (Nikolaisen *et al.*, 2008) is highlighted in the table. The ration of biomass yields at various stocking densities cultivation is based on the experimental data (Bruhn *et al.*, 2011c).

Biomass stocking density	1 kg m^{-2}		2 kg m^{-2}		4 kg m^{-2}		6 kg m^{-2}		8 kg m^{-2}	
	All waters	Shallow nearshore waters	All waters	Shallow nearshore waters	All waters	Shallow nearshore waters	All waters	Shallow nearshore waters	All waters	Shallow nearshore waters
Biomass [10^6 t year^{-1}] (DW)	67,500	591	81,000	710	108,000	946	54,000	473	40,500	355
Ethanol [10^6 t year^{-1}]	2,025–15,525	18–136	2,430–18,630	21–163	3,240–24,840	28–218	1,620–12,420	14–109	1,215–9,315	11–82
Butanol [10^6 t year^{-1}]	2,025–4,050	18–35	2,430–4,860	21–43	3,240–6,480	28–57	1,620–3,240	14–28	1,215–2,430	11–21
Acetone [10^6 t year^{-1}]	675–1,350	6–12	810–1,620	7–14	1,080–2,160	9–19	540–1,080	5–9	405–810	4–7
Methane [10^6 m^3 year^{-1}]	675–6,480	6–57	810–7,776	7–68	1,080–10,368	9–91	540–5,184	5–45	405–3,888	4–34
Protein [10^6 t year^{-1}]	3,375–16,200	30–142	4,050–19,440	35–170	5,400–25,920	47–227	2,700–12,960	24–114	2,025–9,720	18–85
Energy [10^{12} kJ year^{-1}]	1,282,500	11,234	1,539,000	13,481	2,052,000	17,974	1,026,000	8,987	7,69,500	6,740

Source: Adapted with permission from Lehahn *et al.* (2016a).

Table 1.2 Potential for offshore production of biomass and derived products for the 5 near-future deployable biorefinery provinces (NDBP) at the optimum biomass density of 4 kg m^{-2} (Nikolaisen *et al.*, 2008). The notion "All waters" refers to all locations regardless of water depth and distance from the coast, while "Shallow nearshore waters" refers to areas associated with water depths smaller than 100 m and located less than 400 km from the coast.

	All waters	Shallow nearshore waters	EAS[a]	NAT[b]	SAE[c]	SAW[d]	WAS[e]
Biomass [10^6 t year^{-1}] (DW)	108,000	946	435	124	110	62	82
Ethanol [10^6 t year^{-1}]	3,240–24,840	28–218	13–100	4–29	3–25	2–14	2–19
Butanol [10^6 t year^{-1}]	3,240–6,480	28–57	13–26	4–7	3–7	2–4	2–5
Acetone [10^6 t year^{-1}]	1,080–2,160	9–19	4–9	1.24–2.48	1.10–2.20	0.62–1.23	0.82–1.64
Methane [10^6 m^3 year^{-1}]	1,080–10,368	9–91	4,350–41,760	1,240–11,904	1,100–10,560	615–5,904	820–7,872
Protein [10^6 t year^{-1}]	5,400–25,920	47–227	22–104	6–30	6–26	3–15	4–20
Energy [10^{12} kJ year^{-1}]	2,052,000	17,974	8,265	2,356	2,090	1,169	1,558

Notes: [a]EAS — East Asia offshore waters; [b]NAT — North Atlantic; [c]SAE — South America offshore waters — East; [d]SAW — South America offshore waters — West; [e]WAS — West Africa offshore waters — South.

Source: Adapted with permission from Lehahn *et al.* (2016a).

more than 4 orders of magnitude differences in the global production potential (Fig. 1.2(b)). Additional important factors such as wind, waves, and currents will pose additional constraints on the specific location productivity. We found that with the technologies available for offshore cultivation shortly, which require a water depth of less than 100 m for mooring (Fao *et al.*, 2013) (red spot in Fig. 1.2(b)), there is almost no impact on the farm's distance from shore. Importantly, at this depth range practically all the biomass can be cultivated in farms located less than 400 km from the shore, and are thus bounded within the Exclusive Economic Zones (EEZ) (Fao *et al.*, 2013). The global potential of the near-future achievable deployment offshore biomass production (i.e. in regions extending up to 400 km distance from the shore, and with a water depth of up to 100 m) can provide $9.4 \cdot 10^8$ tons DW year^{-1}. This is equivalent to 17.9 EJ year^{-1} of primary energy potential (calculated as LHV). In comparison, the predicted bioenergy potential from agricultural land in 2050 is expected to be 64–161 EJ year^{-1} (Haberl *et al.*, 2011). It is important to point out that the numbers reported here are based on the total potential assessment of ocean areas, and there is no technology available today to utilize these areas.

Almost all of the biomass production potential at a distance smaller than 400 km from shore is concentrated at 13 provinces (Fig. 1.3(a)). Approximately 85% of the production potential at this distance from the shore is associated with 5 regions that are characterized by water depth of up to 100 m (red boxes in Fig. 1.3(a)). These regions, which meet the important water depth criteria for mooring offshore cultivation platforms using near-future technologies, are hereafter defined as near-future deployable biorefinery provinces (NDBP). For each NDBP we extract monthly values of productive surface area (S, Fig. 1.3(b)), defined as the extension of the region allowing biomass production and meeting the 100 m water depth and 400 km distance from shore criteria, and spatially averaged biomass productivity (DW_{mean}, Fig. 1.3(c)). Total biomass productivity (DW_{tot}, Fig. 1.3(d)) is calculated by multiplying S with DW_{mean}. As for the global patterns of biomass production potential (Fig. 1.2(a)), NDBPs are associated with regions of elevated nutrient concentrations at the ocean's surface layer (Fig. 1.3(a)). These elevated nutrient levels result from a variety of dynamical processes as coastal upwelling (e.g. provinces SAW and WAS) and deep wintertime convection (e.g. province NAT), which upwell nutrient-rich waters from below the mixed layer (Williams and Follows, 2002). Monthly variations in surface

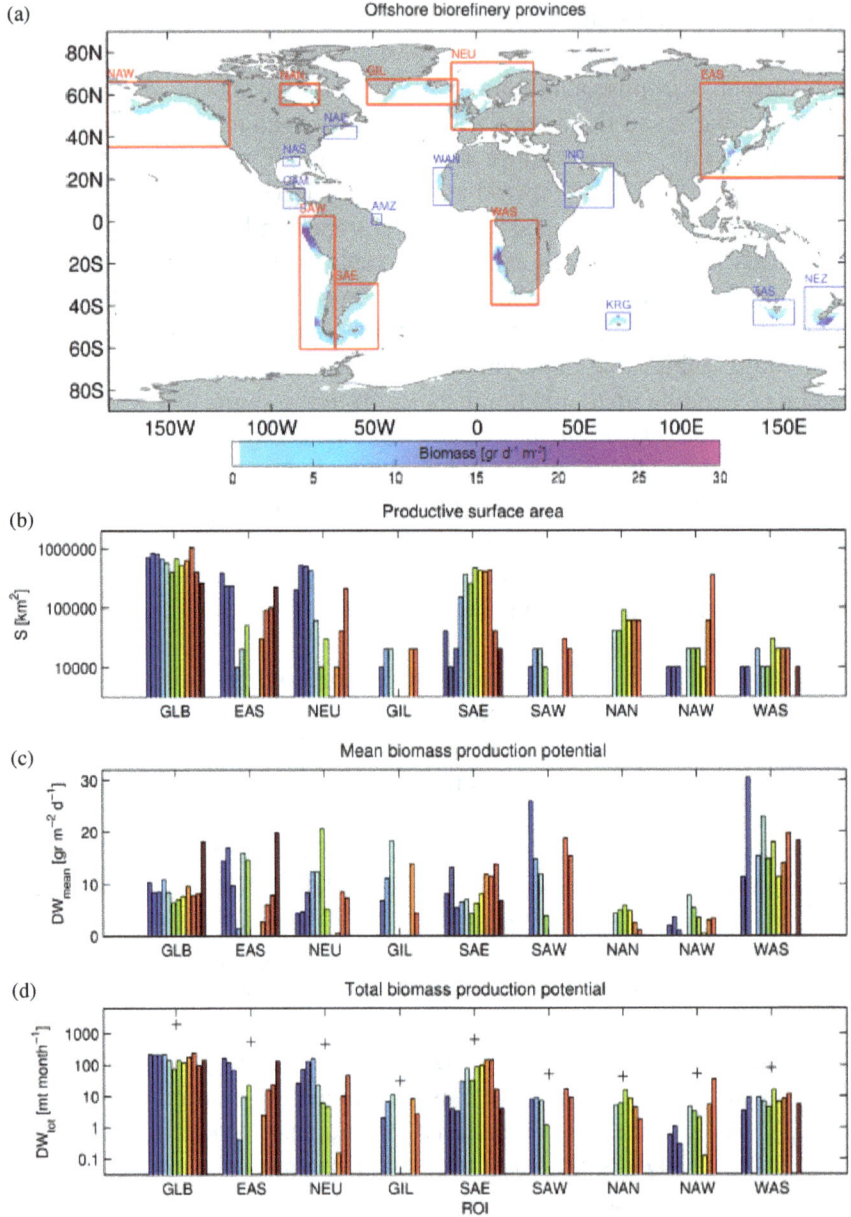

(a)

Offshore biorefinery provinces

(b)

Productive surface area

(c)

Mean biomass production potential

(d)

Total biomass production potential

Fig. 1.3 Regional potential for offshore biorefinery. (a) Potential for biomass production at a distance of less than 400 km from land, taking biomass stocking density of 4 kg m^{-2}. Values in the map will change with changes in stocking density. Boxes delineate major offshore biorefinery provinces, with those permitting biomass production at a water depth of up to 100 m (defined as near-future deployable biorefinery provinces, NDBP) marked in red, and those permitting biomass production only at deeper waters marked in blue. (b-d) Monthly estimates of (b) productive surface area; (c) mean biomass production potential; and (d) total production potential within the 5 NDBP (red boxes and associated abbreviations in panel A) and integrated globally (denoted GLB). Colors denote different months of the year. The analysis is performed over locations associated with a water depth of 100 m or shallower. The + signs mark annually integrated biomass production potential in each region. Assumed biomass density is of 4 kg m^{-2}.

area and potential biomass productivity, which are characteristic to all NDBPs, are driven primarily by seasonal variations in Photosynthetically Active Radiation (PAR) and nutrient availability. The potential for biomass production in the different NDBP is limited for periods of between 8 (provinces SAE and SAW) and 10 (province WAS) months.

1.6 What is the Environmental Cost of Using the Sea? Should We Start Talking About Sea Conversion?

Large scale cultivation can be responsible for positive and negative impacts on coastal and marine ecosystems (Hughes *et al.*, 2012). Therefore, it is necessary to attain balance between biomass for chemicals, food, and fuel production and its environmental cost (Na Wei *et al.*, 2013). Compared to other terrestrial biomass crops, macroalgae have a higher rate of carbon dioxide fixation, which is a potential benefit of large-scale cultivation (Gao and McKinley, 1994). At the same time, there are many environmental risks associated with large-scale offshore macroalgae cultivation.

Many environmental factors such as winds, waves, ocean currents, and rain may adversely impact cultivation, but their effect can be somewhat controlled by selecting favorable sites for cultivation. Natural disasters such as storms, hurricanes, typhoons, and tsunami are generally unpredictable and can destroy cultivation. Type of seabed under the water, the depth of sand, and possible grazers are predictable and can be analyzed before selecting sites. All these factors are avoidable, thus the risks associated with these factors can be reduced.

Risks from factors like light, nutrients, and salinity level of marine water can be controllable by the selection of macroalgal species tolerable to specific environmental conditions. Successful management of the risks related to these factors is essential for the choice of the site for the implementation of the offshore biorefineries with possibilities of many social and environmental benefits (Rui Jiang *et al.*, 2016).

The large-scale cultivation of *Ulva* can strip off all essential nutrients in the immediate vicinity, thus hurting the entire marine ecosystem. Artificial mixing could balance the reduction in nutrient stocks due to biomass cultivation (Pan *et al.*, 2015). In contrast, delay in harvesting, biomass losses, and lack of local grazers could lead to local eutrophication (Yang *et al.*, 2008).

The dense and large quantity of macroalgae cultivation can restrict circulation in the marine system and reduce gas exchange. In addition, the large scale *Ulva* cultivation could create a shadow inside the marine environment at the cultivation site. This can make light penetration difficult and adversely impact the natural ecosystem of the site. In the case of the death of macroalgae, the decay of biomass can result in the depletion of oxygen in the seawater, leading to the mortality of aquatic life in the marine ecosystem. With this, the decay can release hydrogen sulfide gas with offensive odor, thus polluting the air in the nearby environment. Moreover, macroalgae biomass decay could lead to additional environmental hazards such as the formation of toxic concentrations of hydrogen sulfide by sulfate-reducing bacteria, which can affect the sensitive marine ecosystems.

The grazing of *Ulva*, a used model species, is dependent on the presence of herbivore grazers at the cultivation site. There are many possible grazers such as bivalves, ascidians, sponges, amphipods, polychaetes, and gastropods. Even small-sized herbivorous fish can act as grazers that graze on *Ulva* or certain epiphytes on *Ulva*. Grazing can lead to both positive and negative impacts on the production and growth of the biomass. Although high grazing results in an economic loss, it can mitigate the shading impact of macroalgae growth (Worm *et al.*, 2000). The grazing rate depends on the composition of the biomass (Emmett Duffy *et al.*, 2003). It also depends on the other factors such as the presence of dominant grazers, their per capita grazing rate, feeding preferences, and patterns of grazing (McGlathery *et al.*, 2007). For example, species of crustaceans can reduce biomass of macroalgae (Geertz-Hansen *et al.*, 1993) but many invertebrate species generally feed on epiphytes and show a positive effect on macroalgal growth. For example, *Gammarus,* which is one of the most important grazers of *Ulva,* shows higher grazing on epiphytes present on *Ulva* rather than *Ulva* under high nutrient conditions (Kamermans *et al.*, 2002). Epiphytes are generally grown abundantly in the seawater with high nutrient levels and, in such cases, grazers prefer them as food. This could counteract the effects of mild eutrophication (Sand-Jensen and Borum, 1991). Grazers who generally feed on epiphytes can switch to *Ulva* biomass itself if epiphytes are absent. These complex interactions make it a difficult task to quantify the positive or negative impacts of grazing, and monitoring is required at each cultivation site.

An additional environmental risk from large-scale macroalgae cultivation, which requires mitigation strategy, is invasiveness, especially if non-native species are cultivated. For example, in Indian seawater, non-native macroalgae such as *Gracilaria salicornia* and *Kappaphycus alvarezii* are naturalized and show their occupation and spread (Loureiro *et al.*, 2015). The invasive species can adversely impact coral reefs, local species of macroalgae, and other organisms. There are two approaches available to monitor the invasiveness of macroalgae: (1) low-tech approach, which includes the field surveys and morphological identification of invasiveness; (2) high-tech approach, which includes DNA analysis, which shows the genetic structure of the population.

These modeling results show that even using near-future aquaculture technologies, offshore cultivation of macroalgae has the potential to provide some of the basic products required for human society in the coming decades without using any new arable land and freshwater.

This can include one of the following:

1. Displacing entirely the use of fossil fuels in the transportation sector.
2. Providing for 100% of the predicted demand for ethanol, acetone, and butanol.
3. Providing for 5–24% of the demand for proteins.
4. Production of biogas that could displace $5.1 \cdot 10^7$–$5.6 \cdot 10^{10}$ tons of new CO_2 emissions from power generation from natural gas.

In addition to improving offshore cultivation technologies, much attention should be given to the study of the ecological consequences of implementing large-scale offshore biorefinery infrastructures, to ensure their sustainability and to reduce to a minimum their environmental impact. Technological and scientific efforts should be focused primarily on the newly identified near-future deployable biorefinery provinces, where offshore biomass cultivation is expected to be most feasible. Development of sustainable offshore biorefineries infrastructures, if developed carefully, provide a new efficient source for basic products required for human society in the coming decades.

Chapter 2

Blue Economy and Marine Biorefinery

The "Blue Economy" is a relatively new concept that offers a vision of the ocean and coasts as a new source of economic growth, job creation, and investment. The concept of the "Blue Economy" covers three economic forms:

(1) Economy coping with the global water crisis (McGlade *et al.*, 2012).
(2) Innovative development economy (Pauli, 2009).
(3) Development of the marine economy (PEMSEA, 2014).

This extensive interpretation sees a Blue Economy in which economic opportunity is balanced by responsible investment in a sustainable ocean economy — a "win–win" scenario where the private sector, acting through enlightened self-interest, is a catalyst for both economic development and environmental protection. In this vision, the Blue Economy is in itself a source of opportunity, investment, and growth. This chapter examines the Blue Economy concepts as an analytical frame, with a particular focus on macroalgae-based biorefinery as an example of an important sector within the Blue Economy. The policy implications of a rapidly evolving Blue Economy, across multiple sectors, are highlighted.

Blue Economy is a new economy concept for delivering sustainability (Bank, 2015; Mulazzani and Malorgio, 2017; Pauli, 2010). It is one of the potential "sustainable business models", which is important in driving and implementing corporate innovation for sustainability (Bocken *et al.*, 2014). The first Blue Economy concept is developing the need to cope with the global water crisis. Climate change is making some parts of the planet much drier and others far wetter (Kellogg and Schware, 2019). In addition, the demand for water increases with economic growth. Moreover, water "consumption" rises, meaning that the use of water is in such a way that it is not quickly returned to the source from which it was extracted. For example, if it is lost through evaporation or turned into food ("Liquidity crisis", 2016). The big drivers of this are the world's increased desire for grain, meat, manufactured goods, and electricity.

Moreover, around the world, only a few economies price water properly (Baum *et al.*, 2016). Usually, it is artificially cheap, because politicians are scared to charge this essential natural resource. This means that consumers have little incentive to conserve it and investors have little incentive to build pipes and other infrastructure to bring it to where it is needed most. And as the global population rises from 7.4 bn to close to 10 bn by the middle of the century, it is estimated that agricultural production will have to rise by 60%. This will put water supplies under a huge strain as discussed in Chapter 1 in detail and shown in Fig. 2.1.

Water is vital not only for food and domestic well-being. It is fundamental to economic growth as well. Scarcity stalls industrial development by squeezing energy supplies. Electricity generation depends upon plentiful quantities; nuclear power requires water both for cooling turbines and the reactor core itself. The majority of water-intensive industries, such as coal mining, textiles, and chemicals, are found in countries that are particularly prone to water shortages: China, Australia, America, and India

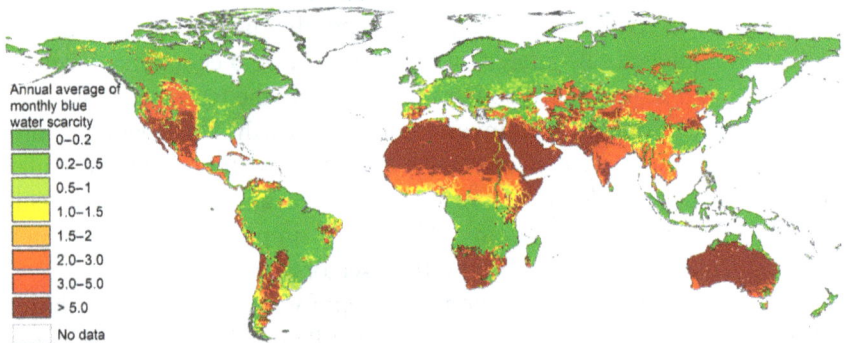

Fig. 2.1   Annual average monthly blue water scarcity at 30×30 arc min resolution. Period: 1996–2005.

Source: Adapted with permission from Mekonnen and Hoekstra (2016).

(Fig. 2.1). The industry can increase strains on supplies too, by polluting water, making it unfit for human use.

Climate change will only make the situation more fraught. Hydrologists expect that a warming climate will speed up the cycle of evaporation, condensation, and precipitation (Cook *et al.*, 2016). Wet regions will grow wetter and dry ones drier, as rainfall patterns change and the rate at which soil and some plants lose moisture increases.

Deluges and droughts will intensify, adding to the pressure on water resources (Kellogg and Schware, 2019). Late or light rainy seasons will alter the speed at which reservoirs and aquifers refill. A warmer atmosphere holds more moisture, increasing the likelihood of sudden heavy downpours that can cause flash flooding across the parched ground. This will also add to sediment in rivers and reservoirs, affecting storage capacity and water quality.

Uncertainty surrounds what this will mean for crop yields, Chapter 1 and (Palatnik *et al.*, 2011). Higher concentrations of carbon dioxide in the atmosphere may make plants use water more efficiently in some parts of the world, but average yields of irrigated wheat — common in countries such as China and India — could drop by 4% and maize harvests would fall everywhere.

There is no single solution to the world's water crisis. But cutting back on use, improving the efficiency of that use, reusing, recycling, and sharing out water more effectively would all help. Here, sea-grown macroalgae-based biorefinery can provide an extremely valuable solution (Lehahn *et al.*, 2016a). Thus, macroalgae cultivation especially offshore could save freshwater by supplying an alternative seawater-based food and materials source.

The second Blue Economy concept is based on the application of mechanisms and principles of nature for the development of humanity, promoting the responsible use of natural resources, the reuse and consequent recovery of waste, and the preservation of the oceans (Mulazzani and Malorgio, 2017). Macroalgae-based biorefinery with significant scope and potential as a sustainable platform can be included as an integral part of the "Blue Economy". Here, Blue Economy represents economic-based activity associated with aquatic biomass focusing on creating innovations, making models, and providing a solution to environmental issues sustainably, with scope to reduce global dependency on the fossils.

Biorefining of macroalgal biomass in a circular loop to maximize resource recovery is being considered as one of the sustainable options that will have both economical and environmental viability. The more the resources recovered as products from the biorefinery, the more will be the economic feasibility of the process. The integration of various unit operations in the macroalgal biorefinery framework enhances resource recovery, process efficiency, and cost-effectiveness; with the primary aim to obtain a range of valuable bio-products. In addition to this process, integration in circular mode promotes self-sustainability and zero discharge to the environment. Macroalgae can be cultivated in seawater polluted by excess nutrients or used to reduce nitrogen and phosphorus in wastewater effluents or desalination brines. By doing this, environmental pollution can be reduced, and resources can be recovered circularly. Nitrogen recycling, for example, can save large amounts of energy invested (and CO_2 emitted) in ammonia production by the Haber–Bosh process. Moreover, phosphorus recycling has major environmental importance as its global reservoirs are rapidly diminishing and therefore its discharge as waste should be highly discouraged. Thus, macroalgal-based biorefineries can contribute to Blue Economy using macroalgae genera for waste recycling into a source of fuels, high-value-added chemicals, and bioactive compounds through systemic production processes.

The third Blue Economy concept seeks to marry ocean-based development opportunities with environmental stewardship and protection. According to the World Bank, the Blue Economy is the *sustainable use of ocean resources for economic growth, improved livelihoods, and jobs, while preserving the health of the ocean ecosystem.* Here, the Blue Economy aims to balance sustainable economic benefits with long-term ocean health, in a manner that is consistent with sustainable development and its commitment to intra- and inter-generational equity. The term has also been used to give greater recognition to the many, though often not priced, ocean values ranging from cultural worth and village-based subsistence economies to commercial and industrial commodities. Under this definition, not all ocean-based activities are consistent with the Blue Economy concept, because many of those activities are not sustainable.

Yet there is a disconnect between the current reality and the ideal. Much economic activity in the ocean remains based on unsustainable over-exploitation or pollution. Governments and businesses alike increasingly recognize the need for a new model. But what that model should look like and how to get there remain unclear. What is widely acknowledged is that solutions designed to simply limit economic activity are not only unrealistic — they are also a missed opportunity for development.

The transition from a conventional economy in the ocean to a sustainable ocean economy could be a tremendous economic and investment opportunity if done right. The risks and challenges are considerable. A new and intensive phase of economic activity in the ocean is getting underway just as the science warns that the seas are facing unprecedented pressures from humans, and that time to salvage them is fast running out.

Blue Economy significantly advances practices beyond existing sustainable development frameworks. A proliferation in terms adds more complexity to an already challenging management space. Nevertheless, the conceptual framework is useful for structuring evaluations of practice and helping to reveal missing ingredients necessary for the sustainable development of oceans.

Nonetheless, there is no shortage of forward-thinking businesses, governments, and ocean advocates exploring sustainable ocean economy. Established marine industries such as shipping, oil & gas, and fishing are being pressured and reshaped by changing consumer and investor preferences, greater regulation and, in some, a perceptible shift towards greater stewardship of the ocean environment and more sustainable business practices. A raft of new activities, such as renewable energy, sustainable aquaculture, ecotourism, and "gray–green" coastal infrastructure, and macroalgae-based biorefineries offer opportunities for sustainable investment that aim to do less or no harm to the ocean or even to profit from reversing damage to ocean ecosystems.

Ensuring greater stewardship and sustainable practices among existing ocean activities, while bringing new endeavors such as macroalgae-based biorefineries to scale, are challenges that require vision, new modes of governance, bold regulatory and behavior changes, and, of course, large investments. In the public sector, the transition will involve sizeable spending on reforms to institutions and laws, and effective monitoring and enforcement. Private-sector investors will need to develop a greater awareness of the opportunities and risks involved in the sustainable ocean economy.

New investment frameworks and financial tools will be essential to develop and govern the new ocean economy in a sustainable way. Continuing with business, as usual, poses great risks. New dialogue on financing the transition to a sustainable ocean economy and the role of macroalgae-based biorefinery in it is vital. Follow-up studies should evaluate how large the opportunity is of the macroalgae-based biorefinery in contributing to the Blue Economy. What are the risks involved? What investment frameworks might be necessary? What capital is available, and how can it be scaled up?

Chapter 3

Marine Macroalgae

3.1 Seaweed Species

Marine macroalgae, generally referred to as seaweeds, are photosynthetic organisms bearing fast growth rates and primarily inhabiting marine environments. Specifically, canopy-forming kelp, rockweed, and red macroalgae provide important ecosystem roles, including primary production, carbon storage, nutrient cycling, habitat provision, biodiversity, and fisheries (Harley *et al.*, 2012). Many seaweeds provide valuable economic benefits for several industries. Worldwide, there are around 25,000 species of marine macroalgae described so far (Algaebase.org), and roughly 45%, 30%, and 25% correspond to the Phylum Rhodophyta ("red"), Ochrophyta ("brown"), and Chlorophyta ("green"), respectively. Nonetheless, these figures are estimates and should be taken with caution due to the on-going process of discovery and description of new species and varieties. The molecular revolution in taxonomy, causing huge changes in the delimitation of taxa, together with a perceived inability of scientific taxonomy and nomenclature to keep up with these changes, contributes uncertainties regarding the actual number of seaweed species (Bolton, 2019; De Clerck *et al.*, 2013).

The three phyla differed in terms of evolutionary patters and physiological signatures. Seaweed size can run from a few mm up to 60 m tall kelps from the brown group. Marine macroalgae have different morphologies and forms and some groups may also have calcifying tissues. Fleshy, filamentous, and sheet-like forms are among the large morphological forms. All seaweeds share chlorophyll "a" as the ultimate photosynthetic pigment, trapping and concentrating solar energy to activate the universal Calvin cycle inorganic carbon fixation pathway. Red seaweeds contain specific pigments, notably phycobilins (phycoerythrin, phycocyanin, and allophycocyanin) to expand the light capture within the visible light (400–700 nm). Brown seaweeds have their pigments as well. In addition, the three different groups have storage compounds resulting from various biochemical pathways, and hence, could offer a large variety of valuable hydrocolloids and active molecules. Seaweeds used primarily dissolved CO_2 present in equilibrium with the atmospheric one, approximately 400 ppm CO_2, on average. But, the major source for photosynthesis is the plentiful ionic carbon form, HCO_3^-, found in seawater about 200 times more concentrated than CO_2. In the intertidal, macroalgae may severely be exposed to air, meaning exposure to desiccation and high temperatures risks, and not many species can stand under these harsh conditions for extended periods.

Seaweeds are typically distributed in the photic zone, between 0–200 m deep, the limit at which there is still enough sunlight for positive photosynthesis. Species adapted to thrive in the intertidal zone present special eco-physiological features allowing them to handle desiccation and excessive sunlight. Most seaweeds present a clear seasonality and they modulate their reproduction cycle accordingly (see below). Marine macroalgae have become the focus of interest in the last decades due to their unique characteristics contributing to several industries worldwide, and offering several ecosystem services largely ignored in the past. Although seaweeds drew renewed interest during the last decades as highly attractive feedstocks, they have been exploited by humans since the early civilizations. The most important seaweed genera in the current aquaculture industry are *Laminaria* (kombu), *Undaria* (wakame), *Porphyra/Pyropia* (nori), *Eucheuma/Kappaphycus*, and *Gracilaria*, which together provide about 95% of the total world seaweed production. Further, most of the seaweed industry is based on an impressively low number of species. For example, only 7 species are economically domi-nant in China (Zhang, 2018).

3.2 Spatial and Chemical Diversity

Most seaweeds are strongly seasonal-dependent, and their growth, repro-ductive strategies, and chemical constituents can be radically dis-tinct throughout the year. These changes are primarily determined by

Fig. 3.1 A field site close to Tel Aviv, Israel, in which wintertime is typically dominated by species of *Ulva* in combination with *Porphyra*. The second species is regarded as a "winter species" and fully retrieves by mid-spring. The upper intertidal in the same rocky platforms show naked towards summertime.

environmental factors such as temperature, light, and day-length, all of which have annual cycles in natural marine environments. Run-off of nutrients during rainy seasons can also significantly affect these cycling. This seasonality can be dramatic sometimes so that one single substrate can present radical views as shown in Fig. 3.1.

Seaweeds are highly diverse, which is reflected in terms of physiological, morphological, ecological, and biochemical traits which are expressed in the algal cellular structure, levels of cell organization, reserve, and structural polysaccharides, their life-history traits, photosynthetic pigments, and different habitats (Chopin and Sawhney, 2019). Generally, algae can be characterized as simple organisms that lack vascular conducting tissue, with no true roots, stems, or leaves as in terrestrial vascular plants. All autotrophic algae have cellular organelles surrounded by membranes, an exception for the nucleus, mitochondria, and chloroplast which possess a double membrane. They have the same metabolic pathways as for the rest of the plants and produce similar protein and carbohydrates. The cell wall usually contains cellulose and additional polysaccharides (Chopin and Sawhney, 2019; Pereira *et al.*, 2013).

As mentioned, generally, seaweed is photosynthetic benthic organisms normally attached to the sea bottom. Sandy coastal habitats have fewer seaweeds since most species cannot efficiently anchor, but there are exceptions such as the green *Udotea* and *Caulerpa* that possess specialized root-like structures. Sandy areas are primarily dominated by seagrasses, particularly in tropical and subtropical habitats (Dawes and Lawrence, 1980).

The vertical and spatial distribution patterns of the different seaweed species in those areas are determined by a variety of abiotic, physical and chemical factors, hydrodynamics and water motion, availability of nutrients, CO_2 and pH (Eggert, 2012; Gil *et al.*, 2008b; Harley *et al.*, 2012; Israel *et al.*, 1999; Ramos *et al.*, 2012). In addition, biotic factors like a competition between species, phenotypic plasticity, epiphytes, and predation also play a fundamental part in the position and boundaries of different seaweeds (Mann, 1991). Survival throughout the seasons in seaweeds is tackled through various ecological strategies. Some seaweeds have a succession of genetically distinct populations that develop one after the other, pending on the environmental pressures. Some species avoid harsh periods by creating spores that may survive long periods. Some even regulate their lifecycle according to the prevailing season from macro- to microscopic forms.

The presence of different species is regulated by the interaction of those different environmental factors. Light and temperature, which are dictated by the seasons, are considered to be the most influential factors in macroalgal biology and ecology. Light is the main component needed for the photosynthesis process, while temperature dictates the metabolic functions and rate of various necessary biochemical and enzymatic processes (Habiby *et al.*, 2018; Harley *et al.*, 2012; Pereira *et al.*, 2013). As the seasons change, the day-length may be shorter or longer and the temperatures can change significantly. Therefore, the seasons can significantly dictate the composition of macroalgae communities in the natural environment.

For typical subtropical environments, and for perennial species, wintertime represents a period of slow growth, but in which significant accumulation of valuable molecules takes place. With increasing temperatures and light in Spring, these seaweeds enhance their growth at the expense of decreased levels of exploitable chemicals in their tissues.

The availability of nutrients is another essential factor for seaweed growth. Nutrients levels largely dictate growth rate, health, and chemical composition of seaweeds (Chemodanov *et al.*, 2019; Neori *et al.*, 2004; Yates and Peckol, 1993). One of the most dangerous disturbances worldwide to marine and coastal habitats is nutrient enrichment or eutrophication phenomena (Nixon, 2009). Excess levels of the nutrient may lead to explosive growth or "blooms" of opportunistic seaweed. Those seaweeds usually exhibit superior nutrient uptake, fast growth rates, and have better reproductive capabilities that allow them eventually to out-compete slower-growing species (Nixon, 2009).

Generally, different species may show a different response to the same environmental disturbance. Usually, more sensitive species will succumb to competitive, tolerate, and more opportunistic species that are favored under the new conditions (Gagnon *et al.*, 2016; Nixon, 2009). Examples of opportunistic seaweed species that are highly successful under the anthropogenic disturbance of nutrient enrichment are the green seaweeds: *Ulva, Chaetomorpha Cladophora* and *Monosroma*, the red seaweeds: *Ceramium, Gracilaria* and *Porphyra*, and the brown seaweeds: *Ectocarpus* and *Pilayella* (Morand, 2005). Contrarily, *Cystoseira* is a good example of a species highly sensitive for such disturbance (Sales and Ballesteros, 2012).

The abundance of intertidal seaweeds depends on their ability to withstand natural hydrodynamic forces in their habitat, such as water motion, currents, and waves action. The different physical properties of each species, its specific morphology, structure and size, tissue strength and resilience, can significantly affect the success of a particular seaweed to face these forces (Carrington, 1990; Williams, 1986). For example, the size of different species of seaweed can be limited in a habitat characterized by strong waves and water motion. Carrington in 1990 showed that the seaweed *Mastocarpus papillate* exhibited a broad range of morphologies and sizes, about different water drag forces (Carrington, 1990). The combination of the different structural properties of seaweeds, alongside the habitat hydrodynamics and other abiotic environmental factors, are responsible for the macroalgal community structure. This interaction, in many cases, will eventually create the typical vertical distribution of seaweeds in intertidal rocky shore habitats (and other benthic organisms) also known as "belts".

In contrast to phytoplankton, seaweeds are multicellular. This important attribute allows the seaweeds to grow larger, and develop vertically away from their holding substrate, bringing them closer to light, and enabling them to harvest nutrients effectively from a greater volume of water (Lobban and Harrison, 1994)). Some of the larger kelps (Fucoids) have special gas-filled structures that help their canopy float and carry their enormous size in the water column (Foreman, 1976). Other smaller species like the green *Codium* have gas (largely O_2) trapped among their filaments for positioning. Most seaweed are sessile, but some are epiphytic, and others may occasionally grow unattached under specific conditions.

The spatial geographical and temporal distribution of seaweed, in contrast to many other organisms, is not uniform, but rather follows a non-random, dynamic way of appearance. That is because different species have varied adaptation for different abiotic and biotic environmental factors. Those factors dictate the seaweed seasonal appearance and their location along the coast (Lobban and Harrison, 1994). Ultimately, the distribution and specific niche in time and space of each species can be concluded as the complex expression of its ecology and life history, effected by different factors operating at different scales and different intensities (Soberon and Peterson, 2005). Another approach of describing seaweed distribution geographically is based on their taxonomical groups. Green, brown, and red seaweeds exhibit distinct distribution patterns around the globe. Contrary to both marine and terrestrial organisms, seaweed richness does not continuously increase towards the tropics and does not follow Rapoport's rule (Bolton, 1994; Pereira and Neto, 2014; Santelices and Marquet, 1998). Generally, brown seaweeds' richness increases, and green seaweeds' richness decreases towards higher latitudes, while the red seaweeds' richness increases from the Arctic to the Tropics and from the Tropics to the Sub-Antarctic region (Witman and Roy, 2013). In the northern hemisphere with much colder waters, brown seaweeds dominate and decrease towards the south warmer waters. Then, southwards while seawater becomes more temperate the red seaweed species will be far more common (Gaspar *et al.*, 2012). The green seaweed distribution may depend less on bio-geographic factors alone, their abundance will usually increase or decrease in response to local environmental conditions. Under good environmental conditions, their number will remain constant and their biomass will be relatively low, while usually late-successional red and brown seaweed species will be widespread. However, when environmental condition and water quality will alter and degrade, opportunistic green seaweed can appear in large numbers and repress the other seaweed species (Bermejo *et al.*, 2012; Gaspar *et al.*, 2012).

Seaweeds are rich in diverse polysaccharides and carbohydrates and have high quantities of essential minerals and vitamins (Stengel and Connan, 2015). For example, a study that quantified the monosaccharide content and diversity of macroalgal species common to the Eastern Mediterranean shores of Israel representing the three major seaweed divisions, namely, Chlorophyta (*Ulva* sp. and *Cladophora pellucida*), Rhodophyta (*Nemalion helminthoides, Galaxaura rugosa,* and *Gracilaria* sp.), and Ochrophyta (*Padina pavonica* and *Sargassum vulgare*) (Fig. 3.2), found that the most abundant monosaccharide was different in 5 out of the 7 investigated species

(Robin *et al.*, 2018), Fig. 3.3. The monosaccharide diversity profile was specific to each taxonomic group, especially in the first two orders of diversity, which correspond to Shannon entropy and Simpson concentration (Robin *et al.*, 2018). The content of monosaccharides released by acid

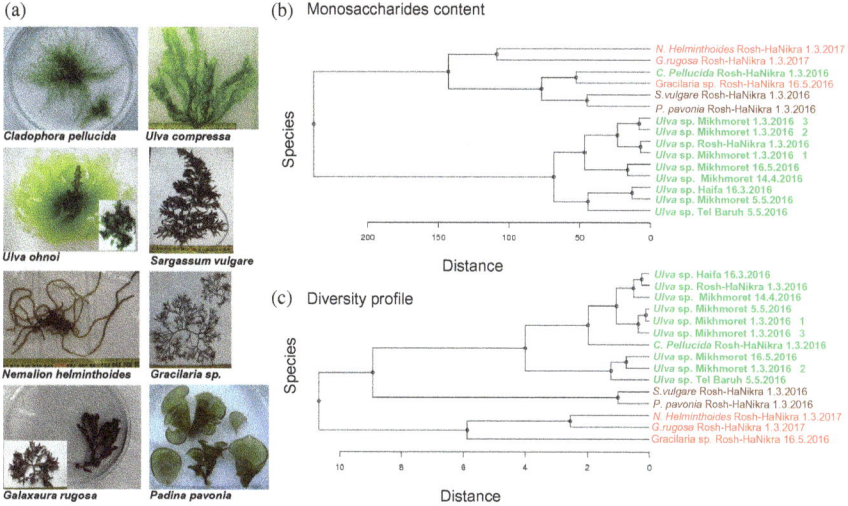

Fig. 3.2 (a) Species identified in the study on Eastern Mediterranean shores of Israel. ((b)–(c)). Sorting of species with a hierarchical clustering algorithm. Sorted by (b). Monosaccharide content (c). Diversities profile of monosaccharides.

Source: Adapted with permission from Robin *et al.* (2018).

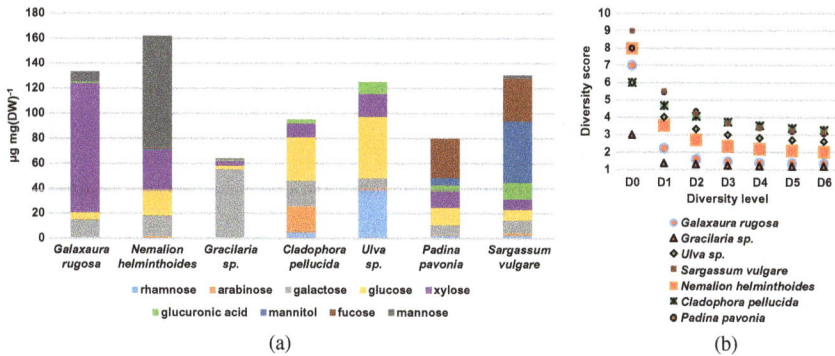

Fig. 3.3 Monosaccharide diversity of macroalgae collected from Rosh-HaNikra reef, Israel. (a) Monosaccharide content released by acid hydrolysis. (b) A diversity profile of monosaccharides. Source: Adapted with permission from Robin *et al.* (2018).

hydrolysis varied by 153% between *Ulva* sp., *C. pellucida*, *G. rugosa*, *N. helminthoides*, *Gracilaria* sp., *P. pavonica*, and *S. vulgare* collected from the same site (Fig. 3.3). *Ulva* sp. collected at different sites at different months showed up to 79% variance in the total released monosaccharides, with up to 270% variance in the content of individual monosaccharides (Robin *et al.*, 2018), Fig. 3.4.

As mentioned earlier, one of the most noticeable differences between the three seaweed Phyla is their pigment composition, which is largely determined by solar radiation's intensity and characteristics. It has been shown that algae that live under low light conditions, such as in great depth, respond by increasing their pigment and accessory pigments content. This shade-algae also exhibits high photosynthetic performance at low light intensities and in contrast lower photosynthetic rates at high light, in comparison to algae or plants that live under high-light

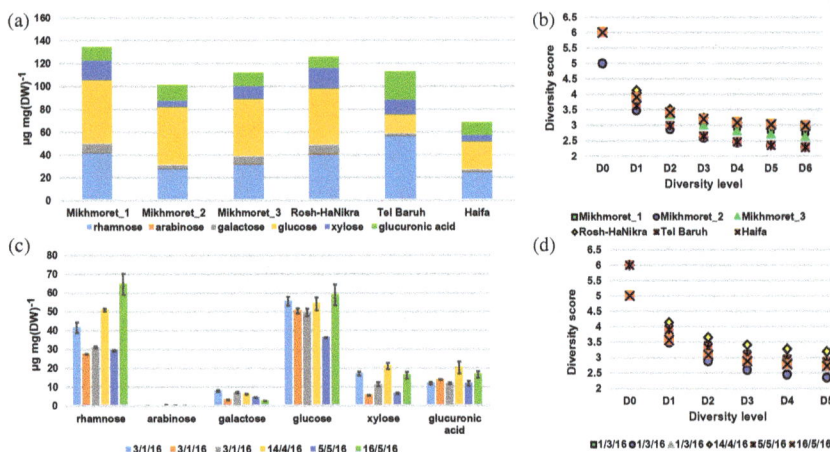

Fig. 3.4 Spatial and seasonal changes in monosaccharide diversity of *Ulva* sp. (a) Content of monosaccharides in species sampled at 4 different sites about 130 km apart on the Eastern Mediterranean shores of Israel. Three sites were sampled at Mikhmoret. (b) A diversity profile of monosaccharides in species sampled at 4 different sites about 130 km apart on the Eastern Mediterranean shores of Israel. Three sites were sampled at Mikhmoret. (c) Content of monosaccharides in algae sampled at different months at Mikhmoret site. (d) A diversity profile of monosaccharides of species sampled at different months at Mikhmoret site. Error bars show ± standard deviation.

Source: Adapted with permission from Robin *et al.* (2018).

environments (Beer and Levy, 1983; Rosenberg and Ramus, 1982). All algae have chlorophyll a and the accessory pigment β-carotene. Green algae also have chlorophyll B. Additionally to chlorophyll A, red algae contain chlorophyll D and phycobiliproteins that are accessory pigments that mask the colors of chlorophyll A and β-carotene and are responsible for the Rhodophyta's red color. Brown algae also have chlorophyll C and A, and a relatively large number of xanthophylls carotenoids that are responsible for their brown color (Kraan, 2013; Maschek and Baker, 2008; Patarra *et al.*, 2011).

In the natural environment, seaweeds experience extreme environmental stresses regularly, and during low tide can be subjected to high irradiance, desiccation, changes in salinity, nutrient unavailability, and more (Collén and Davison, 1999; Hurd *et al.*, 2014; Steen, 2004). Those local conditions of light, nutrient, and temperature regimes have a significant impact on the seaweed metabolite levels and thus on their chemical composition. Furthermore, the seaweed chemical richness can vary according to the seaweed species, lifecycle, development stage, and physical structure. Those facts as well as the understanding of the adaptive mechanism, are important when one's trying to learn about the regulation of concentration of specific compounds in the seaweed tissue (Neori *et al.*, 2004; Stengel *et al.*, 2015).

To acclimate to the stressful and highly fluctuating environmental conditions (e.g. UV radiation, salinity, and temperature stress), and to withstand biological factors such as herbivory, competition, and epiphytes, seaweeds have evolved complex chemical mechanisms, among them, are the formation of special secondary metabolites, accessory pigments, antioxidants, and various protective proteins and molecules. Some important examples of these molecules are accessory pigments like carotenoids and phycobiliproteins, diterpene alcohols, phenolic compounds, terpenes, and MAAs: mycosporine-like amino acids (Karsten *et al.*, 1998; Kumar *et al.*, 2010; Sampath-Wiley *et al.*, 2008; Schmitt *et al.*, 1995). Seaweed-derived secondary metabolisms have been shown to possess a wide range of bioactivities such as antioxidant, anti-inflammatory, antidiabetic, anticancer, antiviral, antimicrobial, antifungal to anti-obesity, and anti-parasitic (Abdelhamid *et al.*, 2018; Patarra *et al.*, 2011; Stengel *et al.*, 2015). For these reasons, seaweeds have a variety of applications in human welfare and are among the most important cultivated marine organisms.

3.3 Knowledge Gaps of the Seaweed LifeCycle for Their Use as Crops

Despite their relative simplicity as compared to terrestrial plants, seaweeds may present complex and sometimes sophisticated lifecycles. The versatility of lifecycles offers an important breakthrough in the development of seaweed aquaculture paving the way for sustainable seaweed biomass in several species worldwide. While the first empirical studies have been carried out and the bulk of seaweed lifecycles have been elucidated, biological processes driving lifecycle, for example for biorefinery, from fertilization to the production of organisms, are the next challenge and fundamental knowledge of macroalgal developmental mechanisms need to be deciphered (Charrier *et al.*, 2017). These gaps can be benefited by the development of cutting-edge technologies in animals and plants, however, many of these technologies need to be adapted to macroalgae because of their specific ecological niche (highly saline) and their biology (in part because of their phylogenetic distance from better-known organisms, (Charrier *et al.*, 2017)).

The algal lifecycle incorporates a variety of reproductive modes: asexual reproduction, sexual reproduction, and regeneration. The lifecycle of most seaweeds is heteromorphic and includes alternation of generations. The thallus can be found in two different multicellular stages, a haploid stage (*n*) called the gametophyte, and a diploid stage (*2n*) called the sporophyte. In some species, both stages look alike and share the same eco-physiological features, while in other species the two stages are so diverse that for a long time they have been treated as different species. As opposed to terrestrial plants, seaweeds usually lack major cell differentiation along the thallus. Entering the reproduction phase involves a developmental process that produces spores through cell differentiation from vegetative cells into reproductive cells. At the end of maturation, flagellated spores are released in high quantity and usually have positive or negative phototaxis features. Fertilization may take place in the water when two spores from different thallus fuse, or within the reproductive cell of the thallus. The lifecycle of most seaweeds include asexual and sexual reproduction routes. Usually, spores that produce from the gametophyte (gametes) can fuse with a gamete from different thallus to form a diploidic sporophyte. However, if a gamete does not fuse with other gametes, it can be developed into a mature gametophyte by an asexual life route. Another asexual reproduction route that is very common in the algaculture industry is a vegetative reproduction using fragmentation.

In general, seaweed blooms in nature have a seasonal cycle. This fact led to the hypothesis that the environmental regime has a strong influence on the development of spores, shifting the cells from a vegetative phase into a reproductive stage. In seaweed aquaculture, sporadic reproduction events triggered by abiotic factors could lead to the development and release of spores, negatively affecting growth rates and yields. A better understanding of the mechanism and the environment regime that are involved with the reproduction phase of seaweeds can improve the cultivation and domestication via crossbreeding and strain selection for desirable features. In the biorefinery industry, the ability to increase specific desirable biomaterials in the raw material using strain selection can dramatically boost the profitability of this industry.

The successful cultivation of *Porphyra* followed by the completion of its lifecycle during the 1940s emphasizes the contribution of research on seaweed reproduction. *Porphyra* is a red alga with top commercial value worldwide and used in the "nori" making international market. In *Porphyra*, the gametophyte bears a blade-like appearance and this is the raw material for nori production. For a long time, farmers have relied on natural swarmer spores that penetrate the gametophyte, fertilize it leading to the forming of the sporophyte (carposporophyte). No one knows where those spores come from. However, this change in 1949 when the microscopic sporophyte conchocelis, which previously were thought to be a separate species of alga, turned out to be another stage in the *Porphyra* lifecycle. This study discovered that the spores that fertilize the gametophyte originated from the sporophyte conchocelis and that this stage can only live in calcareous substrata such as shells of dead bivalves. This insight led to a dramatic progression in *Porphyra* farming techniques.

While in Asia more than 90% of the seaweed production is sourced from cultivation (accounting for 93% of the global production in 2013 (FAO, 2016)), the dominant practice of non-Asian countries is still harvesting natural stocks. Hence, what are the real knowledge gaps in seaweed reproduction that draw back sustainable biomass production, land- or sea-based? There are surprisingly few as most of the reproductive biology of current commercial species is relatively well know. Perhaps the lacking part in the seaweed industry is the small number of species currently cultivated. There should be a wider view and screening of potential species that can be incorporated into the seaweed industry. Perhaps the current potential restrictions to sustainable seaweeds production rely on more efficient cultivation techniques, optimization, and technologies appropriate for both land- and sea-based cultivation.

Chapter 4

What is a Macroalgae Biorefinery?

4.1 Marine Biorefinery as a System

Marine biorefinery adheres to the Merriam-Webster dictionary definition of a *system*: "a set of interacting or interdependent parts forming a complex/intricate whole, whose property is different from the properties of its components". An offshore marine biorefinery is a complex system and the systematic design of its components requires the following:

- Determination of the optimum location
- System design to function in the chosen location
- System integration within the natural environment
- Cultivation platform design — upstream processing
- Conversion processes — downstream processes
- Analysis of social and environmental impacts

A general definition of a biorefinery by Soetaert, as used in the Biorefinery Euroview report, is as follows: "Integrated bio-based industries, using a variety of different technologies to produce chemicals, biofuels, food, and feed ingredients, biomaterials (including fibers), and power from biomass raw material" (Euroview, 2010).

Biorefineries are the manufacturing units of bioeconomies. In a biorefinery, one or several biomass feedstocks (crops, lignocellulosic biomass, seaweed, microalgae, insects, etc.) are processed into a wide range of products as mentioned above. One could also replace "industries" by "system", thus allowing the use of biorefinery for a small-scale system that could not be integrated into the word "industry".

Concerning the actual physical design of a biorefinery, several options exist. The first one is that all the involved processes are integrated into a single location, in the integrated plant. The second option is that the integration is made through different already existing bio-based facilities, and thus, the feedstock of one of the plants is the co-product or waste stream of another one, without requiring the different plants to be physically integrated. This option allows the existence of many biorefineries in the current network of bio-based industries without them being assembled in a single entity with legal status. Finally, a hybrid version of a biorefinery could encompass both the first and the second options, with a central integrated plant escorted by satellites of additional bio-based industries.

Before developing the design and the example of seaweed biorefinery, we must first give the main advantages and incentives to build a biorefinery, answering the "why" questions. In addition to the advantages, we will also underline the potential challenges that arise when designing or operating a biorefinery. We would like to specify that most of the analysis shown here is relevant to any biorefinery and is not specific to seaweed biorefinery.

The main advantage of biorefineries is the capacity to produce multiple products from the same raw materials. Not only is it expected to increase the revenue per mass of feedstock, but it also diversifies the applications thus making the whole system more resilient economically. Moreover, in an ideal biorefinery, there is no waste stream since all biomass constituents find an application. As applications for products from biorefinery go from food to energy, chemical building blocks to biomaterials, it covers the entire spectra of basic commodities for human society. Additionally, several advantages are linked to the use of biomass as feedstock: they are sustainable resources (when managed properly) with far fairer abundance and geographical distribution than fossil resources, and all their constituents are biodegradable. Finally, if we consider biofuel and bioenergy, textile (cotton, etc.), wood (paper, furniture, etc.), food and beverage, cosmetic, nutraceutical and pharmaceutical industries as a subpart of biorefineries since they are bio-based industries, one could say that biorefineries are already a reality. All of those reasons make biorefineries central actors in the current and future sustainable economies.

Concerning more technical details, the co-production of multiple products usually refers to the multiple subsequent processing of the biomass feedstock during several stages. Each stage leads to the production of one or several different products. Since each step removes a fraction of the biomass, one that could negatively impact the following process step, the subsequent steps are becoming simpler and more efficient. For example, in the case of seaweeds, by extracting protein and water-soluble molecules in a first stage, the content of residual water-soluble polysaccharides becomes considerably higher, while the extraction product obtains a higher purity (i.e. by having less protein and water-soluble materials such as salt) than if the polysaccharides would have been extracted first. This is critical since the processing is a major economical and technological hurdle of biorefineries. More precisely, biorefineries facilitate the purification steps of biomolecules by removing potential "contaminants" (which are co-products) beforehand, and as the purification downstream processes are one of the major costs in biomass processing, substantial benefits can be made. Also, certain processes such as crushing, milling, and other cell and tissue disruptions will usually benefit most of the subsequent processes by improving the contact surface of the biomass and the accessibility of intracellular components. Thus, the energy and cost of such a process benefits the entire processing chain instead of a single product.

Concerning the challenges of building and operating biorefineries, there are uncertainties on the real economical, environmental, and social benefits. For example, since the exergy return on investment (EROI) of biorefineries is low, if not negative, most biorefineries require a considerable amount of energy input, which today origins from burning fossil fuels and thus negating some of their sustainability advantages. Another example is the competition between food and first-generation biofuels. Many other aspects, such as integration in the current manufacturing system, feedstock supply chain, environmental regulations, trade agreement, competitions with fossil products and water or land scarcity are all potential hurdles to the development of biorefineries. We will thus focus on the technological aspect. Indeed, if processes in petroleum refineries are well understood and mastered, including at an industrial scale, this is not the case for biorefineries. If technologically it is possible to produce anything from biomass, the efficiency and commercial viability of most processes, specifically on a large scale, has yet to be proven.

The next challenge resides in designing the biorefinery, i.e. to integrate the processes into one optimized process flow (see the following Sections 4.3 to 4.6). This is a challenge because biomass processes were usually designed for the sole production of one product and not the preservation of the leftover for subsequent processing. Their integration required the understanding of the effect of each process on each constituent of the biomass, which is the production target of the subsequent processing steps. Thus, specific and non-destructive processes should be located upstream of the biorefinery chain while the more severe and destructive ones need to be kept as downstream as possible. As of today, there is no widespread and successful standard for biorefinery design.

Although a considerable effort is being deployed to answer those two technological gaps (better bioprocesses and integration of those processes) at the research level (see Section 4.6 for example), there are still gaps to fill.

4.2 General Thermodynamic Considerations in Marine Biorefinery Design

The determination of the optimum marine biorefinery location is predicated by environmental and socio-economic conditions. Environmental conditions are numerous and include the conditions for macroalgae growth and risks. Assessment of offshore locations for cultivation suitability is an expensive and time-consuming task. For the initial assessment, the following model has been recently suggested to assess the biomass growth rate at a specific geographical location (Lehahn *et al.*, 2016a):

$$\mu \; = \; \mu_{\max} \cdot f\left(I, T, S, N, P\right) - r_{\text{resp}} \qquad (4.1)$$

where macroalgae biomass growth rate (μ) is calculated based on its maximum possible growth rate ($\mu_{\max}$), as a function of light intensity (I), temperature (T), salinity (S), nutrients (N and P for nitrate and phosphate, respectively), and respiration rate (r_{resp}).

In this model, we assume that each of the factors has a separate impact on the biomass growth rate. However, following Liebig's law of the minimum, we formulate the growth rate as a function of the limiting factors, which is the scarcest resource to the macroalgae's requirement. This formulation is based on a simplifying assumption that there are no interactions between these factors and thus no co-limitations. However, temperature and salinity effects were excluded from the minimum law, as they constitute background conditions for metabolic efficiency, and are not considered as growth resources. Therefore, the approximated function for biomass growth rate appears in Eq. (4.2):

$$\mu = \mu_{max} \min\{f(I) \cdot f(N) \cdot f(P)\} \cdot f(T) \cdot f(S) - r_{resp} \qquad (4.2)$$

where μ_{max} is the maximum growth rate, r_{resp} the respiration rate (d^{-1}) defined as

$$r_{resp} = r_{resp\_ref}\theta^{T-T_{ref}} \qquad (4.3)$$

where r_{resp_ref} is the maximum respiration rate at reference temperature (T^0C), θ is the empirical factor. The function $f(I, T, S, N, P)$ is defined as follows in Eq. (4.4)–(4.8):

$$f(I) = \frac{I}{I_{opt}}e^{\left(1-\frac{I}{I_{opt}}\right)} \qquad (4.4)$$

where I is the light intensity at the time (t) and I_{opt} is the optimum light intensity for specific species biomass accumulation. I is affected by stocking density, which determines light availability per unit of biomass (Nikolaisen *et al.*, 2008; Oca *et al.*, 2019)

$$f(T) = e^{-a\left(\frac{T-T_{opt}}{T_x-T_{opt}}\right)^2} \qquad (4.5)$$

$T_x = T_{min}$ for $T \leq T_{opt}$ and $T_x = T_{max}$ for $T > T_{opt.}$

$$\text{For } S > S_{cr} f(S) = 1 - \left(\frac{S-S_{opt}}{S_x - S_{opt}}\right)^m \qquad (4.6)$$

where $S_x = S_{min}$ and m_{min} for $S < S_{opt}$ and $S_x = S_{max}$ and m_{max} for $S \geq S_{opt}$ for $S < S_{cr}$

$$f(S) = \frac{S - S_{min}}{S_{opt} - S_{min}}$$

$> N_{min}$) and ($P > P_{min}$):   (4.7)

P and N: $P < h$:

$$f(N, P, C) = 1$$

$< 1 \, f(N, P, C) = f(N)$:

$$f(N) = \frac{N_{int} - N_{int\_min}}{keq + N_{int} - N_{int\_min}}$$

$> hh \, f(N, P, C) = f(P)$:
P_{int_max}:

$$f(P) = \frac{P_{int}}{P_{int\_max}}$$

P_{int_max}:

$$f(P) = 1$$

where μ_{max}, r_{resp_ref}, T_{ref}, θ, I_{opt}, T_{min}, T_{max}, T_{opt}, S_{cr}, S_{min}, S_{max}, N_{min}, P_{min}, l, h, hh, keq, N_{int_min}, P_{int_max} are species-specific parameters. Example of the use of this model for *Ulva* growth can be found in (Lehahn *et al.*, 2016a). Daily production per m² of FW biomass is derived by multiplying calculated growth rate (μ) with biomass density (σ) (kg m⁻²) (Nikolaisen *et al.*, 2008), Eq. (4.8):

$$BM_{FW} = \mu \cdot \sigma \qquad (4.8)$$

The example of the model used has been demonstrated on the macroalgae *Ulva* species, to assess its global cultivation potential (Lehahn *et al.*, 2016a). The work identified potential provinces where *Ulva* species can be cultivated for the maximum biomass yields offshore.

System engineering design that will fit the chosen location in the harsh marine environment is a challenge. Attaching seaweeds to ropes, lines, or nets is a traditional way of cultivation since installation and maintenance costs are very low. One of the cultivation methods is growing seedlings directly on the ropes (Peteiro and Freire, 2012).

Another approach is via transplantation: seedlings are grown indoors, then cultured in greenhouse tanks, and finally, the small fronds are transplanted onto ropes in the sea. The ropes should be tensed to reduce diffusion boundary layers surrounding the thalli, thus increasing water exchange for efficient nutrient availability and fast growth (Hurd, 2000; Neushul *et al.*, 1992). Horizontal and vertical ropes systems were also discussed by Peteiro and Freire (2012). A system with concentrically horizontal ropes was tested in Buck and Buchholz (2004). Seabed planting can be cited as well as a cage system (open-topped cage anchored to the seafloor) for near-shore cultivation (Hanisak, 1987b). Co-management with other offshore systems like wind farms, for instance, was considered (Buck *et al.*, 2008; Jung *et al.*, 2013; Klimakrise, 2009; Michler-Cieluch *et al.*, 2009) and discussed as an environmentally and economically advantageous approach (Reith *et al.*, 2005a).

Optimizing downstream processing of the biorefineries is a complex task, which depends upon the species' choice, their chemical composition, available technologies, human resources, and economic and environmental impacts. Exergonomics, introduced by Yantovsky for the analysis of energy systems, links invested and operational exergy expenditures and allows one to find optimal exergy efficiency of production systems (Yantovski, 2000; Yantovski, 1989). Yantovsky also suggested that "for more reliable decision-making, the simultaneous optimization of three target functions: exergy, money, and pollution, is needed" (Yantovski, 2000). However, the designer should reduce not only the pollution from the system but must also consider the multiple complex effects that an energy system — especially a large-scale renewable energy system — has on the ecosystem services of the surrounding environment. In the Section 4.3, we will examine the use of environmental ergonomics, a recently developed tool for energy and production system analysis (Golberg, 2015), for the design and analysis of offshore marine biorefineries. Environmental exergenomics includes mechanical (technological efficiency of system), capital efficiency of the system, and environmental efficiency, measured by the eco-exergy (exergy contribution of ecosystems to the biorefinery).

4.3 Environmental Exergonomics for Marine Biorefineries

Let's consider the marine biorefinery system on the production scale. The system converts solar and mechanical energies into concentrated energy

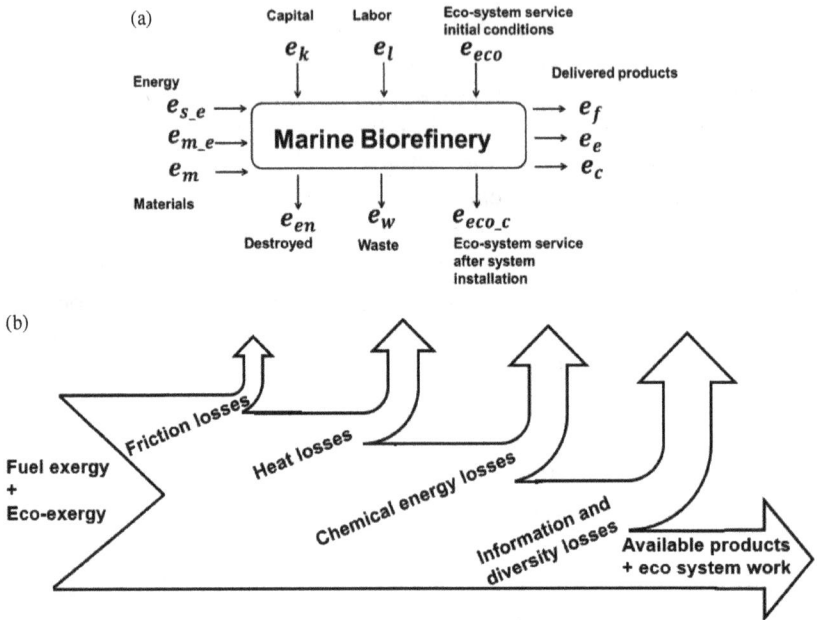

Fig. 4.1   Exergy flow in marine biorefineries. (a) Exergy currents. (b) Environmental exergonomics exergy diagram for marine biorefineries.

products, such as food, energy, and chemicals. If the system performance is measured in the units of exergy, as in the most general case (Fig. 4.1(a)), the inputs to the process are represented by an exergy stream of solar energy supply (e_{s_e}), mechanical energy supply (e_{m_e}), materials (e_m), capital inflow (e_k), human labor (e_l), and information, represented by eco-exergy (e_{eco}). The outputs are the delivered exergy contained in food (e_f), useful energy, such as biofuels (e_e), and platform chemicals (e_c), exergy rejection to the environment (e_{en}), materials waste (e_w), and eco-exergy, information, loss (or gain) ($e_{eco\text{-}c}$). As both physical and information exergies are conserved in these systems, the system will experience continuous physical and information exergy losses. The Sankey exergy diagram for the exergy flow in marine biorefineries, which includes the eco-system's eco-exergy losses, is shown in Fig. 4.1(b).

4.3.1 *Marine biorefinery system boundary*

The biorefinery system includes the physical, financial, and environmental components. The physical boundaries for the analyzed energy systems are the boundaries of the physical territory (allocated ocean area and required onshore facilities) where the system is constructed and operated. The capital boundaries include the capital invested in system construction, maintenance, and deconstruction. The environmental boundaries include the ecosystems that are located in the area that is occupied by the cultivation site and processing facility or that are affected by the facility construction. The system also includes produced products, wastes, jobs, and local ecosystems. The exergy time history of energy unit — including construction exergy current ($\dot{e}_c$), operation exergy current ($\dot{e}_{con}$), and deconstruction exergy current ($\dot{e}_d$), as well as the exergy currents changes in the surrounding ecosystem ($\dot{e}_{eco}$) — are summarized in Fig. 4.2.

4.3.2 Marine biorefinery system efficiency

Under the assumption that each of the input and output factors can indeed be described using physical and informational exergy functions, the efficiency of a system using the environmental exergonomics method is described in Eq. (4.9):

$$\eta_{env} = \frac{1}{\frac{1}{\eta} + \frac{1}{K} + \frac{1}{E}} \tag{4.9}$$

where η_{env} is the total sustainable energy system efficiency, or the main criterion for environmental exergonomics;

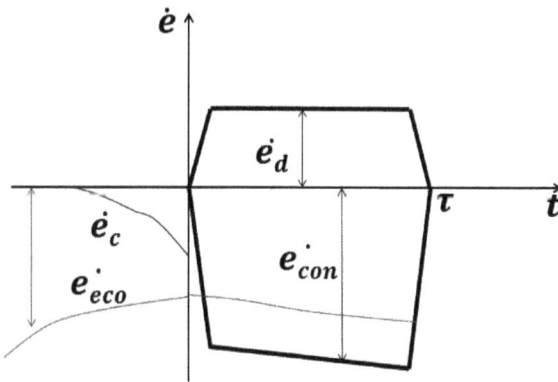

Fig. 4.2 The history of a biorefinery system, including its effects on the environment.
Source: Adapted with permission from Golberg (2015).

η is the technological/mechanical system efficiency, based on the operational exergy flow (Yantovski, 2000):

$$\eta = \frac{\delta_0}{\delta_i} \tag{4.10}$$

where δ_i is the inlet exergy current and δ_0 is the output exergy current.

K is the net exergy financial coefficient, the ratio of delivered exergy to invested exergy (money) (Yantovski, 2000):

$$K = \frac{\frac{de_d}{dt} \cdot \tau}{e_c + e_l} \tag{4.11}$$

where e_d is the delivered exergy, e_c is the invested exergy needed for system construction, e_l is the invested labor, and τ is system operation time.

E is the ecological or ecosystem efficiency of the energy conversion system based on the eco-exergy flow. We will use a ratio of eco-exergy before and after energy system construction and use:

$$E = \frac{\frac{de_d}{dt} \cdot \tau}{e_i} \tag{4.12}$$

where e_i is the consumed eco-exergy, which is described by the reduced ability of the ecosystem to perform work:

$$e_i = e_{eco\_0} - e_{eco\_\tau} \tag{4.13}$$

The Environmental exergonomics main criterion function is defined as

$$Z_{env} = \frac{1}{\eta} + \frac{1}{K} + \frac{1}{E} \tag{4.14}$$

Assuming that K and E are independent, for the arbitrary functions $K(\eta)$ and $E(\eta)$:

$$\frac{dZ_{env}}{d\eta} = -\frac{1}{\eta^2} + \frac{-\frac{dK}{d\eta}}{K^2} + \frac{-\frac{dE}{d\eta}}{E^2} \tag{4.15}$$

For $\frac{dZ_{env}}{d\eta} = 0$:

$$\eta^2 = \frac{K^2 E^2}{-E^2 \frac{dK}{d\eta} - K^2 \frac{dE}{d\eta}} \qquad (4.16)$$

and thus

$$Z_{min} = \frac{\left(-E^2 \frac{dK}{d\eta} - K^2 \frac{dE}{d\eta}\right)^{1/2} + K + E}{KE} \qquad (4.17)$$

and

$$\eta_{opt} = \frac{KE}{\left(-E^2 \frac{dK}{d\eta} - K^2 \frac{dE}{d\eta}\right)^{1/2}} \qquad (4.18)$$

where η_{opt} is the optimum efficiency of the system. The correlations between K and η, and between E and η, are study specific. If a correlation between these efficiencies is found, as suggested in the early works of Szargut (Szargut, 1971), further functional analysis using a single parameter, for example, η or monetary cost is possible (Yantovski, 2000).

4.3.3 Calculation of exergy currents

Physical exergy is defined as the maximum amount of reversible work that can be produced by getting the temperature, pressure, velocity, position within a gravitational field, and chemical composition to equilibrium with the defined reference state. Equation (4.19) describes the physical exergy of the system in the most general form (Simpson and Edwards, 2011):

$$\delta = \left[h - h_0 - T_0 \left(s - s_0\right)\right] + \frac{\left(V - V_2\right)^2}{2} + g\left(z - z_0\right) + \sum_i \left(\mu_i c_i - \mu_0 c_0\right) \quad (4.19)$$

The first term of the equation includes the classical thermodynamic properties — enthalpy (h), temperature (T), entropy (s) — known for many substances and mixtures in a wide range of states. The second and third terms are a result of measured position (z) and velocity (V) relative to the reference state, and their exergy and energy contents have the same numerical value as proposed in Hermann (2006). The fourth term is the chemical exergy of basic system elements (μ), and is the chemical potential. For all properties, subscript "0" stays for the value of the property at standard conditions.

The capital exergy currents can be divided into monetary and labor currents. This subdivision and separation of the labor current from the monetary investment proposed by Sciubba (2012, 2003) emphasize the important impact of energy systems on workers and society. The detailed analyses of capital exergy currents can be found in Sciubba (2012, 2003).

For simplicity, in this work, the capital exergy current is defined as the exergy required to build the unit and the exergy equivalent of working hours invested by the system staff during the system's lifetime:

$$e_{c+l} = e_c + e_l \tag{4.20}$$

where e_c is the exergy required to build the unit, and

$$e_l = K_{\text{labor}} \cdot n_{\text{workers}} \cdot WH \tag{4.21}$$

where K_{labor} is the exergenic equivalent of labor (Dai *et al.*, 2014), and WH is the annual work hours.

4.3.4 *The emergence of Eco-exergy currents*

The term eco-exergy was developed in ecology (Jørgensen, 2007). The concept of eco-exergy was first applied to ecology in 1970s (Jørgensen, 1990; Jørgensen and Mejer, 1977) and the last four decades led to the formulation of the "maximum exergy principle in ecology", which described the formation of biodiverse communities in terms of thermodynamics (Xu *et al.*, 2002) Eco-exergy has been used in ecology to express emergent properties of ecosystems arising from self-organization processes in the evolution of their development (Zhang *et al.*, 2010). Exergy has also been used as an objective function in ecological models to assess the changes and concentrations of various species in the ecosystem under stress (Xu *et al.*, 2002).

Eco-exergy is a measure of the maximum amount of work that an ecosystem can perform when it is brought into thermodynamic equilibrium with its environment (Jørgensen, 2007) Eq. (4.22):

$$e_{eco} = RT \sum_{i=0}^{n} \left[C_i lnln \left(\frac{C_i}{C_i^{eq}} \right) + \left(C_i - C_i^{eq} \right) \right] \qquad (4.22)$$

where R is the gas constant, C_i is the concentration of species i in the system, and C_i^{eq} is the concentration of species i in the reference environment. The term eco-exergy evolved from the use of entropy in the information theory, where entropy is the average amount of information contained in each message received (Harte and Newman, 2014). In application to ecosystems, the information is coded in the genetic load of the organisms. In early works, Jørgensen *et al.* (2007) proposed the following equation for the calculation of an ecosystem's exergy (Xu *et al.*, 2002):

$$e_{eco} = RT \left(\mu_1 - \mu_1^{eq} \right) \sum_{i=1}^{N} RT \sum_{i=2}^{N} \left(C_i ln \left(P_{i,a} \right) \right) \qquad (4.23)$$

where $\mu_1 - \mu_0$ is calculated from standard chemical potentials of the organic matter, C_i is the concentration of the species in the environment, and $P_{i,a}$ is the probability of producing the component i at thermodynamic equilibrium. $P_{i,a}$ can be found from the number of permutations among the characteristic amino acid sequence for the considered species. Since living organisms use 20 different amino acids and each gene determines on average 700 amino acids, P can be evaluated using Eq. (4.24) (Xu *et al.*, 2002):

$$P_{i,a} = 20^{-700G} \qquad (4.24)$$

where G is the number of genes (for the standard table).

Eco-exergy has been used as an ecological indicator to assess the ecological condition and ecosystem health (Dalsgaard *et al.*, 1995). The most recent definition of eco-exergy is (Jørgensen, 2015)

$$e_{eco} = f \sum_{i=1}^{n} \left(B_i \beta_i \right) \qquad (4.25)$$

where f is the work-energy per unit of biomass (Jørgensen, 2015), which on average is 18.7 kJ gr^{-1}. B_i is the biomass weight of the species, i (gr), and β_i is the weighting factor available in tables in Appendix A (Jørgensen *et al.*, 2005). β_i is equal to RTK, where R is the gas constant, T is absolute temperature, and K is Kullback's measure of information based on information embedded in the genes of the species (Svirezhev and Steinborn, 2001). Kullback's measure of information defines the incremental changes in the system information as a result of the transition from a "reference state (i_o)" to a current one (i) as follows (Svirezhev and Steinborn, 2001):

$$K = \sum_{i=1}^{n} p_i ln \frac{p_i}{p_{i0}} \tag{4.26}$$

The impacts assessment of the large-scale biorefinery systems on the local and global ecosystems are rarely found in the literature. The installation of a large-scale biorefinery requires the deployment of large territories of the sea. This, in turn, may cause change to local biodiversity and may affect even larger ecosystem services (Hernandez *et al.*, 2014). These novel uses of the sea affect the habitat, food, and water availability, and preying strategy in animal species. It can also lead to the introduction of invasive species that decrease natural biomass biodiversity (Crowl *et al.*, 2008). The above-mentioned examples of ecological changes in areas with biorefinery installations can affect biodiversity and thus the exergy of the ecosystem. The change in the eco-exergy in the area in which the biorefinery system is installed can be calculated using Eq. (4.27):

$$e_i = f \sum_{i=1}^{n} \left(B_i \beta_i\right)_o - f \sum_{i=1}^{n} \left(B_i \beta_i\right)_\tau \tag{4.27}$$

where the first term (subscript "0") stays for the eco-exergy of the ecological system before biorefinery construction, and the second term (subscript "τ") stays for the eco-exergy of the ecological system after the biorefinery deconstruction.

4.4 Determination of Optimum Scale and Serviced Area for Marine Biorefineries

Previous studies on the cost function of agricultural processing systems (French, 1960) and onshore macroalgae for biofuel biorefinery energy efficiency analysis (Golberg *et al.*, 2014) show that feedstock transportation costs limit the size of the biorefinery. Transportation costs limit the maximum possible distance of the cultivation site to the processing facility. However, different from the near-shore facilities, where the costs on biomass transportation from the sea to the onshore processing facility are known (~30% of the macroalgae cost (Valderrama *et al.*, 2013)), the real monetary transportation costs from the open ocean offshore biorefineries can be only estimated. A more realistic approach is to estimate the energy expenses required for transportation that will limit the distance of an offshore cultivation site from the processing facility.

The maximum economic distance from the processing facility of an offshore cultivation area is calculated using Eq. (4.28):

$$D_t = \frac{\varepsilon \sum_{p=1}^{n} E_p}{2E_t} \qquad (4.28)$$

Where D_t (km) is the maximum economic transportation distance, and ε is the ratio of the energy embedded in the final products that can be used for transportation of the feedstock to keep the process economically viable. Here we assume that the transportation vessel travels only one direction with cargo and is empty on its way back.

To exemplify the estimation of the transportation energy constrains of off-shore cultivation, following (Lenstra *et al.*, 2011), we assumed that the transportation will be done with an Aframax ship tanker. The tanker capacity is 100,000 tons and the average fuel consumption (between full and empty cargo) is 25.4 gal km^{-1} (4019.55 MJ km^{-1}) of heavy ship oil. Previous extensive studies in the bioethanol industry showed that for profitability and positive net energy balance, the energetic cost of transportation should be not more than ~1.8% of the total energy embedded in the final products, distributed equality between biomass transportation and final products distribution (USDA, 2008). Therefore, we constrained the total energy expenditures (ε) on transportation to 0.9% of the energy embedded in the potential products of the transported macroalgae biomass. The recent work on offshore cultivation analysis of *Ulva* sp. showed the D_t of the farm varies from 115–690 km, depending on the moisture content of the macroalgae during transportation.

4.5 Practical Implications and Processes Selection

The main challenge in the design of the macroalgae biorefinery is the process selection. Today, they are numerous different processes that can be applied to biomass, including macroalgae, to extract and create additional value from it. Making an exhaustive list of them will not help understanding how they can be integrated so we will spare the reader of such exercise. Instead, we will try to answer two questions: how to choose processes steps for seaweed biorefinery and what should be considered when integrating them?

A first step in the process design choosing what product to produce and which processes to use. These two decisions are coupled, as products are chosen based on both their cost of production, derived from process efficiency and economic value, which determines how expensive can the production process be. We propose three different approaches for product and process selection, based on answering three different questions: What do you want? What is there? What do you get?

The first approach, "what-is-there?", is to fractionate the seaweed biomass into its constituents, as pure as possible. This approach follows a basic logic: each constituent has different properties that can be used to separate them from one another (for example, by different solubility in a solvent) and then different applications can be found for each of them. The main constituents belong to different biomolecule groups: protein, ash, lipid, and polysaccharides in seaweed. However, in many cases, specific molecules among the different groups are the targets due to their specific properties compared to the other molecules of the same type. For example, we can mention enzymes or pigments for protein; starch, cellulose and water-soluble polysaccharides for polysaccharides; and other specific bioactive molecules (phenolic compounds, plant biostimulants (i.e. auxin), etc.). Enzymatic processes could fit in this category as their substrate is usually unique: amylase target starch, protease protein, and so on. However, they do not help to extract their substrates, but rather removing them by hydrolysis. Processes extracting specific constituents are notably chosen by many researchers to analyze the isolated biomass constituents. However, technologies used in this approach are sometimes not scalable, and the extensive purification steps to obtain high purity extracts are often done for sole analytical purposes. For example, many researchers have reported on the difficulty to produce protein concentrate and isolate from seaweed (Tamayo Tenorio *et al.*, 2018). Note that it is not possible to perfectly fractionate most seaweed into their pure constituents.

The "What-do-you-want?" approach is based on which product you are interested in. This approach is similar to the first one, however, in some cases, the product is not a pure compound, but a mixture of several of them that together bring the required properties for a chosen usage. Examples of such situations encompass raw extracts serving as bio-stimulants, animal feed and food products that contain various amounts of nutrients (protein, carbohydrates, fibers, lipid, microelement), feedstock for bioenergy productions, biomaterials, etc. In other cases, the target product can be directly synthesized from the biomass. For example, the direct trans-esterification of microalgae fatty acids to biodiesel or simul-taneous saccharification and fermentation for bioethanol productions are "product-focused" processes that do not require the extraction of specific constituents. Nevertheless, designing a process to obtain a certain product can prove to be difficult, notably because evolution did not make seaweed to meet the needs of Human society, thus it usually requires an extensive transformation of the mixture of biomass constituents before it can be called a "product".

Another approach is "what-do-you-get?". Under this approach, tech-nological limitation becomes the central backbone of the process choice. The idea is that with current technological progress, only certain extracts or products can be obtained and thus used, and only certain technologies are available at the industrial scale. The choice of process is thus based on the advantages of a certain technology over others and what is obtained afterward is a consequence of this choice. This takes origin in the idea that if a certain process is simple, cheap, scalable and/or efficient, then the only barrier of its commercial use is to find a relevant application for its product. For example, several cell disruption processes, which are a cru-cial upstream treatment, are currently available. The choice of the cell disruption method will, therefore, be based on the cost of the process, its efficiency, its simplicity and scalability, its sustainability, and/or its effect on biomass constituents. The result of the treatment is not necessarily a product by itself. Besides, some emerging technologies (supercritical car-bon dioxide extraction, ultrasonic treatment, microwave treatment, pulsed electric field treatment, etc.) with numerous benefits over "classical" processes, have been tested on seaweed biomass because of different technical advantages. The obtained effect or product is studied only on the second step, assessing how this process can be implemented in a biorefinery. This is a common approach in research but can have relevance to industrial uses too, notably as new technologies can offer a technological advantage over competing technologies, and in some cases may be innovative and thus "patentable". In general, the "what-you-get?" approach is also practical: since implementing the process is relatively easy, the main concern becomes finding an application for the obtained product, rather than developing a process for the desired product.

Finally, it is possible to use two of the three approaches to choose the different stages of a biorefinery design, alternating between extracting specific components, generating target products or implementing effective processes.

Concerning the integration of the processes, as we mentioned before, "specific and non-destructive processes should be located upstream of the biorefinery process chain while the more severe and destructive ones need to be kept as downstream as possible". By destructive and severe, we are referring to processes that permanently affect the chemical structure, nature, and properties of the vast majority of biomolecules in the biomass. For example, hydrothermal liquefaction degrades protein, lipid, polysaccharide and most of the organic molecules in the initial biomass feedstock. Thus, it is crucial to extract valuable molecules before any process that could lead to their degradation or a loss of activity due to the physicochemical properties applied (pH, temperature, etc.). This can be done by enzymatic and microbial activities, for example. The choice of the order of extraction is critical. One needs to ask himself what to extract first? What to extract next? Moreover, the residue after each step might carry unwanted characteristics from the previous step such as residual solvent, high or low pH or high temperature, thus adding extra steps to adjust the process parameter (neutralizing the pH, washing, cooling, etc.). This has to be taken into account when designing the biorefinery process. For example, a molecule that would be easily extracted using simple water extraction can be extracted first, since it would have minor effects on the residual biomass. Another example is using solvent recovery steps as a pre-heating stage for the following treatment that requires heating. At the same time, upstream treatment such as drying and storage can affect all molecules, including polysaccharides present in the cell wall which are usually not as sensitive as other molecules (for example proteins, pigments, or lipids). Therefore, an ideal "smart" approach needs to be considered to first optimize the efficiency of each process, and second, reduce energy and other resource input (chemical, water, another solvent, etc.)

while keeping the whole process as simple as possible (quickest, the smallest amount of steps, cheapest technology, scalable). This smart design needs to be done on a case-by-case basis since there is no consensus on seaweed biorefinery design. Moreover, it is expected that each biorefinery will have a unique design due to unique feedstocks, unique choices of products or process technologies, or the uniqueness of the local/regional/national situation.

4.6 Examples of Co-production

A wide range of products can be made from seaweed, ranging from functional food, animal and fish feed, energy including biofuel, biomaterials for biomedical or other purposes and additional applications including water treatment, paper making, bioelectronics, and plant bio-stimulants. However, most published works focus on only one of those products, and few in comparison consider the biorefinery/integration/co-production approach. We examined a sample of 21 published works on the co-production of valuable outputs from seaweed biomass that encompasses most of the recent literature as far as we know (Table 4.1). Most published co-productions achieved the extraction of a salt-enriched aqueous fraction as fertilizer, structural water-soluble polysaccharides (Ulvan, Alginate, Agar, Carrageenan and Fucoidan), pigment and lipid hydrophobic extracts, and a residual fraction used as a soil conditioner, a protein-enriched extract (animal feed), a bioenergy feedstock (biogas, combustion fuel or biochar), or bio-sorbent. Some include the extraction of other polysaccharides (cellulose or starch), or the hydrolysis of some of the carbohydrates for the production of bioethanol, biodiesel, or platform chemicals. Interestingly, there is a gap in the introduction of certain original but promising applications of seaweed biomass into the biorefinery scheme. Indeed, there are many more works done on a single product than on co-products, thus, many potential processes and applications were not explored as biorefinery steps. We have included a similar amount of works for each type of macroalgae (Brown, Green, and Red) as well as some studies evaluating a combination of all of them together (only 3/21). The major part (76%) of the studies were experimentally based, while the remaining was knowledge-based (simulation-based on previous experimental works or hypothetic conditions). The full usage of the biomass without waste or residue was prevalent in all knowledge-based and

Table 4.1 Examples of proposed seaweed biorefineries.

Algae type	Products	Experimental/ Knowledgebase	Scalability	Full biomass utilization	Sustainability assessment	Economical assessment	Integrated process assessment	References
B	Biodiesel/Alginate	Exp.	x	x	x	✓	1	Kim et al. (2019)
B	Ethanol extract/Fucoidan/Alginate/ Sugar/Biochar	Exp.	x	✓	x	✓	0.5	Yuan and Macquarrie (2015)
B	Ethanol/Protein/Liquid Fertilizer	Kn. based	✓	✓	✓	x	0	Seghetta et al. (2016a)
B	Ethanol/Liquid fertilizer/Animal feed	Kn. based	✓	✓	✓	x	0	Seghetta et al. (2016b)
B	Alginate/Fucoidan/Bioactive liquid extract/Ethanol	Exp.	x	x	x	x	0	Kostas et al. (2017)
B	Ethanol/Alginate/Energy	Kn. based	✓	✓	x	✓	0.5	Konda et al. (n.d.)
B	Succinic acid/Fertilizer/Bioactive phenolic extracts	Exp.	x	✓	x	✓	1	Marinho et al. (2016)
B+G	Bio-sorbent/Lipids/Animal feed	Exp.	✓	✓	x	✓	0	Masri et al. (2018)
B+G+R	Nutrients/Minerals/Salts/Cellulose microfibrils/Platform chemicals/ Protein/Structural polysaccharides	Kn. based	✓	✓	✓	✓	0	Sadhukhan et al. (2019)
B+G+R	Bio-crude/Biochar/Liquid extract/ Gaseous phase	Exp.	x	✓	x	x	0	Raikova et al. (2017)
G	Aqueous extract/Ulvan/Protein/ Biogas	Exp.	x	✓	x	x	0.5	Mhatre et al. (2019)

Type	Products	Study basis	Scalability	Full Biomass Utilization	Sustainability Assessment	Economical Assessment	Integrated Process Assessment	Reference
G	Animal feed/Biofuel ABE/1,2-propanediol	Exp.	x	v	x	x	0	Bikker et al. (2016c)
G	Aqueous extract/Lipids/Ulvan/Cellulose/Ethanol	Exp.	x	x	x	x	0.5	Trivedi et al. (2016)
G	Aqueous extract/Starch/Lipids/Ulvan/Proteins/Cellulose	Exp.	x	v	x	v	0	Prabhu et al. (2020b)
G	Aqueous Extract/Lipids/Ulvan/Proteins/Cellulose	Exp.	x	v	x	x	0	Gajaria et al. (2017)
R	Liquid Fertilizer/Carrageenan/Ethanol/Biogas	Kn. based	v	v	v	x	0	Ingle et al. (2017)
R	Liquid Fertilizer/Pigments/Lipids/Agar/Cellulose/Ethanol	Exp.	v	x	v	x	0	Baghel et al. (2015)
R	Ethanol/Agar/Solid fertilizer	Exp.	x	v	x	x	0	Kumar et al. (2013)
R	Aqueous extract/Lipids/Carrageenan/Residue	Exp.	x	x	x	x	0.5	Peñuela et al. (2018)
R	Pigments/Liquid fertilizer/Lipids/Agar/Ethanol/Soil conditioner	Exp.	x	v	x	v	0.5	Baghel et al. (2016)
R	Levulinic acid/Biochar	Exp.	x	v	x	v	0	Cao et al. (2019)

Note: B = Brown, G = Green, R = Red. Scalability: "was industrial-scale considered in the study?". Full Biomass Utilization: "was there leftover, residue, or waste stream that were not considered?". Sustainability Assessment: "where the environmental, energy or element cycle (such as carbon cycle) considered?". Economical Assessment: "Where the cost of the biorefinery, the price of the products, any other economic study was undertaken?". Integrated Process Assessment: "were the co-production compared to single-product (0.5) and/or the different co-productions combinations and orders (0.5 or 1) considered or not (0)?". "v" = Yes; "x" = No.

most (69%) of experimentally based works. Less than half (44%) of the experimentally based articles included an economic assessment, which was more prevalent than scalability (12.5%) or sustainability assessment (12.5%). On the contrary, knowledge-based works always considered scalable processes and included a sustainability assessment in 60% of the studies outstripping economic consideration (40%).

An example of the biorefinery process design for co-production of starch, salt, lipids, ulvan, protein, and cellulose from a green seaweed *Ulva ohnoi* is shown in Fig. 4.3 (Prabhu *et al.*, 2020b). The following

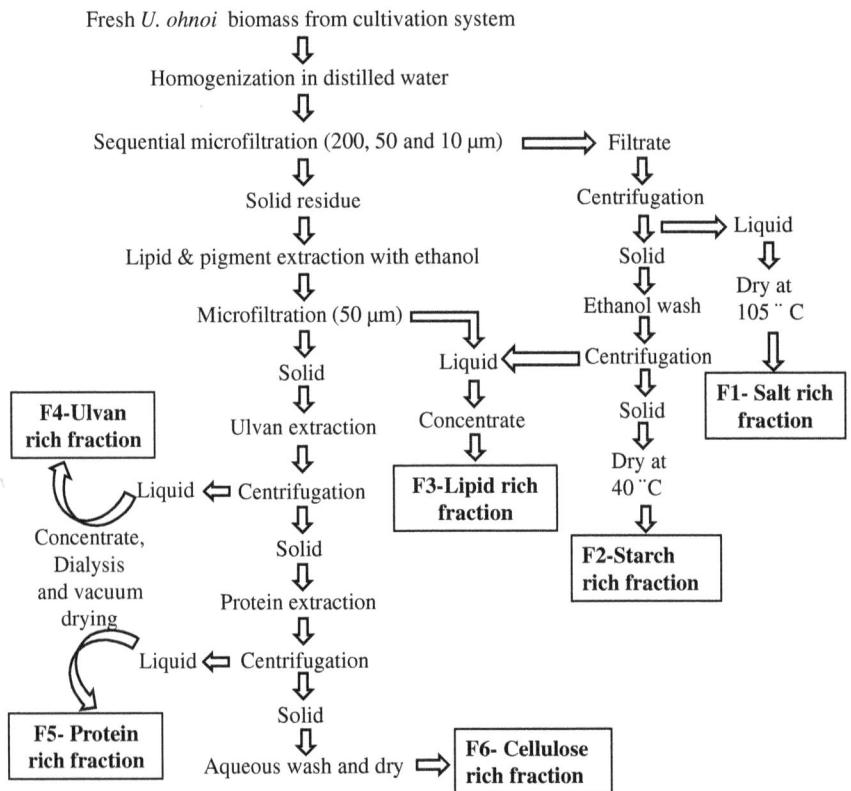

Fig. 4.3　The scheme of *Ulva ohnoi* biorefinery process. The integrated production of a wide range of valuable products, fractions F1-F6. F1-Salt fraction, F2-starch fraction, F3-lipid fraction, F4-ulvan fraction, F5-protein fraction and F6-cellulose fraction.

Source: Adapted with permission from Prabhu *et al.* (2020b).

steps show the example of the co-extraction process applied to this specific algae species.

Step 1: Starch and mineral salt extraction: Starch was extracted as reported in the earlier study (Prabhu *et al.*, 2019a). Shortly, 200 g FM *Ulva ohnoi* was mixed with approximately 2800 ml distilled water and homogenized to obtain slurry with the help of homogenizer (HG-300, Hasigtai Machinery Industry Co., Ltd., Taiwan). The slurry was sequentially filtered through nylon filters having pore size of 200, 50, and 10 μm, to obtain the filtrate. The filtrate was centrifuged at 5000 rpm for 10 min. The supernatant was dried at 105°C to recover the salt fraction. The lipids and pigments in the starch pellet were removed by washing three times with excess absolute ethanol (total 600 ml). The off-white pellet left behind was dried at 40°C until the constant mass was recorded.

Step 2: Lipid extraction: Lipid extraction was carried using the absolute ethanol method (Glasson *et al.*, 2017). Solid *Ulva ohnoi* biomass after starch extraction (residue left in the 200, 50, and 10 μm nylon filter) in step 1, was suspended in absolute ethanol (500 ml) at room temperature and mixed using a magnetic stirrer. After 3 h of mixing, the mixture was filtered through 50 μm pore size nylon filters using a glass vacuum filtration unit. The above extraction procedure using ethanol was repeated thrice. The ethanol fraction was combined, concentrated and dried using a rotary evaporator. The ethanol fraction recovered in the starch purification step was added to this fraction before it was concentrated. The fraction was finally dried at 40°C until a constant mass was reached and the final mass was recorded.

Step 3: Ulvan extraction: Ulvan was extracted using the oxalate salt method as described by (Robic *et al.*, 2008). Solid biomass left after filtration in step 2 was dried at 23°C for 5 h and then suspended in 1 L $(NH_4)_2C_2O_4$ (0.05 M). The mixture was incubated at 80°C for 2 h with gentle mixing every 15 min. After the incubation the mixture, while it was still warm (~45°C), was subjected to centrifugation for 10 min, at 1811 g. The supernatant was concentrated down to about 300 ml and then dialyzed using 8 kDa MWCO dialysis membrane for 24 h, with 3 changes of distilled water. The dialysate was finally freeze-dried and weighed to estimate the mass of the ulvan fraction.

Step 4: Protein extraction: Solid biomass residue remaining after centrifugation in step 3 was used for protein extraction using alkaline treatment method, with 1 L of a 0.25 M NaOH solution at 80°C for 2 h (Mhatre *et al.*, 2018). Then the mixture was cooled at room temperature and then subjected to centrifugation for 10 min, at 1811 g. The supernatant was collected and neutralized using 6 N HCl. The neutralized liquid was concentrated to 300 ml, dialyzed, lyophilized and weight was measured to record the mass of protein fraction.

Step 5: Cellulose extraction: The residue left at the end of protein extraction (step 4) was washed with excess deionized water to attain the neutral pH, followed by filtration through a nylon filter of 200 μm pore size. The solid residue was dried at 40°C and the mass of the cellulose fraction was recorded.

The example of recovered six fractions appears in Fig. 4.4. Using this protocol, 90.31 ± 1.94% of the initial biomass was recovered in separated products. The fraction of the recovered products from initial dry weight biomass was 45.42 ± 1.91% salts, 3.67 ± 1.38% starch, 3.81 ± 1.26% lipids, 13.88 ± 0.40% ulvan, 14.83 ± 1.06% proteins, and 8.70 ± 1.87% cellulose (Prabhu *et al.*, 2020b).

However, most of the studies did not completely evaluate the integration of co-processing by comparing it to single-product cases, different combinations among the selected products, or the order in which the

Fig. 4.4   Image of the end products (dry) extracted in an integrated biorefinery. F1-salt fraction, F2-starch fraction, F3-lipid fraction, F4-ulvan fraction, F5-protein fraction and F6-cellulose fraction.

Source: Adapted with permission from Prabhu *et al.* (2020b).

products are recovered from the seaweed biomass. Indeed, only two of those works included a thorough study of the order in which the processes should be integrated, while the big part followed a knowledge-based approach based on destructive/non-destructive and specific/nonspecific processes, as described in the Section 4.5.

These observations suggest that there are considerable gaps in the scientific literature on seaweed biorefinery and researchers are still working on the proof of concepts at a laboratory scale. The gap between laboratory and pilot or commercial scale could explain why most of the studies do not investigate sustainability or economic aspects. Indeed, when simulations were used, they all considered a fully industrial process, which is a condition for relevant environmental or economic assessment. Thus, in addition to a proper process integration assessment, the systematic use of scalable technology (even at a "high" laboratory scale (1 kg or 1 L) or pilot scale) would be a major milestone for the development of knowledge required for economic and sustainable analysis. Those analyses are key for commercial trials or policymaking, and thus the long-term development of seaweed biorefineries. Aiding scientists to achieve such milestones, for example by easing the access and affordability of pilot-scale testing, would fasten the development of seaweed biorefinery.

The common outputs from each co-production design of Table 4.1 were split into the three categories defined in the previous see Section 4.5: "what is there?", "what do you want?" and "what do you get?". Their process order, economical value, and the destructiveness were ranked according to their position in the cascading biorefinery (from upstream to downstream, see Table 4.1 for Sadhukhan *et al.* (2019), and general knowledge of the authors on process severity, respectively (Table 4.2).

The choice of the order (Table 4.2) was quite similar to the one proposed above (see Fig 4.3). This table does not take into consideration the post-harvest processing (storage, drying, etc.), the pre-treatment (milling, crushing, etc.), and the purification steps of solid/liquid separation (filtration, centrifugation, etc.) and product isolation, as most of the works in Table 4.1 consider non-scalable processes. To obtain further information on those key processing steps, refer to the relevant chapter such as in Chapter 7. Interestingly, most of the processes are fitting in the category "what you get?", as authors seemed to preferably investigate the outcome of certain processes on seaweed biomass. The "what is there?" approach was followed for the obtention of rare materials found in seaweed

Table 4.2   Analysis of process outputs from co-production examples.

	Process outputs	Process order	Economic value	Destructiveness
What is there?	Seaweed water-soluble polysaccharides (alginate, ulvan, fucoidan, agar, carrageenan)	2	H	•
	Other polysaccharides (cellulose, starch)	1/4	M	•/••
	Pigments (phycobiliproteins)	0	H	—
What do you want?	Biofuel (ethanol, biodiesel)	4	L	••
	Platform chemicals (levulinic acid, succinic acid, ABE, 1, 2-propanediol)	4	M	••
	Biogas	5	L	•••
What do you get?	Bio-crude	5	L	•••
	Biochar	5	L	•••
	Solid fertilizer	5	L	—
	Liquid fertilizer	0	L	—
	Animal feed	5	L	—
	Protein extract	3	M	•
	Organic solvent extracts	1	L-M	—

Note: "ABE" = Acetone, Butanol, Ethanol. Process orders: from upstream ("0") to downstream ("5"). "Economical value:" = low, "M" = medium, "H" = high. Destructiveness (for the obtention of the related outputs): "•" = low, "••" = medium, "•••" = high, "—" = none or as-coproducts of another process.

(unique polysaccharides, rare red pigments) while the product-focused approach "what do you want?" was used mainly for energy outputs such as biofuel. The outputs that can be rank in "what you get?" can be considered in two categories. Firstly, extraction with a specific solvent such as alkali or organic solvents leads to the extraction of a wide range of molecules, mainly protein, and lipid and pigments, respectively. Pigments are a significant fraction of biomolecules in seaweed that can impact the visual aspect of all fractions and are thus removed upstream, where the extraction is the most efficient. Secondly, downstream processes have a residue or waste stream that is a mixture of "left-over" of non-extractable molecules such as cell wall components, and those are thus obtained per default. Thus, little efforts were made to find applications for those left-overs: such as animal feed, soil conditioner, combustion fuel, or HTC feedstock, all of these being of low value. Generally, all publications include the production of necessary outputs such as energy carrier (biofuel, biochar, etc.), food, animal feed, and fertilizer, all of them being of low values, underlining the will of the authors to make sustainability of basic human need a priority of seaweed biorefinery.

4.7 Additional Considerations and Future Perspectives

Current hurdles to seaweed biorefineries such as seasonal feedstock availability (Milledge and Harvey, 2016a), un-sustainability of mono-culture (Borines *et al.*, 2011), competition with other biomass feedstock, large energy requirements for processing and high capital cost for equipment, could be overcome by several means. For example, by integrating different feedstock from marine (fishery waste, etc.) (Kerton *et al.*, 2013) or land-based (agricultural waste, etc.) sources to be processed in the same biorefinery, together with seaweed. Some additional considerations to seaweed biorefinery and future perspectives are presented in this section.

To improve the sustainability of biorefinery in general, certain authors suggested using some outputs from seaweed processing as inputs required for the same biorefinery system. This is particularly relevant for energy (Ingle *et al.*, 2017; Konda *et al.*, 2015) and certain seaweed cultivation input (CO_2, nutrients). Usually, the energy obtained from combustion, biogas or other energy products would not suffice for the total energy input of the seaweed biorefinery, but it will improve its sustainability and lower the utility expenditure. Similarly, CO_2 or nutrients from various processes can be recycled in a seaweed cultivation system.

The capacity of seaweed to trap nutrients from its surrounding water is an ecosystem service that could be taken into account when assessing the sustainability and Life Cycle Assessment (LCA) of seaweed biorefin-eries (Neveux *et al.*, 2018; Seghetta *et al.*, 2016a). In an area where the excess of nutrients is released into ocean water, algal bloom (AB) occurs, a major problem in the polluted marine area around the globe, leading to ecosystem disruption or even collapse, economic loss, and health hazard.

Seaweed farming could alleviate such AB by buffering the nutrient released into natural marine habitats and removing those nutrients after harvesting. It could also serve as a carbon sink, reducing marine dissolved CO_2 and reducing the effect of ocean acidification on the marine ecosystem. In a more general aspect, seaweed can be used for bioremediation of heavily polluted marine areas (Sadhukhan *et al.*, 2019). Certain processing such as hydrothermal carbonization, liquefaction, anaerobic digestion, or combustion can process seaweed with pollutants (organic pollutants, heavy metals, etc.) that would not be suitable for other purposes such as human or animal consumption. This concept can be pushed forward by the concept of IMTA: Integrated Multi-Trophic Aquaculture, where seaweed is used as a nutrient recycler for fish effluents while serving as feed for fish or shellfish (Nardelli *et al.*, 2019; Troell *et al.*, 2009).

Interestingly, if biorefineries are usually advocating for co-production against the single-product economy, it is peculiar to notice that most of the published work on biorefineries, including seaweed biorefineries, focus on single feedstock. Feedstock availability, cost, and quality are paramount factors for the successful implementation of a biorefinery. Allowing flexibility in the feedstock choice is thus not just an advantage, it should be a necessity. This approach is process-dependent and feedstock-dependent, but has already been reported for bioenergy production, as those processes are usually feedstock-flexible (Appels *et al.*, 2011; Borines *et al.*, 2011; Cesaro and Belgiorno, 2015; Ghatak, 2011; Goh and Lee, 2010; Milledge *et al.*, 2014a; Park *et al.*, 2012; Sanchez and Cardona, 2008; Tabassum *et al.*, 2016).

Similarly, the outputs of seaweed biorefineries can be blended or further processed with the ones from another biomass processing. For example, the seaweed-based animal feed can be blended with other feeds such as oilseed meals. Similarly, cellulose or starch can find applications together with their equivalent from land plants. Ash can be used in construction materials. Recently, seaweed was hydrothermally processed together with marine plastic wastes (Raikova *et al.*, 2019). In the same idea, it is possible to refine seaweed-based bio-crude with other bio-crude or its fossil counterpart. Sugar-rich liquor obtained from seaweed can also be used with others to produce any sorts of chemicals such as biofuel, platform chemicals, or bioplastics (Baghel *et al.*, 2015; Ghosh *et al.*, 2019; Kostas *et al.*, 2017). This may open a new era of inter-biomass integrated biorefinery systems.

Finally, co-production of valuable outputs, if they are technologically feasible, are still market-dependent. Thus, producing alginate alongside ethanol from brown seaweed, for example, might not be a commercially relevant choice as the scale of the ethanol market is several orders of magnitude higher than the one of alginate (Konda *et al.*, 2015). The capacity of biorefineries to overcome this by having flexible process flows and output portfolios is crucial. In the example mentioned above, the alginate can be hydrolyzed into sugars and fermented to platform chemicals or other chemicals.

Chapter 5

Environmental Impacts of Seaweed Aquaculture

Because of the limitation of terrestrial resources (land area and freshwater), there is a globally increasing demand for marine resources to address the many challenges of the expanding human population, climate change, population aging, and food security (Pereira *et al.*, 2013; Stengel and Connan, 2015). Seaweed biomass is a palliative solution to these concerns thanks to its ability to utilize sea- or brackish water and non-fertile lands, and thanks to its high productivity. On-going developments in cultivation technologies and methods and numerous breakthroughs in the research regarding seaweed sourced functional and bioactive compounds (Neori *et al.*, 2004) promote this resource towards increasing commercialization. The demands for seaweed biomass have also been driven by the globally increasing awareness and acceptability for seaweed in the Western world.

With more than 90% of the seaweed biomass being provided to the industries via cultivation, it is a consensus that seaweed aquaculture is a safe practice for the environment. Cultivation of algae delivers considerable environmental benefits coupled with social and economic assets, contributing to the general welfare. As discussed in this chapter, there are, however, some drawbacks that need to be taken into account. The impacts of seaweed farming may not be as destructive as some other human activities, but they should still be considered when establishing new farms or managing existing farm sites. The location of the farm has a crucial effect on its environmental impact. This impact, positive or negative, is usually more significant in coastal areas which are highly influenced by human activity and thus more sensitive than offshore areas. For example, establishing seaweed farms above seagrass beds may lead to reduced productivity and shoot density, or altered meiofaunal abundance. Seaweed farming in the vicinity of coral reefs has beneficial effects of local ocean acidification mitigation (see Section 8.23), and on the other hand, may cause an overgrowth of corals. Other studies suggest changes to herbivorous fish communities in adjacent areas because seaweed farms changed the environment. Some seaweed groups, for example, the Green *Ulva* sp., tend to thrive in disturbed and polluted ecosystems. Therefore, coastal regions that suffer from excess nutrient input and consequent eutrophication and biodiversity losses may benefit from controlled seaweed farming (see Section 8.8). Otherwise, striving to minimize the negative effects of seaweed farming on local ecosystems, there is a consensus that seaweed farms should be shifted to deeper, sandy-bottom areas rather than being placed in shallow, enclosed environments. This location-focused approach might also be relevant for inland cultivation, although land-based is regarded as a much safer practice, as mass exchange with the environment is much lower.

5.1 Ecological Relevance of Seaweeds in the Marine Environment

In coastal marine ecosystems, seaweeds are considered the main primary producers and make an important contribution to the ecosystem's health, both locally and globally. In addition to their notable high productivity among phototrophic organisms (ca. an order of magnitude more productive than planktonic), seaweeds have a significant ecosystem function as nutrients and carbon sinkers (Chopin *et al.*, 2012) and play multiple ecological roles. These include providing space and substrate for marine microorganisms, supplying a nursery ground for fish, and maintaining the overall biodiversity structure (Satheesh *et al.*, 2017).

5.2 Invasive Species Through Seaweed Cultivation

Increasing discoveries of marine organism invasion hot-spots have underlined the environmental threats arising from seaweed invasion. There is a consensus regarding the major vectors of the introduction of alien species. The increasing maritime transportation traffic on a global basis (primarily ballast waters and hull fouling) and aquaculture practices have significantly contributed to the displacement of seaweed species from distant geographical sites, and are responsible for seaweed invasions. Additional invasion vectors are artificial routes that communicate otherwise disconnected marine environments, for example, the Suez Canal, in the Levant basin of the Mediterranean Sea (Israel and Einav, 2017). Here, marine organisms that have made their way from the Indo-Pacific Oceans via the Suez Canal are collectively called "Lessepsian invasion" and this term also applies for seaweeds. In the last decades, an increasing number of macroalgal species have been exposed, including cryptic species not described before. Some of these species may create intensive algal blooms in subtidal areas. Altogether, there are about 130 non-indigenous seaweed species (NISS) for the whole Mediterranean Sea, some also of Atlantic origin. Following explosive growth, large biomass drifts are becoming common onshore. Later, growth slows down, but in many cases, the invaders become a significant component in the local seaweed assemblages thereafter. At present, our very poor knowledge of the physiological tolerance range of exotic seaweeds, the attributes of their life histories, and the genetic make-up of their populations result in the inability to predict some of the biotic consequences of NISS in association with warming and other globally changing marine environments. In this context, explaining the disappearance of species is generally much harder than their appearance as invasive ones. Once settled and following evaluation for its potential domestication, NISS could safely be adopted for seaweed aquaculture in the invaded region. As a rule, seaweed farms choose species that are local to the specific site, while cosmopolitan species could be considered when environmental evaluations are carried out. This is to prevent potential ecological threats derived from using non-native cultivation practices. As indicated below, sea-based culture is far more prone to potential environmental impairment than land-based culture.

5.3 Potential Threats of Offshore Cultivation

Offshore seaweed cultivation will eventually be feasible when technological barriers will be resolved, especially for exposed areas. As discussed extensively in recent years, offshore constructions like wind farms can serve as a platform for developing seaweed cultivation grounds far from the shoreline. These platforms will in short time allow for the establishment of the new-born artificial ecosystem, with marine biodiversity of a yet unknown ecological value. In offshore regions, where seaweed is usually non-native, their environmental effects are for most cases unknown and based on assumptions and theoretical assessments. However, as offshore areas are vast and open, the potential negative impact is assumed to be relatively small. Nonetheless, one known environmental risk associated with offshore farms is biomass loss due to fierce water motion, which may cause consequent harmful seaweed blooms. Therefore, the selection of seaweed species to be cultivated is critical since few species will likely be able to remain intact and maintain stable high growth in exposed environments. Further, importing non-native species from far geographical sites could lead to unwanted ecological consequences, and thorough risk assessments should be performed before a location is chosen.

An example of the invasion of commercial seaweed species into the local ecosystem is the bioinvasion of *Kappaphycus alvarezii* on corals in the Gulf of Mannar, India. *Kappaphycus alvarezii* (Doty) Doty ex. P. Silva (*Rhodophyta, Solieriaceae*) is a Philippine-derived rhodophyta which has been intensively introduced into the coastal ecosystems in the tropics for commercial production of carrageenan. It was first introduced into the Gulf of Mannar Marine Biosphere Reserve (GoM), South India, for commercial cultivation in 2002 despite warnings from the local scientific communities. In 2008, it was reported to successfully escape the cultivation areas and establish colonies in the coral reefs in the Gulf of Mannar (Chandrasekaran *et al.*, 2008) Fig. 5.1.

Since the report of its invasion in 2008, the removal of *K. alvarezii* from the reefs has been started using manual removal. This was, however, an unsuccessful attempt that led to a negative impact on the eradication program. Regrowth of *K. alvarezii* from removal points and drifting broken fragments resulting during removal have led to the further establishment in the reef environment as reported in 2014 (Kamalakannan *et al.*, 2014). The same species is expanding in the Caribbean in the coasts of Costa Rica, 12 years after the beginning of its commercial cultivation in the area (Cabrera *et al.*, 2019).

Probably the most known examples of the negative impact of seaweed farming on the environment are the *Ulva prolifera* green tides in the Yellow Sea, China, which have been occurring since 2007 (Zhang *et al.*, 2019). From 2007 to 2017, the direct economic losses caused by the macroalgal blooms (Fig. 5.2) were as high as 1.3 billion RMB (~185M USD) (Ye *et al.*, 2011).

Fig. 5.1 (a) and (b) Invaded colonies of *Kappaphycus alvarezii* in Krusadai Island as observed from the water surface during low-tide conditions; (c) and (d) overgrowth of *K. alvarezii* on top of colonies of *Acropora* sp., as a green mate; (e) growth of *K. alvarezii* on lateral sides of colonies of *Acropora*; (f) and (g) ridges and valleys of coral colonies invaded and doomed by *K. alvarezii*. h, Complete covering of coral surface by rubber-like major axis (smothering) of *K. alvarezii*.

Source: Adapted with permission from Chandrasekaran *et al.* (2008).

Fig. 5.2 Photos showing the development process of the green tide in the Yellow Sea in 2012. (a) Initial floating algae (black arrows) in Subei Shoal on 26 April 2012; (b) Aggregated floating algae with a closer image of the individual floating patch (c) on 6 May; (d) A long band of floating algal mat in June. (e) A large-scale floating mat in the open water of the Yellow Sea in June of 2012.

Source: Adapted with permission from Wang *et al.* (2015b).

Besides eutrophication, it is thought that the rapid expansion of *Porphyra* mariculture is another important anthropogenic factor that contributes to the development of green tides providing substrata for the *Ulva* blooms. *Porphyra* mariculture in 2012 was approximately $4.1 \cdot 10^4$ ha with about 4500 m of ropes per ha being used for cultivation, which is thought to provide substrata for *U. prolifera* (Li *et al.*, 2015). When *Porphyra* is harvested from mid-April to mid-May, *U. prolifera* fronds that are attached to the rafts fall into seawater, together with the *U. prolifera* fragments removed from the ropes by local farmers (Fig. 5.3), both of which provide the most direct and initial biomass supply for the green

Fig. 5.3 Photographs of attached and disposed green macroalgae from the *Porphyra* aquaculture rafts in Subei Shoal. (a) Structure of *Porphyra* aquaculture rafts; (b) Macroalgae attached on the rafts after the *Porphyra* net was removed, a closer image of the abundant macroalgae attached on the connecting ropes (e); (c) On-site cleaning of the green macroalgal epiphytes from the connecting ropes by a metal hook attached to a tractor; (d) Macroalgae clumps leftover on the muddy flat of Subei Shoal.

Source: Adapted with permission from Wang *et al.* (2015b).

tides (Zhang *et al.*, 2019). Annually, approximately 6500 tons of *U. prolifera* fall off the *Porphyra* rafts, 62% of which float in the surface water, spread out with wind and ocean currents, and finally bloom into massive green tides (Wang *et al.*, 2015b; Zhang *et al.*, 2019).

Additional sources of concern are the potential facilitation of disease and alteration of population genetics and the local physiochemical environment (Campbell *et al.*, 2019). Other potential negative effects of offshore cultivation are pollution via use of marine vessels and infrastructure, the use of unsustainable materials, the risk of plastic and oil pollution due to spills, and wave damage, and the excessive use of fuel for ongoing operation. These can be minimized by wise planning and managing but cannot be eliminated. Finally, offshore cultivation may yield seaweed of lower productivity and lower protein content due to lower environmental nutrient concentration, and these effects on the value of biorefineries should be taken into account.

5.4 Advantages of Land-based Practices

Seaweed aquaculture in semi-controlled environments such as land-based systems is highly environmentally sound. As mentioned, there are economic constraints in tank cultivation as it is a relatively expensive practice compared to other cultivation approaches. Nevertheless, the advantages of land-based settings are numerous and are mainly reflected by sustainable growth and biomass production with desired attributes for biorefineries. Using tank cultivation, it is possible to domesticate attractive species and strains with commercial value. Further, assuming there is full control of seaweed effluents before excessive seawater is recycled back into the sea, one can consider the cultivation of foreign species. High productivities are achieved by water agitation, usually by aeration, which reduces boundary layers and enables more efficient use of nutrients (Israel *et al.*, 2006). Also, by aeration, all the algae receive equivalent illumination for optimal photosynthesis. When sunlight radiance is too strong and inhibits the growth, it can be reduced by artificially shading the cultivation tanks. By regulating seawater exchange rates and seaweed density in the cultivation tank, the negative effects of epiphytes may be prevented. Another significant advantage of pond cultivation is the ability to control mineral nutrition, resulting in the manipulation of organic and inorganic components of the seaweed biomass. At the same time, land-based cultivation may require high energy inputs for pumping water and aeration and significant labor expenses, and result in a large areal footprint in coastal regions where free land is becoming scarce and expensive.

Tank cultivation of seaweed can also be used for bioremediation of polluted or eutrophicated coastal waters, thus contributing to a nutrient reduction in regions that suffer from excess, unbalanced, nutrient accumulation (see Section 8.8). Coastal waters eutrophication and resource depletion have become serious issues in many coastal regions in the world. Seaweeds can remove nutrients that cause eutrophication (for example, the removal of about 75,000 t of nitrogen and 10,000 t of phosphate from coastal waters in China, Zheng *et al.* (2019)). Seaweed can also efficiently remove carbon in addition to CO_2 and release O_2 into coastal waters (see Section 8.12). Seaweed cultivation can save the use of chemical fertilizers, pesticides, and farmland by replacing conventional terrestrial vegetable cultivation.

A significant additional advantage of using seaweeds in land-based bioremediation is that biomass can be conveniently harvested with mesh, nets, or ropes due to their large size, thereby physically removing contaminants or excess nutrients captured in the biomass from seawater. Then, the resultant seaweed biomass can be used for bio-products, either as raw biomass or as feedstock for the extraction of bio-based chemicals. These features provide, in many cases, an economically viable bioremediation process, generating benefits from both the cleanup of wastewater effluent discharge and the valorization of biomass. Subsequently, any remaining biomass is typically used for applications in lower-value product applications. Similarly, raw biomass that is not suitable for the extraction of high-value products can be used directly for lower-value products in animal feed, bioenergy, and fertilizer industries. In all cases, however, seaweeds have the potential to sequester contaminants from polluted seawater, through land-based and coastal bioremediation, and recycle these contaminants into useful bioproducts through the harvest and processing of biomass.

5.5 Spatial Management

Seaweed species, as well as coastal water qualities, vary by region and therefore spatial management and water quality monitoring in coastal seaweed farms are necessary. This is to avoid the overloading of futuristic seaweed farms and potential environmental issues that may arise.

For example, competition between species for nutrient availability, allelopathy, and related ecological factors that may negatively interact between farms. Global planning of maritime areas dedicated to aquaculture is advisable on local and global scales for optimal productivity.

5.6 Environmental Benefits of Integrated Multi-Trophic Aquaculture (IMTA)

These systems have been under development/improvement since first implemented a few decades ago, and they are designed to use seaweeds (or aquatic angiosperms in freshwater systems) as mitigation agents for excessive nutrient loads caused by several forms of fed aquaculture, namely fish cultivation, in both fish-ponds or fish-cages in the sea. One of the main environmental problems caused by fed monocultures is the surplus discharge of organic matter and dissolved nutrients into the environment. The nutrient-rich effluents can negatively affect natural ecosystems, causing habitat modification, water quality degradation, coastal eutrophication and other adverse effects (Chopin and Sawhney, 2009). A potential solution to avoid the pronounced degradation of coastal ecosystems and their services is to implement an ecologically sound approach that minimizes the impact on the environment.

Integrated Multi-Trophic Aquaculture (IMTA) advocates the integration of fed species like finfish, with inorganic and organic extractive species like seaweeds (primary producers) and shellfish (filter feeders). The seaweeds use wastes of the fed organisms which are rich in dissolved ammonia and phosphate, to form new biomass. By doing so, the seaweeds can clean and treat the water, preventing nutrients and waste from spreading and harming the environment, while stabilizing the levels of oxygen, pH, and CO_2. This creates ecological balance and provides long-term sustainability while adding another valued marine crop. It is believed that the cultivation of seaweeds in their preferable nutrient-rich environment will establish a healthy basis to enhance desirable natural materials by the seaweed industry. Co-cultivation of valuable marine crops is a more restricted form of IMTA and may offer even enhanced economic benefits. For example, the co-cultivation of mussel and seaweed are fast-growing sectors worldwide. Both mussels and various seaweed species can be grown on similar basic longline structures in an integrated manner, offering several benefits such as better space utilization of limited permitted sites, shared use of the capital costs of expensive anchors, lines, and buoys, and better risk management via crop diversification. The additional benefits of using multiple complementary nutrient bio-extractive crops are improved ecosystem services such as improved water quality, provision of the structure resulting in nursery and foraging habitat for other species, and sustainable seafood supply. Co-cultivation of seaweed and shellfish is often mentioned as a multi-use approach to efficiently use space in offshore wind parks.

Land-based systems in many countries since the 1980s have had the most success using green algae in the genus *Ulva* and red algae in the genus *Gracilaria*. Species of these widespread genera grow well in sheltered waters, and strains that grow without sporulating under high-density conditions can be found. A benefit of *Gracilaria* compared with *Ulva* is that there is a long-established industry extracting agar from the former, whereas the world use of products from *Ulva* biomass is more limited, nonetheless, has been rapidly expanding in recent years. However, the areal production of *Ulva* is generally greater than *Gracilaria* in many systems, and *Ulva* has become the seaweed of choice in land-based integrated aquaculture around the world. *Ulva* has often been chosen in the past despite the lack of established uses for the biomass produced.

Chapter 6

The Marine Biomass Feedstock

6.1 Natural Stock Collection History

Collection of seaweed natural stocks in coastal communities has been performed for centuries, with early documentation dating back to the Neolithic period, some 14,000 years ago (Buschmann *et al.*, 2017; Dillehay *et al.*, 2008; Erlandson *et al.*, 2015). Rare archeological evidence suggests that ancient hunter-gatherers in coastal regions used seaweed as food and medicine (Dillehay *et al.*, 2008), while later, post-Neolithic revolution (the first agricultural revolution) applications include also feed, soil conditioning, and fertilization (Buschmann *et al.*, 2017; Delaney *et al.*, 2016; Monagail *et al.*, 2017).

The most prominent example of seaweed utilization for human consumption is found in East Asia (i.e. China, Japan, and Korea), where it became a luxury product and even a tax alternative (Yang *et al.*, 2017). Seaweed such as *Porphyra* is mentioned not once in the Chinese literature, including the earliest written record of their human usage, from about 1700 years ago (Buschmann *et al.*, 2017; Yang *et al.*, 2017). However, the tradition of eating natural seaweed exists also in other coastal areas, such as in Ireland, Brittany, Iceland, Maine, Nova Scotia (Mouritsen *et al.*, 2013), and the South Pacific Polynesian and Melanesian Islands (South, 1993). In Ireland, evidence shows that seaweed was gathered for food at least since the Middle ages, while fertilization and feedstock applications date back at least several centuries (Morrissey *et al.*, 2001). The brown fucoid *Ascophyllum nodosum*, which dominates the rocky coast of the North Atlantic, has been harvested for hundreds of

Fig. 6.1   Jeju, South Korea, Haenyeo female divers who dived into shallower areas of the sea, compared to men, who usually went to the sea by boat to fish. Due to a lack of men, a lot of women had to dive into the deep sea to get seafood. As sea diving became a female-dominated industry, many of the haenyeo subsequently replaced their husbands, as the primary laborer.

Source: https://www.visitjeju.net/.

years, as well as other species such as *Laminaria digitata*, *Chondrus crispus*, and *Palmaria palmata* (Monagail *et al.*, 2017). Maërl, which is composed mostly of the calcareous seaweeds *Lithothamnion coralloid* and *Phymatolithon calcareum*, has a long history of harvest along the Atlantic coast of France for soil conditioning and as a replacement for lime in agriculture (Monagail *et al.*, 2017). In some cultures, such as in Brazil, Japan, Hawaii, Portugal, South Africa, British Columbia, Alaska, South Korea, and Ireland, seasonal gathering or harvesting of seaweed, and the consequent drying, have been traditionally led by women (Monagail *et al.*, 2017), Fig. 6.1.

6.2 From Domestic to Industrial Use

Around the 17th century, with the introduction of the seaweed-based potash into the glass and soap industries, seaweed harvesting in some regions exceeded the domestic, local community level, and became more intensive (Delaney *et al.*, 2016). Potash production was later replaced by Iodine production and finally by the hydrocolloid (gelling) industry, which is active till these days (Buschmann *et al.*, 2017; Delaney *et al.*, 2016). However, even though seaweed harvesting has been intensified, natural stocks remained stable as long as harvesting techniques and cautious local management traditions had not changed (Monagail *et al.*, 2017). Traditional sustainable harvesting includes, for example, the harvesting of *Laminaria digitate* in Ireland which is done by hand during low tides, and consists of cutting only the upper three-quarters of the frond by knife, leaving the stipe and lower part of the frond intact to regenerate quickly (Morrissey *et al.*, 2001). Harvesting practices for *Ascophyllum nodosum* in Ireland have barely changed for centuries. Since the 19th century, family foreshore patches were harvested in a regulated rotational manner, marking the margins of each patch by rocks called "mearing stones" (Monagail *et al.*, 2017). Another example of a local harvesting management comes from the Kombu kelp harvesting villages of the Hidaka District in Japan. In these villages, after a severe competition among harvesters caused resource depletion, community regulations were set, and further sustainable harvesting has been instructed by the hatamochi, the person authorized to define harvesting times and periods (Iida, 1998).

Only in the 1960s, when the demand for seaweed started to rise, mechanized harvesting techniques came into use (Monagail *et al.*, 2017). Two mechanized harvesting solutions are the Norwegian "seaweed trawler" with highly efficient suction capabilities, and the French "Scoubidou trawl" that harvests *Saccharina latissima* using a crochet-hook-like implement, which rotates around the fronds and uproots them to be pulled on board, both have played a major role in the seaweed industry (Monagail *et al.*, 2017). However, the use of similar suction-based mechanical harvesters in the Canadian Maritimes resulted in uncontrolled over-harvesting and had to be stopped in 1994 (Monagail *et al.*, 2017). Another destructive harvesting mechanism is the mechanical mining of maërl beds via sea-floor dredging, which led to a decline in its natural status and abundance till it had to be included in national inventories of sites of conservation interest (Barbera *et al.*, 2003). At the same time, less destructive solutions were developed. One example is the Canadian boat and rake technique from the 1970s. The long-handle rake, with its specially designed cutting head, is deployed by the harvester from a suitable boat, cutting the floating seaweed canopy where the majority of the bio-mass is, while leaving behind some meristematic tissue to allow for regrowth of the canopy within a year or two (Monagail *et al.*, 2017; Sharp *et al.*, 2006). Another example is the harvesting of *Macrocystis pyrifera* in the Pacific Ocean water off Baja California and California with highly mechanized large vessels, capable of hauling 300–550 tons of wet weight in each load. These vessels remove the surface canopy of the seaweed (33–50% of the plant biomass) at a depth of 1.2 m, leaving the sporophylls and the frond producing meristems untouched, thus allowing regeneration (Barilotti and Zertuche-Gonzalez, 1990).

6.3 Preventing Over-Harvesting

Seaweed constitutes an important role in intertidal and subtidal ecosystem structures. Large-scale and repeated removal of marine macroalgae has a direct influence on marine biodiversity. Uncontrolled harvesting decreases the abundance of megafaunal invertebrates, fish, and apex predators, reduces contributions to the marine carbon cycle, and weakens the ecological services these habitats provide to coastal areas, such as remediating eutrophication by removal of dissolved nutrients and protection from erosion and hazardous waves (Monagail *et al.*, 2017). The increased demand for seaweed and the introduction of mechanized harvesting techniques required the implementation of suitable management programs. The risk of potential over-harvesting was generally known, and in many regions harvesting was managed and natural stocks were continuously monitored. Sustainable wild seaweed exploitation has been carried out for decades in some areas in Chile, Portugal, South Africa, Norway (Buschmann *et al.*, 2017), Ireland (Morrissey *et al.*, 2001), and other countries.

A good example is the commercial harvesting of kelp in central California, in which previous knowledge was applied, restricting the harvesting frequency to two cuttings per year since it first began (Barilotti and Zertuche-Gonzalez, 1990). In addition, aerial photographs were used between 1971 and 1979 to examine the long-term stability of the kelp beds, without revealing any stability changes (Barilotti and Zertuche-Gonzalez, 1990). Furthermore, concerns regarding the effects of the harvesting were followed by a series of studies to determine the effects of harvesting on survivorship in this locale and to determine the maximal sustainable harvesting (Barilotti and Zertuche-Gonzalez, 1990). When comparing the possible harvesting effects of *Macrocystis pyrifera* to *Eucheuma uncinatum*, the latter was found to be much more sensitive, due to the inability of the practiced harvesting techniques to keep the source of spores for the next generation untouched (Barilotti and Zertuche-Gonzalez, 1990).

At the same time, many other coastal regions suffered from poor management and consequent over-exploitation. As the hydrocolloid agar became a major seaweed product globally, agarophyte seaweeds have been widely harvested, leading to over-harvesting in many places. During the 1970s, *Gracilaria* was over-harvested in central Chile as a consequence of a high market price of agarophyte and poor economic situation (Monagail *et al.*, 2017). Similarly, unregulated agarophytes harvesting in Brazil in the 2000s led to declining populations and a prolonged decrease in productivity (Marinho-Soriano *et al.*, 2006; Monagail *et al.*, 2017). In Japan, anthropogenic activities causing water stagnation and increased sedimentation, followed by natural events such as a volcano eruption and a long rainy season, greatly diminished *Gelidium* spp. beds (Fujita *et al.*, 2006). Also in Morocco *Gelidium* spp. beds deteriorated, potentially harming the global production of microbiology-grade agar (Buschmann *et al.*, 2017). However, other species have also suffered from local over-harvesting effects. Along the European-North African Atlantic coast, over-harvesting of *Gelidiales* was experienced (Buschmann *et al.*, 2017), while on Prince Edward Island and Nova Scotia in Canada, long-term extensive rake harvesting of Irish moss (*Chondrus crispus*) beds transformed the domination to *Furcellaria lumbricalis* (Monagail *et al.*, 2017; Sharp *et al.*, 2006). *Himanthalia elongate*, which has been harvested in Europe for centuries and used as food, fertilizer, and hydrocolloid source, is currently harvested in France, Ireland, Portugal, and Spain mainly for human consumption (Stagnol *et al.*, 2016). As seaweed collection for personal consumption is not regulated or managed, *Himanthalia elongate* popularity caused reduced local abundance and even local extinction (Monagail *et al.*, 2017).

Finally, sustainable harvesting of natural stocks of seaweeds is a matter of combined management and suitable techniques. Harvesting techniques that do not harm the base of the plant and do not leave bare areas allow rapid regeneration and minimize ecological effects (Morrissey *et al.*, 2001). Also, harvesting should be managed in a way that allows sufficient regeneration time, which can range between 3 and 5 years

(Monagail *et al.*, 2017; Morrissey *et al.*, 2001), depending on the specific species and environmental conditions. Poor resource management, such as opportunistic harvesting, excessive removal of holdfast material, trampling, and enhanced grazing by herbivores (Monagail *et al.*, 2017), all place additional stresses on the resource, potentially leading to complete stripping of the seaweed bed (Monagail *et al.*, 2017).

6.4   Future Perspectives on Natural Stocks Collection

Present sustainable seaweed harvesting has become possible thanks to knowledge and experience accumulated in coastal communities, followed by scientific-based ecological models and management plans (Buschmann *et al.*, 2017; Monagail *et al.*, 2017). The success of these management plans, as demonstrated with the 540 million dollar worth kelp beds in Northern Chile, depends on the collaborative efforts of all involved stakeholders, including fishers, industry, government, and scientists (Monagail *et al.*, 2017; Vásquez *et al.*, 2014). Although current management plans are mostly local, some initiatives to develop global certification programs, or useful practice guides for seaweed harvesting, have been proposed (Monagail *et al.*, 2017). However, the scale to which an industry based solely on natural stock harvesting can grow is rather limited (Buschmann *et al.*, 2017). Therefore, increasing demands for seaweed for food products and other applications can be delivered only by seaweed agricultural cultivation (Buschmann *et al.*, 2017; Monagail *et al.*, 2017). The transformation from natural stock harvesting to controlled cultivation has started many years ago in East-Asia (Delaney *et al.*, 2016; Lindsey Zemke-White and Ohno, 1999; Yang *et al.*, 2017), but will inevitably penetrate also to the European and American seaweed industries, which still rely on natural resources (Buschmann *et al.*, 2017).

Today, 32 countries actively harvest seaweeds from wild stocks, harvesting annually more than 800,000 tons, which are only a small fraction of total global seaweed production (Monagail *et al.*, 2017). Looking forward to the future of seaweed harvesting suggests that it is likely to remain mostly a cultural-traditional habit (Monagail *et al.*, 2017; Morrissey *et al.*, 2001). Therefore, future large-scale biorefinery and industry feedstocks are not expected to be based on natural stocks' harvesting, but mostly on cultivated seaweed (Buschmann *et al.*, 2017).

6.5 Choice of the Cultivated Species

Seaweed culture is an attractive option to attain sustainable biomass production with valuable chemicals and molecules that may fit the biorefinery needs. A key decision in the biorefinery design is selecting a suitable species of seaweed. Like in other business-oriented initiatives, the farmer/innovator should identify a relevant target product(s) according to a market survey. Next, after matching the target products with a suitable biochemical composition of a specific species, the economic feasibility of its commercial cultivation should be examined. The economic feasibility depends on the balance between revenue from selling the biomass and the costs of the process, which include one-time establishment expenses (infrastructure, equipment, etc.) and on-going operation and maintenance costs (energy, chemicals, labor, rental, fees, distribution, and more).

The cost of the process also depends on the chosen species, as different species require different conditions, including different water exchange and aeration rates, different harvesting and treatment frequencies, etc. Due to the high costs of tank seaweed cultivation, most likely low valued seaweed products will not be suitable, but rather specific targeted molecules that may sell at high prices in the local and international markets.

Monoculture aquaculture requires strains that maintain high productivities, preferably year-round. When year-round cultivation of a single species is not possible, for example, due to climate limitation (cold winter, hot summer, etc.), several species can be cultivated seasonally, thus increasing the system efficiency. In some cases, high yields of the target product will require various cultivation manipulations. For example, protein content in many species can be increased by controlling the fertilization regime, light intensity, and harvesting timing. Therefore, the selected strain should naturally contain the target molecule and respond positively to the enrichment manipulations. In addition, the selected species will preferably reproduce vegetatively, by continuous growth from filaments or blades cut out of the same adult individual. In this kind of reproduction, it is possible to maintain over time the genetic value of the seaweeds. However, the lack of sexual reproduction through a genetic exchange might debilitate biomass production and increase vulnerability to diseases, predators, and extreme environmental conditions. Indeed, similar to terrestrial crops, economically important seaweeds are susceptible to disease and parasitism. Disease outbreaks may occur not only in natural populations but also during cultivation, which can result in significant economic losses. In particular, the recent intensification of seaweed cultivation with larger monocultures has led to some disastrous disease outbreaks, and one example is recently referred for *Saccharina* aquaculture in China (Wang *et al.*, 2019).

6.6 Offshore Cultivation

Looking towards sustainable large-scale biorefineries, the macroalgal feedstock cannot be based on the harvesting of wild stocks or cultivation in onshore or near-shore farms. Wild-stock harvesting leads inevitably to over-exploitation while on- or near-shore farming competes with food crops or coastal uses (Buschmann *et al.*, 2017) and is limited by decreasing available areas (Forster and Radulovich, 2015; Möller *et al.*, 2012). Two main solutions withstand the conditions above. One, envisioning to construct vast seaweed farms in coastal unfertile deserts, was presented by Guillermo Garcia-Blairsy Reina (Buschmann *et al.*, 2017). The second, with wider potential for global implementation, is the offshore cultivation, which is further discussed in what follows (Golberg *et al.*, 2020).

6.6.1 *Developing offshore concepts*

Early reports of the offshore algae cultivation concept proposed to release juvenile *Sargassum* sp. 500 miles offshore the USA–Canada border and harvest them offshore the USA–Mexico border, for the production of methane in onshore anaerobic digesters (Szetela *et al.*, 1976). In the 1960s, this proposal inspired Howard Wilcox from the San Diego Naval Undersea Center to envision and develop the first multi-product floating seaweed farm called the "Ocean Food and Energy Farm Project" (Roesijadi *et al.*, 2008). Due to the energy crisis of the 1970s, this project was stopped in favor of prioritized biofuel production programs. The "Marine Biomass Program", in which offshore kelp cultivation structures were tested off the coast of Southern California, operated during the 1970s and the early 1980s in California (Roesijadi *et al.*, 2008, 2010). This program has made significant advances in understanding the complexity of the marine biological system and in enhancing growth data but failed to overcome the difficulties of working in the open ocean, especially the stability of the cultivation systems and the attachment of the algae to the systems (Golberg *et al.*, 2020; Roesijadi *et al.*, 2008).

Following the beginning of the new millennium, with increasing awareness of the environmental effects of the industrial era (Suutari *et al.*, 2015), scientific engagement with offshore biomass cultivation, which developed during the 1960s through to the 1980s (Roesijadi *et al.*, 2008, 2010), has become significant again (Buck *et al.*, 2004; Buck and Buchholz, 2005, 2004; Hughes *et al.*, 2012; Korzen *et al.*, 2015a; Reith *et al.*, 2005; Roesijadi *et al.*, 2008, 2010b; Suutari *et al.*, 2015; van den Burg *et al.*, 2013). Although previous techno-economic assessments were not favorable of offshore algae cultivation (Golberg and Liberzon, 2015a; Roesijadi *et al.*, 2008, 2010), four decades of technological evolution, cast into the current political-environmental context, has led to a re-examination of this idea (Feinberg and Hock, 1985). This technological evolution includes experience gained through oil and gas exploration, advancements in the oceanographic and atmospheric sciences, major improvements in both tensile strength and weight of materials that can be used at sea, and improved understanding of seaweed life cycles (Roesijadi *et al.*, 2008; Santelices, 1999). One example is the development of flexible and submersible offshore aquaculture structures, such as the SUBFLEX, which is being operated offshore Israel since 2006 (Drimer, 2016). Simultaneously, the establishment of offshore wind farms (Reith *et al.*, 2005) and the inevitable distancing of aquaculture facilities from the coast (Troell *et al.*, 2009), facilitated an additional potential reduction in cultivation costs via integration of infrastructure and operations (Buck and Buchholz, 2004; Reith *et al.*, 2005; Zollmann *et al.*, 2019a).

Furthermore, the improved knowledge regarding the lifecycle of different species enables a better design of a complete cultivation cycle. Therefore, macroalgae cultivation systems may include multiple cultivation steps, combining intensive on-land tanks or ponds and extensive open-sea systems (Buschmann *et al.*, 2017; Santelices, 1999). In addition, a hatchery/nursery may be used as a preliminary stage before sea cultivation, enabling continuous cultivation with lower dependency on seasonality effects and lower susceptibility to biomass degradation, diseases, and pests (Gupta *et al.*, 2018). For example, by controlling the timing of germination of *Ulva* sp. or by preserving its sporelings in a hatchery (Gao *et al.*, 2017), cases of sudden sporulation of adult thalli, which are common for this species (Niesenbaum, 1988), may be decreased (Zollmann *et al.*, 2019a).

6.6.2 *Present near-shore cultivation*

Notwithstanding, current global macroalgae cultivation is still strongly dominated by near-shore cultivation. Furthermore, only several genera (*Saccharina*, *Undaria*, *Porphyra*, *Eucheuma*, and *Kappaphycus* and *Gracilaria*) and countries (mostly South-East Asian) produce more than 98% of the macroalgal biomass (Buschmann *et al.*, 2017). In 2014, the leading macroalgae cultivating countries were China and Indonesia with over 10 million tonnes dry weight each, followed by the Philippines and the Korean Republic with over 1 million tonnes each, North Korea, Japan, Malaysia, and Zanzibar with over 100,000 tonnes each, and Chile with 12,836 tonnes (Buschmann *et al.*, 2017; Fao, 2016). East-Asia, which is leading in macroalgae cultivation, is also where cultivation originated. Early cultivation of *Porphyra* spp., also known as *Nori*, in Japan of the 17th century, was performed by placing camellia branches or bamboo into shallow coastal water and allowing the algae to grow on it naturally (Delaney *et al.*, 2016). Only some 70 years ago, the discovery of the lifecycle of the Nori in the early 1950s by Dr. Kathleen Drew, which was a significant breakthrough, enabled the commencement of commercial cultivation (Delaney *et al.*, 2016). Consequently, cultivation has expanded to neighboring countries, and cultivated macroalgae in Japan and China surpassed wild harvest about 40 years ago (Ricardo *et al.*, 2015). Since then, a big portion of the growth accounts for the introduction of *Kappaphycus* and *Eucheuma* farming for carrageenan throughout the tropics (Ricardo *et al.*, 2015).

Understanding the lifecycle of a species appears to be an important prerequisite for its commercial cultivation (Santelices, 1999). The lifecycle of Nori, for example, includes a stage in which spores adhere to oyster shells before they are released into the water column (Delaney *et al.*, 2016). Similarly, some macroalgae species, such as kelp, are unitary and require a hatchery/nursery as a preliminary stage before sea cultivation. Other algae species, such as *Gracilaria* and *Kappaphycus*, are clonal, and can be fragmented vegetatively and propagated directly for growth in culture systems (Buschmann *et al.*, 2017; Santelices, 1999). Therefore, macroalgae cultivation systems may include one or multiple cultivation steps, combining intensive on-land tanks or ponds and extensive open-sea systems (Buschmann *et al.*, 2017; Santelices, 1999).

Open-sea cultivation systems include ropes, lines, nets, rafts, and cages, which are all popular due to inexpensive installation and maintenance (Fernand *et al.*, 2016; Ricardo *et al.*, 2015), though more suitable for near-shore use rather than for offshore conditions. Fernand *et al.* (2017) collected data reported from offshore algae cultivation systems between 1980 and 2015 (Table 6.1). A more advanced system is the offshore-ring that was developed by (Buck *et al.*, 2004; Buck and Buchholz, 2005) after testing various carrier constructions and mooring systems for *Laminaria saccharina* in the rough conditions of the North Sea. This carrier resists current velocities of 2 m per second and wave heights of 6 m and can be equipped with culture lines that can be collected offshore or transported for onshore harvesting. The modularity of this ring enables its future integration in aquaculture systems located in or attached to offshore wind farms. Olanrewaju *et al.* (2017) have designed a moored multi-body floating seaweed farm, which according to simulations and model tests, should withstand the monsoon prone conditions offshore Malaysia (Golberg *et al.*, 2020).

6.6.3 *Future offshore systems for biomass production*

Different approaches have been suggested regarding the future design of offshore cultivation systems. The commonly used extensive approach allows the algae to grow without adding nutrients or applying external mixing. The main advantage of this approach is decreased labor, technology, and energy inputs, thus improving energy balance, while the main disadvantage is decrease in biomass yields, leading to a large area demand (Buck *et al.*, 2008). Extensive cultivation can be performed on anchored platforms or free-floating enclosures (Roesijadi *et al.*, 2008; Zollmann *et al.*, 2019a).

Free-floating enclosures can be released in areas with predicted currents, or followed with tracking devices, and finally collected and harvested when time and location are suitable, and biomass weight is

Table 6.1　Biomass yields of seaweeds cultivated offshore as reported from 1980 to 2015. Collected by Fernand *et al.* (2016).

Cultivation system	Species	Location	Yield	Units	Reference
Rope, vertical	Undaria pinnatifida	North Western coastal bay of Spain	8.3	kg ww/m/139d	Peteiro and Freire (2012)
			21[a]	t ww/ha/year	Peteiro and Freire (2012)
Rope, horizontal	Undaria pinnatifida	North Western coastal bay of Spain	5.9	kg ww/m/147d	Peteiro and Freire (2012)
Rope (concentrical)	Laminaria saccharina	German North Sea	4	kg ww/m/6months	Buck and Buchholz (2004)
Ropes farm, horizontal	Laminaria japonica	Hokkaido, Japan	10^6	t ww/41.2km^2/year	YOKOYAMA and S. (2007)
Rope, horizontal, transplanted[b]	Saccharina latissima	Northern Spain, Bay of Biscay	7.8	kg ww/m/106d	Peteiro et al. (2014)
			45.6	t ww/ha/106d	Peteiro et al. (2014)
Rope, horizontal	Laminaria saccharina	British Columbia, Canada	3–8	kg ww/m/8months	Druehl et al. (1988)
Rope, horizontal	Laminaria groenlandica	British Columbia, Canada	2.6–20.5	kg ww/m/18months	Druehl et al. (1988)
Rope, horizontal	Cymathere triplicata	British Columbia, Canada	1.1–2.7	kg ww/m/7months	Druehl et al. (1988)
Rope, vertical[c]	Palmaria palmata	Northwest Scotland	1	kg ww/horizontal meter of top rope/year	Sanderson et al. (2012)
Rope, vertical[c]	Saccharina latissima	Northwest Scotland	28[d]	kg ww/horizontal meter of top rope/year	Sanderson et al. (2012)

Method	Species	Location	Value	Units	Reference
Rope, horizontal	*Alaria esculenta*[e]	Ard Bay, Carna, Co. Galway, Ireland	45.7	kg ww/m/year	Kraan and Guiry (2001)
Rope, horizontal	*Saccharina latissima*	Isle of Man, Irish Sea	2.8	kg dw/m/year	Holt (1984)
Cage[f]	*Gracilaria tikvahiae*	Indian river lagoon, Florida	9.7[g]	g dw/m²/day	Hanisak (1987)
Cage[f]	*Gracilaria tikvahiae*	Hutchinson Island, Florida	22.4	g dw/m²/day	Hanisak (1987)
Nylon line attached to stakes fixed in sea bottom	*Eucheuma spinosum* (Bohol)	Zanzibar Island, Tanzania	5.4–7%[h]	daily growth rate	Lirasan and Twide (2013)
Floating raft with rope	*Sargassum naozhouense* (Tseng et Lu)	Liusha Bay, Xuwen, Guangdong, China	1750	kg ww/km/95d	Xie et al. (2013)
Rope net with two bamboo poles	*Ulva prolifera*	Jiangsu coastline, China	198.6[i]	kg ww/ha/5months	Liu et al. (2010)
Rope net with two bamboo poles	*Ulva intestinalis*	Jiangsu coastline, China	89.2[i]	kg ww/ha/5months	Liu et al. (2010)

Notes: ww: wet weight; dw: dry weight.
[a]Estimated value.
[b]Transplants were 2.1 kg fresh wt m⁻¹ rope.
[c]Droppers 1 m apart, with one 10 cm section of seeded string for every 1 m of dropper to 7 m depth.
[d]Highest mean yield obtained for a long line.
[e]High yielding strain.
[f]2 cm plastic mesh on 2.5–5.0 cm diameter PVC pipe frames measuring 1 m × 1 m × 0.25 m deep (0.6 m² cage).
[g]Average between two stations (7.8 and 11.6 g dw/m²/day).
[h]Minimum and maximum of five test plants at different locations.
[i]Average of six stations.

satisfactory. This concept mimics the growth of the naturally free-floating *Sargassum* in the Sargasso Sea or of "green tides" in nutrient polluted water (Ricardo *et al.*, 2015). A theoretical large-scale system of this kind was suggested by Notoya (2010), who proposed to grow seaweed beds on 100 km² rafts, floating away from shipping lanes. This concept, even on much smaller scales, is very challenging due to the need to design robust and self-sustained cultivation systems which can withstand harsh offshore conditions. Furthermore, this concept may be suitable only to several specific species, such as brown macroalgae in the genus *Sargassum*, which are stiff and have internal floating mechanisms (Radulovich *et al.*, 2015; Zollmann *et al.*, 2019a).

Anchored platforms, which are already in practice, can be sited in areas that are favorable for cultivation, aiming for optimal temperatures and sunlight, water motion which is sufficient to break down diffusion barriers, and natural supply of nutrients, for example, in natural upwelling zones (Roesijadi *et al.*, 2008). Furthermore, cultivation platforms can be located in eutrophicated regions, combining environmental bioremediation with biomass production (Fei 2004; Xu *et al.*, 2011). When environmental concentrations of nutrients are low, nutrients may be provided by artificial upwelling of deep nutrient-rich water as suggested already in the 1970s in the "Marine Biomass Program" (Roesijadi *et al.*, 2008, 2010). The main obstacle for applying this concept is the high energy requirement of pumping large volumes of water from depths of hundreds of meters. Therefore, artificial upwelling may become feasible only when combined with offshore, self-sustained power sources. An interesting venture can be the integration of deep seawater pumping for nutrient supply with Ocean Thermal Energy Conversion (OTEC) technology, which utilizes deep seawater for power generation based on temperature difference with surface water (Roels, 1979). This technology, which is currently in early stages of implementation, is relevant only for regions where the temperature difference between surface and deep water is high enough (i.e. above 10°C), such as in tropical regions (Roels *et al.*, 1979; Zollmann *et al.*, 2019a, 2019b). Another solution for supplying nutrients offshore is the polytrophic aquaculture, also known as Integrated Multi-trophic Aquaculture (IMTA) (Ashkenazi *et al.*, 2018; Neori *et al.*, 2004). This approach, which is already used in large-scale near- and onshore seaweed cultivation facilities, can significantly increase system sustainability. The underlying theory of IMTA is that waste nutrients from higher-trophic-level species can be recycled into the production of lower-trophic-level crops of commercial value, such as macroalgae (Troell *et al.*, 2009). Theoretically, co-cultivation of different species can increase productivity by increasing the light-harvesting efficiency. This can be done, for example, in a layered seaweed cultivation system, employing typical light absorption characteristics of green, brown, and red macroalgae, respectively, thus improving light use (Reith *et al.*, 2005; Zollmann *et al.*, 2019a).

In contrast to the extensive approach, the intensive approach emphasizes the importance of achieving maximal biomass yields, even at the expense of energy costs. Golberg and Liberzon (2015), for example, have modeled smart mixing regimes to improve biomass productivity by enhancing light harvesting and carbon fixation. Mixed water cultivation is commonly applied to onshore reactor cultivation of free-floating algae (Chemodanov *et al.*, 2017b; Forster and Radulovich, 2015) and has been shown to improve the yield and prolonged cultivation period in near-shore pilot experiments (Chemodanov *et al.*, 2019) (Figs. 6.2 and 6.3). However, applying free-floating algae cultivation offshore, mixed or non-mixed, is challenging due to strong ocean currents and increased loss risks, which may lead to uncontrolled macroalgal blooms (Liu *et al.*, 2009). Furthermore, the energetic and technical feasibility of mixing seaweed in offshore cultivation farms is yet to be assessed (Zollmann *et al.*, 2019a).

Lehahn *et al.* (2016) analyzed the global potential of far offshore *Ulva* biorefineries to provide food, chemicals, and energy. Also, this analysis located suitable cultivation areas, defined distance and depth limitations, and analyzed environmental risks and benefits of large-scale offshore macroalgal cultivation. Finally, although offshore cultivation of macroalgae is regarded as a sustainable alternative biorefinery biomass source, it faces significant challenges before it can meet the requirement for a consistent supply of high-volume feedstock. These challenges include rough offshore conditions that increase construction and maintenance costs (van den Burg *et al.*, 2013), a need for mechanized harvesting solutions for non-linear algal morphologies (Roesijadi *et al.*, 2010b), scarce nutrients and expensive fertilization (Reith *et al.*, 2005; Roesijadi *et al.*, 2008), losses and pests that may decrease productivity (Ingle *et al.*, 2018a; Rocca *et al.*, 2015), incomplete understanding of lifecycle of some species (Gupta *et al.*, 2018), and potential ecological effects that require further research (Lehahn *et al.*, 2016a; Roesijadi *et al.*, 2010b; Zollmann *et al.*, 2019a).

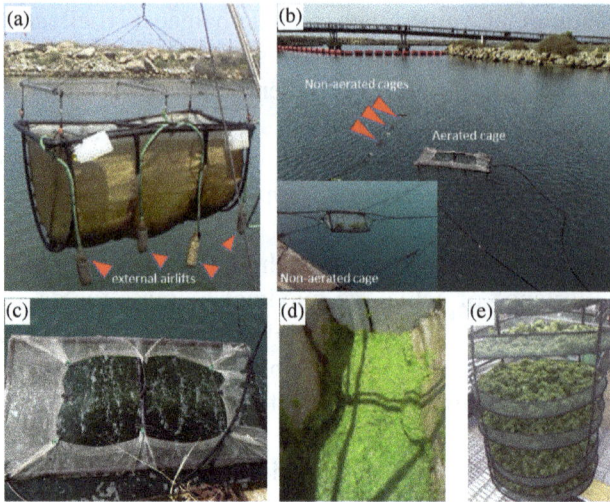

Fig. 6.2 (a) Digital image of the cultivation reactor with external airlifts. (b) Deployment of the reactor with algae to the cultivation site. (c) Tumbling with air and mixing of *Ulva* sp. biomass in the reactor. (d) Harvested *Ulva* biomass after water removal with gravitation. (e) Solar dried *Ulva* biomass.

Source: Adapted with permission from Chemodanov *et al.* (2019).

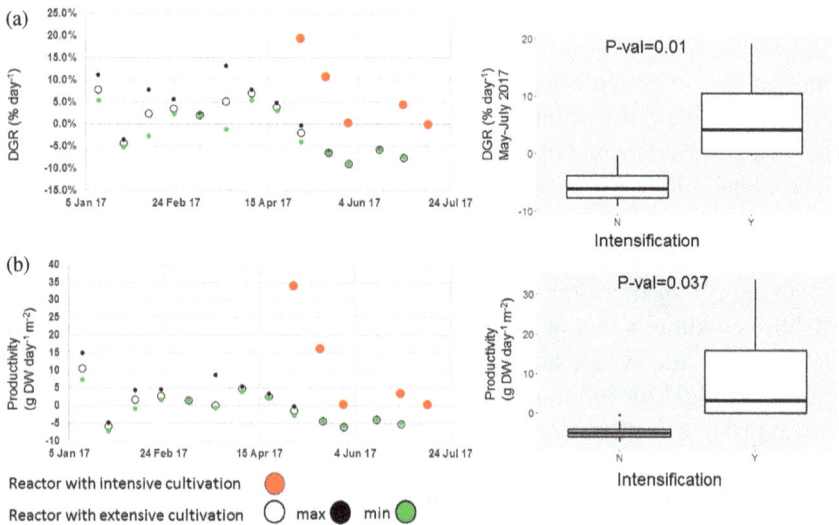

Fig. 6.3 Daily growth rates (a) and productivity (b) of *Ulva* in the intensified and extensive cultivation systems in the sea ($n = 3$). Minimum, maximum and average measurements are shown.

Source: Adapted with permission from Chemodanov *et al.* (2019).

6.7 Integrated Multi-Trophic Aquaculture

At present, aquaculture is one of the fastest-growing animal–food production sectors in the world, reaching an average growth rate of 6.9% y^{-1}. According to FAO, aquaculture production attained an all-time high of 90.4 million tons (live weight equivalent) in 2012 (worth US$144.4 billion). Today, aquaculture provides about 42% of the world's seafood supply and is expected to expand by 50% by the year 2030. FAO suggests that fisheries and aquaculture assure the livelihoods of 10–12% of the world's population (Bostock *et al.*, 2010). Marine seaweeds aquaculture sector represents 45.9% of the biomass and 24.2% of the value of the world total mariculture production, worth US$ 28.1 billion (Chopin and Sawhney, 2009). A wise approach for effective productivities refers to Integrated Multi-Trophic Aquaculture (IMTA) which, among other merits, looks after large biomass production. IMTA addresses the integrated farming of several species from different trophic levels, in close proximity. The concept has long been in use in Asia and contributes significantly to the sustainability of aquaculture, driving ecological efficiency together with environmental acceptability, product-diversity, profitability, and other benefits to society.

6.7.1 *The IMTA approach and systems*

IMTA is a concept which promotes the co-cultivation of species from different trophic levels so that the effluents of one species feeds other species. Co-cultivation has both environmental and economic advantages as it reduces the environmental effects and increases the potential production of the system. This process is based on the need of the primary producers for mineral nutrients, hence decreasing the amount of phosphate and ammonia discharged to the environment, depending on their uptake and storage abilities. When IMTA is applied in the ocean, seaweeds are cultivated downstream fish cages, preventing the accumulation of excess nutrients in the environment.

IMTA operations may be designed for land-based, onshore, shallow waters, or sea-based (near- or offshore) systems, marine or freshwater, and may comprise combinations of several species (Chopin *et al.*, 2010a). Land-based and onshore IMTA involve phototrophic, primary producers, species like seaweeds, water plants, or moss cultivated in tanks, ponds, or reactors connected in serial with fish ponds or downstream of fish pond effluents. Trapping of dissolved chemical compounds by seaweeds is termed bioremediation. Seaweeds are promising biofilters and bioremediation for various forms of dissolved nutrients efficiently removing between 35–100% of the dissolved nitrogen produced by adjacent intensive fish culture (Troell *et al.*, 2003). Generally, one can estimate that the excess nitrogen from the culture of 1 kg of fish could nourish the culture of over 5 kg of seaweed. Excess phosphorus can also be removed in smaller proportions than nitrogen (Buschmann *et al.*, 1996; Neori *et al.*, 2004). In land-based IMTA, most studies have used species of *Ulva* and *Gracilaria* because they offer reliable experimental models and have an attractive economic market value (Martínez-Aragón *et al.*, 2002). Long-term productivity of an IMTA platform growing fish, sea urchins, and *Ulva* on a semi-commercial scale has recently been demonstrated (Shpigel *et al.*, 2018), Fig. 6.4. This system allowed improved economic viability, including a shorter than usual growth period of *Paracentrotus lividus* to commercial size, enhanced gonadal development and quality, reduced food conversion ratio, and reduced effluent treatment cost.

On offshore IMTA, *Ulva* (Prabhu *et al.*, 2020) and other species have been adopted (see Fig. 6.5), such as the red *Kappaphycus* and several brown kelps including *Laminaria, Macrocystis,* and *Saccharina*

Fig. 6.4 The IMTA concept applied to land-based aquaculture involving 3 major steps where seawater flows from the reservoir to seaweed tanks by gravity to reduce energy expenses.

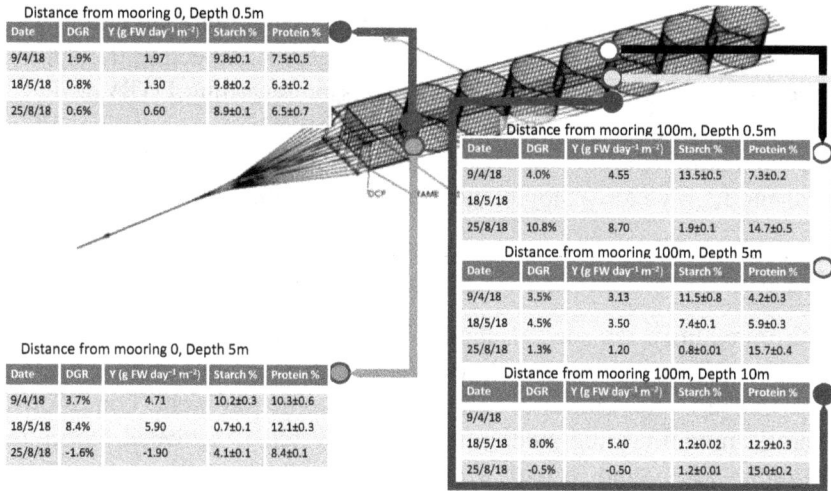

Distance from mooring 0, Depth 0.5m

Date	DGR	Y (g FW day^{-1} m^{-2})	Starch %	Protein %
9/4/18	1.9%	1.97	9.8±0.1	7.5±0.5
18/5/18	0.8%	1.30	9.8±0.2	6.3±0.2
25/8/18	0.6%	0.60	8.9±0.1	6.5±0.7

Distance from mooring 100m, Depth 0.5m

Date	DGR	Y (g FW day^{-1} m^{-2})	Starch %	Protein %
9/4/18	4.0%	4.55	13.5±0.5	7.3±0.2
18/5/18				
25/8/18	10.8%	8.70	1.9±0.1	14.7±0.5

Distance from mooring 100m, Depth 5m

Date	DGR	Y (g FW day^{-1} m^{-2})	Starch %	Protein %
9/4/18	3.5%	3.13	11.5±0.8	4.2±0.3
18/5/18	4.5%	3.50	7.4±0.1	5.9±0.3
25/8/18	1.3%	1.20	0.8±0.01	15.7±0.4

Distance from mooring 0, Depth 5m

Date	DGR	Y (g FW day^{-1} m^{-2})	Starch %	Protein %
9/4/18	3.7%	4.71	10.2±0.3	10.3±0.6
18/5/18	8.4%	5.90	0.7±0.1	12.1±0.3
25/8/18	-1.6%	-1.90	4.1±0.1	8.4±0.1

Distance from mooring 100m, Depth 10m

Date	DGR	Y (g FW day^{-1} m^{-2})	Starch %	Protein %
9/4/18				
18/5/18	8.0%	5.40	1.2±0.02	12.9±0.3
25/8/18	-0.5%	-0.50	1.2±0.01	15.0±0.2

Fig. 6.5 *Ulva* spp. cultivation far offshore attached to the fish farm. Fish farm design, details seaweed cages in relation to fish cages are shown. *Ulva* biomass growth and biochemical composition.

Source: Adapted with permission from Prabhu *et al.* (2020).

(Buschmann *et al.*, 2017). The two main important groups of extractive organisms for future IMTA designs are bivalve mollusks and seaweed. Both extract their nourishment from the surrounding aquatic environment. The bivalve mollusks collect and filter the suspended organic materials (uneaten feed, phytoplankton, and bacteria) to grow, while the algae assimilate the dissolved inorganic nutrients (Neori *et al.*, 2004).

Offshore IMTA requires inspiring challenges and several approaches have been recently described by Buck *et al.* (2018). Altogether, for European IMTA to become a reality shortly, solutions for biological, operational, market, and legislation aspects of IMTA will be needed (Kleitou *et al.*, 2018). Yet, proposed areas that are likely to benefit from focused research and development of IMTA include environmental mitigation and enhanced production of marine food (Kleitou *et al.*, 2018). It is believed that in the future more advanced IMTA schemes will consist of a larger number of species that will occupy additional ecological functions, in a way that will further reduce potential environmental harm and increase profitability (Chopin and Sawhney, 2009). Selecting the appropriate species is a considerably important issue regarding IMTA designs. Optimally, the utilized species should have several key traits, including efficient bioremediation capabilities, high economic value, high growth rates, and high yields (Chopin and Sawhney, 2009).

6.7.2 *Environmental and economic views on IMTA*

Environmentally, the species should be local to avoid species introduction, should not have a negative interaction with other native organisms and fit the climatic and oceanographic ambient conditions. Overall, modern integrated aquaculture systems, seaweed-based in particular, are bound to play a major role in the sustainable development of coastal aquaculture (Neori *et al.*, 2004). Although IMTA allows an aquaculture farm to produce and sell additional products with minimum investment, this option is not always realized. In some cases, seaweeds are used only to satisfy environmental regulation, and are not further processed to products of economic value. IMTA enables the creation of sustainable self-providing "limited ecosystems", composed of man-engineered food chains. Still, the integration of different species, as described, presents economic advantages. The extractive species' biomass can be used as another valued marine crop that can further increase the final farm profits. Furthermore, the different species are adding economic diversification and stability, securing the farm net gain if one of the farmed species' market value declines (Chopin *et al.*, 2010b).

The economical viewpoint has high significance since it is an integral part of the IMTA concept alongside the ecological scope (Chopin *et al.*, 2001; Neori *et al.*, 2004)). In recent years, there have been increasing numbers of studies dealing with wide information about the IMTA's economic implication, from the revenues of the integrated species to exclusive cost details of the assembly phases for the potential marine aqua farm facility (Bixler and Porse, 2011; Korzen *et al.*, 2015b; Krishnan and Narayanakumar, 2013). Ecosystem services attained by the incorporated extractive species need to be considered as part of the gross aquafarm profits (Chopin *et al.*, 2010b). When external benefits such as ecosystem services would be translated into actual monetary value, IMTA is expected to become a more attractive option for aquafarmers.

6.7.3 *Gaps and challenges in seaweed IMTA*

The IMTA concept is currently between the stages of advanced research and experiment and industrial implementation. Although much work was done to validate the integration of seaweeds into the aquaculture industry, there are still many questions and much information that require verification. Among those questions are factors that can influence the design of seaweed culture and their functional capabilities in a potential commercial system (Troell *et al.*, 2003). A suggested model should be fitted to the local habitat, after evaluating location-specific parameters such as temperature, irradiance, and seasonality that can determine the seaweed performances. The species/strains should be indigenous and to be integrated they should not indicate allelopathy when they are in proximity to one another. Furthermore, the design should present cost-effective evaluation, considering the value of the incorporated seaweeds, ecologically and economically. Offshore, IMTA implementation is even more challenging, due to major technological and engineering limitations and proof of seaweed growth.

A few studies from recent years have identified the main limitations for commercial implementation of IMTA systems. These limitations include lack of financial and governmental support, legislation bottlenecks, general lack of R&D, process complexity, biological issues such as bio-fouling and seasonal production, negative public perceptions, lack of cost-effective system designs, high operation costs, and questionable profitability (Kleitou *et al.*, 2018; Sheng Lee, 2018). This wide array of limitations can be overcome only by multi-disciplinary and cross-industry collaboration, preferably also with governmental support (Kleitou *et al.*, 2018).

6.7.4 *IMTA examples globally*

Currently, only a few countries have implemented IMTA on a commercial scale, and globally, most seaweed culture is performed as open water monoculture (Nazar *et al.*, 2019). The most prominent example of commercial, large scale IMTA is Sanggou Bay in Shandong, China. Farmers in Sanggou Bay have intentionally cultured species from multiple trophic levels in combination since at least 1996. Thus, the bay is divided into a few zones: Finfish are farmed in the inner bay, scallops and oysters are farmed in the mid-bay, bivalves and macroalgae are farmed in combination in the outer-mid-bay and macroalgae culture are dominant towards the mouth of the bay, where hydrodynamic conditions are optimal. Consequently, Sanggou bay has been well researched in an IMTA context, and successful nutrient transfer between the various trophic levels was shown. This region is continuously used as a ground for experiments and assessments of various managing schemes and mariculture combinations (Wartenberg *et al.*, 2017). Hitherto, most countries did not regulate or enforce aquaculture industries to treat their effluent water (Troell *et al.*, 2003). Denmark, however, has limited the amounts of wastes released to coastal waters from aquafarms, recommending the coupling of biofilters in the form of seaweed or mussels to the existing finfish mono-aquafarms (Holdt and Edwards, 2014). The approximate cost of treating 1 kg of nitrogen is estimated at $44 (Holdt and Edwards, 2014). The recycling service of carbon, nitrogen, and phosphorous is to be recognized as environmental benefits and financial advantage by both government and industry. In accordance, some of the few examples of commercial IMTA come from Denmark. One example is the cultivation of *Saccharina latissimi* by Hjarnø Havbrug in the vicinity of Horsens Fjord, approximately 100 m from a blue mussel SmartFarm™ and 500 m from rainbow trout (*Oncorhynchus mykiss*) farm cages. A study by Marinho *et al.* (2015) described the seasonal variations in the amino acid profile and protein content of the seaweed, and pointed out a few considerations for harvesting timing, including biomass yields and composition and epiphytes interferences. In Spain, the cultivation of macroalgae in integration with mussel rafts was examined, offering promising results. A study showed that cultivating *S. latissima* in Northwest Iberian Peninsula is fully viable. The integration of this culture with mussel rafts has presented an ability to produce protein-rich algal biomass with higher commercial value as feed or food (Freitas *et al.*, 2016).

6.8 Pest Management in Seagriculture

Though the occurrence of epiphytes and fauna is the inevitable marine environment of seagriculture, for example, Fig. 6.6 and (Yamamoto *et al.*, 2013), increases in their abundance can negatively impact seaweed crop yields. For example, decreases in biomass production and product quality generated from seaweed cultivated in Malaysian and Filipino seawater were linked to epiphytic outbreaks (Vairappan *et al.*, 2007). In The

Fig. 6.6 Epibiota associated with seaweed. Blue arrows: host seaweed. Red arrows: epibiotic organism. (a). **Host-** *Porphyra* sp. from Taiwan, from tank cultivation. **Epiphyte-** young *Porphyra* of the same species, scale bar: 45 mm. (b). **Host-** *Spatoglossumsolierii*, from Rosh-Hanikra reef, Israel. **Epiphyte-** *Porphyra linearis*. Scale bar: 43 mm. (c) **Host-** *Ulva* sp., tank cultivation, destruction of thalli after sharp warming. **Epiphyte-** bacterial destruction. Scale bar: 26 mm. (d) **Host-** *Ulva rigida* from Haifa, Israel, grown in the offshore system in a net cage. **Epiphyte-** *Ulva compressa*. Scale bar: 362.5 mm. (e) **Host-** *Porphyra linearis*, **Epiphyte-** the bacterial destruction of thalli at the end of the cultivation season (due to water warming), scale bar: 41 mm. (f) Magnification of host and epiphyte same as in **D**, scale bar: 0.59 mm. (g,h,i) **Host-** same as in photo **Epiphyte-** Nematode and Copepods. Scale bar: 0.62 mm, 0.22 mm, and 71 μm, respectively. **Host-** *Ulva rigida*. **Epiphyte-** *Alteromonas* sp. grown on an agar petri dish, around the cut *Ulva* sp. fragment. Scale bar: 40 *μ*m.

Source: Adapted with permission from Ingle *et al.* (2018).

Philippines, between 2011 and 2013, the "ice-ice" disease in *Kappaphycus alvarezii*, caused by a bacterial infection (Hurtado *et al.*, 2006; Largo *et al.*, 1995), led to a heavy economic loss of almost $310 million (Cottier-Cook *et al.*, 2016). However, in contrast to terrestrial agriculture, the control and management of marine pests have been neglected and there is a gap in the methods addressing and managing them. Based on this data on crop losses from the current largest world producers, it is necessary to consider the problem caused by pests and develop appropriate risk management processes for seagriculture. One approach to tackle these problems is to understand the macroalgae interactions with associated epibiotic species. Because of the extreme sensitivity of marine ecosystems to chemical treatment, only ecosystem-based pest management strategy, with a focus on biological control that is friendly to the local ecosystem, would be applicable.

The Integrated Pest Management (IPM), developed for terrestrial agriculture, is the decision-based process used to optimize the control of all types of pests economically and ecologically (Ehler, 2006). In terrestrial agriculture, the IPM strategy not only shows effective management of pests and diseases but also reduction of production costs for farmers (Ahuja *et al.*, 2015). IPM integrates multidisciplinary methodologies for the development of agri-ecosystem management strategies while reducing negative impacts on public health and the environment (Smith *et al.*, 1976).

A recent study proposed a Marine Integrated Pest Management (MIPM) framework, the key aspects of which we discuss below (Ingle *et al.*, 2018). According to the International Plant Protection Convention (IPPC), the term *pest* means "any species, strain, or biotype of plant, animal, or pathogenic agent, injurious to plants or plant products" (FAO, 2004). As seagriculture is mostly done in rural areas by low-income farmers, the problem of pests in this industry is not widely addressed. According to the IPPC definition, pathogenic bacteria, fungal infection, epiphyte types that penetrate the tissues of host seaweed and seaweed grazers can be considered as pests in seagriculture.

The study suggests to define the term "*pests*" in seagriculture as, "any species, strain, or biotype of plant, animal, or pathogenic agent, potentially harmful or injurious directly or indirectly to cultivated seaweed or its products". The word "harmful" implies responsible for any adverse impact on the growth rate, quality, and quantity of seaweed crop, and the word "injurious" means any physical damage to seaweed plant which get injured by pests. The word "directly or indirectly" refers to whether the seaweed is the direct target of certain pests or not. Data in Fig. 6.7 map the potential pests in seagriculture according to the suggested definition. Table 6.2 describes various seagriculture pests as appear in the literature.

To address the problem of pest management in seagricultre, we propose the following MIPM approach. The proposed framework is divided into six sections as shown in Fig. 6.8. The continuous monitoring is expected in every stage of the proposed MIPM cycle. The establishment of a knowledge base (Section 1) is the primary task in the field of MIPM, followed by an inspection and risk assessment primarily to the cultivation

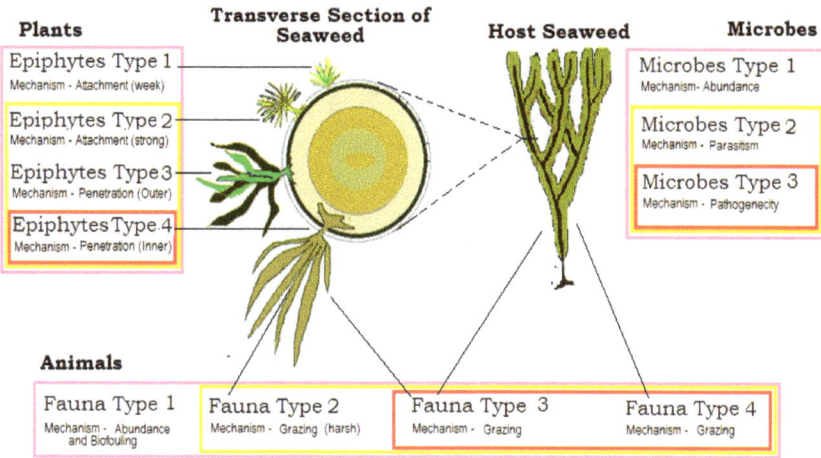

Fig. 6.7 Categorization of pests in seagriculture. Red boxes show the types of epibiota that can be responsible for the direct damage to seaweeds. These can be considered as pests as per IPPC's definition. The yellow boxes show the possible pests after modification in the IPPC definition by adding the word "directly or indirectly". Finally, the pink brackets show the possible pests after the modification in the word "injurious" to "potentially harmful or injurious" in the IPPC definition of pest. Microbes type 1 are responsible for the adverse impacts on the growth of seaweed competition. Microbes type 2 show parasitism with seaweed and become responsible for harm to seaweed. Microbes type 3 show pathogenicity to seaweed and can be responsible for injuries and diseases in seaweed. Fauna type 1 show biofouling or abundance in the seaweed ecosystem and may be responsible for the possible harm indirectly. Fauna type 2 show harsh grazing of epiphytes may be responsible for the indirect injuries to seaweed. Fauna type 3 show grazing of epiphytes and seaweed crops both as well as being responsible for direct injuries to the seaweed. Fauna type 4 show grazing of seaweed crop. The cross-section of host seaweed shows the level of epiphytic penetration in the tissues of the host plant. Epiphyte type 1 attaches weakly to the surface of the host plant but does not show any injury but can be harmful indirectly. Epiphyte type 2 attaches strongly to the surface of the host and can be responsible for the indirect harms and injuries to the host plant. Epiphyte type 3 bleaches the deck-lamella, and penetrating the outer layer of the host's cell wall, is surely responsible for injuries to host indirectly. Epiphyte type 4 penetrates deck-lamella and outer layer of the host's cell wall with a direct injury.

Source: Adapted with permission from Ingle *et al.* (2018).

(Section 2), followed by prevention (Section 3), control (Section 4), intervention and mitigation (Section 5), and evaluation and record-keeping (Section 6).

Table 6.2   The injuries to seaweed due to potential pests.

Type of pest	Basiphyte seaweed	Pests	Type of damage	Study area	Study year	References
Phytoplankton (microalgae)	*General seaweeds*	Epiphytic diatoms	Cover the seaweeds	Iceland	2005	Totti *et al.* (2009)
Pathogenic microalgae	*Kappaphycusalvarezii*	Filamentous red algae	Biofouling leads to diseases	Brazil	2008–2009	Marroig and Reis (2015)
Fungi	*Kappaphycusalvarezii* and *K. striatum*	*Aspergillusochraceus, A. terreus* and *Phoma* sp.	Potential for the ice-ice disease	Philippines	2007	Solis *et al.* (2010)
Microalgae and bacteria	*Kappaphycusalvarezii*	Filamentous red algae, *Alteromonas* sp., *Flavobacterium* sp. and *Vibrio* sp.	Goose-bump like structure, formation of pit, infection can lead to disease	Philippines, Indonesia, Malaysia, Tanzania	2005–2006	Vairappan *et al.* (2007)
Epiphytes type 1	*Laminariahyperborea* (*Gunn.*) *Fosl.*	*Pulmariapulmatu*(*L.*) *Kuntze, Ptilotaplumosa,* (*C. Ag.*) *Huds., Membranopteruulata* (*Huds.*) *Stackh.,* and *Phycodrysrubens* (*L.*) *Batt.*	Covers the seaweed totally	United Kingdom	1968–1969	Whittick (1983)
Epiphytes type 2	*Gracilariagracilis*	*Chaetomorpha* sp., *Antithamniondensum, Acrochaetiumsp*	Strongly attached epiphytes	Argentina	2006–2008	Martin *et al.* (2013)

Epiphytes type 3	*Acrochaeteheteroclada*	*Choundruscriospus*	Penetrates the host but not harmful directly	Canada	1990	Correa (1991)
Epiphytes type 4	*Gracilariatikvahiae*	*Ulvalactuca*	Penetrates in the cortex and harm to cortical tissues	United States	2000	Dawes et al. (2000)
Epiphytes type 4 acts as endophytes	*Laminariasaccharina*	*Laminariocolaxaecidiolax* and *Laminarionemaelsbetiae*	Enter in the medulla of seaweed and damage to host	Germany	1993–1995	Heesch and Peters (1999)
Surface grazers	*Laminaria species*	Sea-snail *Lacuna vincta*, sea urchin *Strongylocentrotusdroebachiensis*	Surface grazing, not specific harm but attract to other animal	Canada	1985	Johnson and Mann (1986)
Harsh grazer of epiphyte	*Macrocystispyrifera*	*Oxyjuliscalifornica*	Damage to blades of seaweed due to harsh grazing	United States	1979	Bernstein and Jung (1979)
Grazer to epiphyte and seaweed	*Ulva* spp.	*Gammarus*	In low nutrients condition Gammarous prefer to feed on *Ulva*	Netherlands	1995	Kamermans et al. (2002)
Grazers of seaweed	*Chondrus Crispus*	*Gammarusoceanicus, Idoteabaltica, Lacuna uincta*	Damage to tissue	Canada	1981	Shacklock and Croft (1981)

Source: Adapted with permission from Ingle *et al.* (2018).

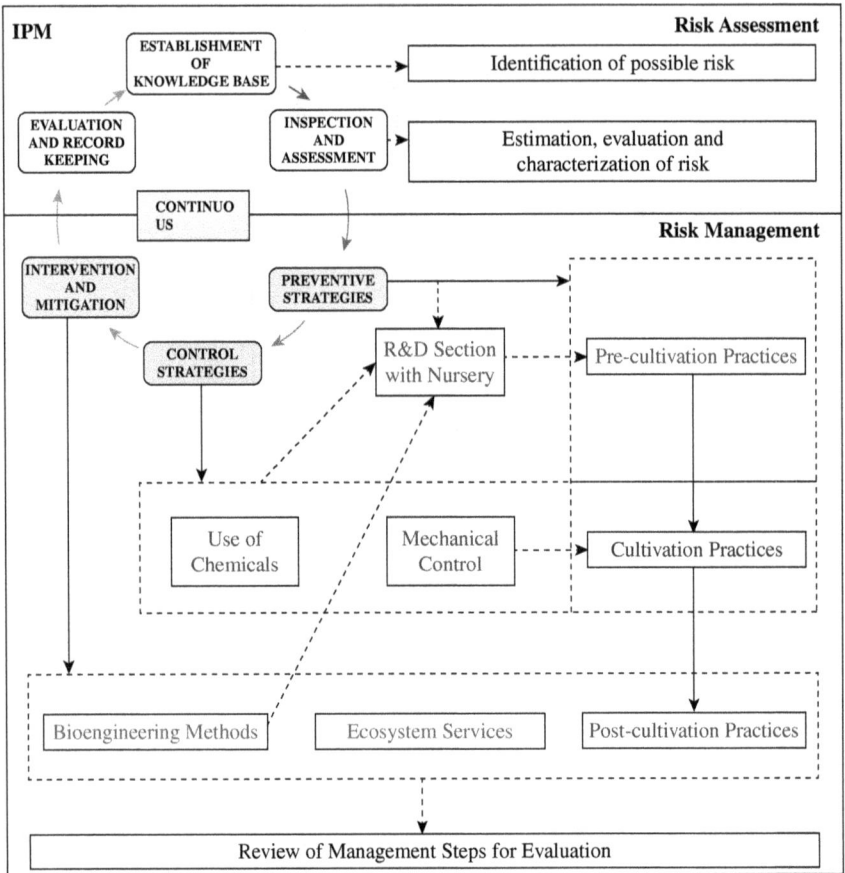

Fig. 6.8   Marine Integrated Pest Management (MIPM) framework for the seagriculture. *Source*: Adapted with permission from Ingle *et al.* (2018).

Section 1: Knowledge base

The establishment of a knowledge base is a crucial task that includes the knowledge of the seaweed crop, the seasonal production, favorable conditions, possible environmental risks, and possible pests available in the cultivation area. It should also include the information and knowledge about the natural enemies of the pests, their availability and abundance, crop lifecycle and life span, all known risks to the cultivated crop, including the market and possible environmental impacts due to the cultivation.

To establish a knowledge base, it is necessary to obtain the following information for each cultivation site:

- The selection of suitable species. The species should be resistant to seasonal changes and known diseases.
- The lifecycle of a specific crop, its genetic sequences, methods of cultivation, possible association with other species, and general practices that are in use to minimize pests and possible diseases.
- The current ecosystem in that area, dominant species of epiphytes and grazers.
- Accepted level of pests and their impacts on cultivation and environment.
- Target and non-target species, their lifecycles, dependency on each other, possible links to diseases.
- Possibility of pests and disease forecasting techniques, the use of non-chemical methods and management practices.
- Possible risk identification to crops from environmental factors and from crops to the environment.

Section 2: Inspection and assessment
Inspection is the primary step of the assessment of any new cultivation set up. Within this step, the possible risk of pests should be characterized by its potential and frequency. The base of this step is setting the goals and defining the management options to tackle the pest problem. An inspecting team, a monitoring team, and an assessment team should know the best environment required for seaweeds and the actual environment at the site, and suggest the required actions. Assessment of possible pest problems should be performed after the inspection. It should consist of observed types, nature, availability, and distribution of pests. It should also include the length of the crop exposure to pests, crop doze response to the pests, and end-point effect. The success of this step depends on the availability of data, resources invested to complete the assessment, and the accuracy of the risk analysis.

Section 3: Prevention
Importantly, the successful removal or reduction of epiphytes during the nursery stage can prevent further growth and large-scale infection at the cultivation site. For example, in *Gracilaria* tank culture, epiphytes can be inhibited by controlling the pH (Friedlander and Ben-amotz, 1991).

A wise selection of cultivation sites and suitable seaweed species, followed by inspection and monitoring of seawater quality and seasonal changes, can prevent pest damage. Routine cleaning and mechanical removal of debris from the bottom of the cultivation site can limit the number and diversity of epiphytic species. The use of healthy seedlings and their maintenance in clean tanks can reduce the epiphytic outbreaks (Vairappan *et al.*, 2007).

Crop rotation, fallow, availability of a trap crop and intermediate crops are all possible strategies generally used in IPM of terrestrial agriculture, which, if adapted to seagriculture, may be effective in reducing pests without harming the ecosystem. Trap crops are plants that are cultivated with the main crops to attract and divert the pests from the main crop (Shelton and Badenes-Perez, 2006). For example, in terrestrial agriculture, marigold or cucumber is cultivated as a trap crop in tomato farming (Kambrekar *et al.*, 2008). A dense cultivation system can minimize the risk of epiphytes; however, dense cultivation can shadow large areas of the sea, thus producing adverse environmental impacts. At the same time, the common practice of attaching seaweed to ropes at the sea surface is favorable for epiphytes to grow and for many animals to move freely in the cultivation area.

Section 4: Control
There is a lack of commercially available pesticides that can be used in the seaweed cultivation, particularly in nearshore or offshore environments. Although multiple environmental limitations must be taken into consideration, some chemicals can be applied for the onshore cultivation or nursery level. It is reported that 4–6% of hypochlorite solutions kill 80% of *Ectocarpus* and *Enteromorpha* epiphytes (Ugarte and Santelices, 1992). Low levels of copper can be useful for controlling certain epiphytes (Fletcher, 1995).

Pest control in early colony-forming stage at the nursery is the most important control step, as without it contamination can propagate to a large scale. Indeed, nurseries can serve as the collection units or species banks. For example, in *Kappaphycus* cultivation, pollution prevention is suggested for *polyphonic* epiphytes and bacterial infections (Anicia and Alan, 2006).

Applying a suitable cultivation method also impacts the quality of seaweed products and increases its growth rate, while limiting the potential damage from pests. For example, vegetative fragmentation and tissue culture methods were compared (Ganesan *et al.*, 2015) for growth rate and quality of agar of *Geliadiella*. In *Gracilaria*, it was found that the cultivation period (Alejandro and Patricio, 1993) and the spore culture method (Alveal *et al.*, 1997) are the most important factors for control of epiphytes.

Mechanical control is a simple and effective but labor-intensive method to control large pests. There are few reports related to mechanical control of pests in tank seagriculture. These include intensive water filtration and multiple water exchange (Ugarte and Santelices, 1992). Besides, nutrient control, for example, nitrogen limitation (Friedlander and Ben-amotz, 1991) is another strategy to prevent pests. Also, avoiding farming at the dominant gazer breeding season can prevent large-scale damages to the seaweed crops.

Section 5: Intervention and mitigation
The overall objective of the intervention and mitigation step is to nullify (1) the impacts of seaweed farming on the environment and (2) pest impact on seaweed. This could be achieved by the development of supporting ecosystem services. This step should provide the knowledge for better management, conservation, and rehabilitation of the ecosystem while mitigating the resource degradation. For example, the dominant epiphytes can be managed with amphipods and grazers, which feed on them. In *Sargassum* farming, the amphipods *Ampithoemarcuzii, Caprellapenantis,* and *Jassafalcata* reduced epiphytic growth without any damage to the crop (Duffy, 1990). Another example is the development of associated bacteria with symbiotic relation with seaweed that could enable better growth and development (Wichard *et al.*, 2015). The use of such bacteria in the nursery level of seaweed can improve the quality of the crop.

The collaboration with local fishermen and fishing agencies will be helpful to get the knowledge of any initial outbreaks of epiphytes. This can encourage the monitoring team and save further yield losses.

Weather conditions and forecasts should be followed and monitored to mitigate possible damage due to any sudden environmental risk such as hurricane, cyclone, high-tide, hailstorm, etc.

Section 6: Evaluation and record-keeping
The review of all steps followed in risk management is the primary step of evaluation and record keeping. The evaluation process should include

the success and failure of each step of MIPM and all management practices to tackle the pest problem with definite objectives. For example, while evaluating the term "site selection" for cultivation in "pre-cultivation practices", the success and failure of the selected site should be discussed to establish the "criteria for the site selection".

There are a few options to convert the evaluation records into a knowledge base:

- Establishing a discussion forum to share the research updates and latest knowledge in this area by the researchers, research institutes, and governmental bodies working in this area.
- Establishing a network or website to share ideas, views, suggestions, prevention-control and mitigation measures with all farmers and stakeholders. We are here suggesting to develop webpages on social media devoted to this work, which can be utilized as knowledge-sharing platforms. The shared records should include information about the behavior of specific pests in specific seasonal and environmental conditions.
- The industries that use seaweed as the raw material can take efforts in implementing the MIPM by promoting farmer awareness. This can be done by regular visits, seminars, and workshops, and can promote the sustainability, consistency, quality, and quantity of the raw material.
- Establishing large-scale cultivation facilities, including the preparation of seedlings in large-scale farms, and developing seaweed breeding programs and seaweed gene banks can be useful backup techniques for cases of failure.

6.9 Monitoring of the Cultivation Process

The growth rate and chemical composition of macroalgae can vary substantially depending on physical, chemical, and biological parameters. These parameters include macroalgae species, thallus size and age, light (Chemodanov *et al.*, 2017b; Rosenberg and Ramus, 1982) (stocking density (Friedlander and Levy, 1995), temperature (Friedlander, 1991; Zollmann *et al.*, 2018), water motion (Gallagher *et al.*, 2018; Harrison and Hurd, 2001; Koehl, 1999; Msuya and Neori, 2008; Posten, 2009; Stewart and Carpenter, 2003), salinity (Friedlander, 1992; Hanisak, 1987), dissolved oxygen, carbon dioxide, pH (DeBusk and Ryther, 1984) (nutrients availability (Korzen *et al.*, 2015b; Luo *et al.*, 2011; Shahar *et al.*, 2020), and bacteria and pest's existence (Alsufyani *et al.*, 2020; Ingle *et al.*, 2018; Lehahn *et al.*, 2016a; Øverland *et al.*, 2019). Therefore, it is crucial to monitor both the environmental conditions and the algae status (chemical composition and growth) during cultivation and alter possible conditions in order to maintain and achieve the desired metabolite production.

One of the first steps to achieve this goal is developing an experimental verified metabolic model specified for each species and desired product, that will describe the impact of nutrients, salinity, temperature, and other environmental parameters on the algae growth rate and chemical composition. In recent years, multiple cultivation experiments have examined and monitored the effect of these parameters. Usually, the measurements are carried out manually using sensors (for temperature, turbidity, PAR, salinity, water velocity, etc.) or by water and algae sampling (for algae growth, composition, diseases, pests, etc.) followed by laboratory analysis. These methods are labor-intensive, costly, and can be done in resolutions of hours to few days (onshore systems) or even several days to weeks (offshore systems), depending on the measurements required.

Monitoring provides useful information on the growing phase and estimates of the final biomass yields. Using the information, several studies developed models for estimating the net primary productivity (NPP, defined as g DW m^{-2} day^{-1}, a measurement of the efficiency of conversion of solar to potentially useful chemical energy) of selected seaweeds for ecological and environmental applications. Duarte and Ferreira (1997), for example, published a dynamic model, which predicted NPP for *Gelidium sesquipedale*. The physiological model was useful for the management of natural stocks of this species as raw material for multiple industries. The model considered productivity as a function of light intensity, temperature and rates of respiration, product exudation, frond breakage, and mortality and included a conversion factor between carbon and dry weight (Buschmann *et al.*, 2017). Another model was developed for the optimization of harvesting *Ascophyllum nodosum* from natural populations (Seip, 1980). The model allowed for the determination of the optimum stocking density and the rate and frequency of harvesting, allowing only 7% of the estimated standing crop to be harvested, to produce a sustainable harvest for several decades of exploitation (Buschmann *et al.*, 2017a).

However, regardless of how helpful these models might be, they cannot be used directly for modeling macroalgae farming sites yet, as environmental conditions in those cultivation sites are different from those occurring in nature (Buschmann *et al.*, 2017a). Furthermore, metabolic reactions occur in shorter time scales, of hours to days, than the actual measurements. Therefore, there is a need to monitor and model the cultivation process in higher resolutions, which will allow making real-time decisions to control cultural practices such as harvest, fertilization and herbicide, and pesticide application. Also, with the enlargement of macroalgae field sites, manual measurements and sampling are no longer able to fulfill the requirements for cultivation monitoring. Automated sensors can be used to monitor environmental parameters in higher time resolutions; however, their use in offshore sites is limited due to possible harsh conditions, such as waves and storms.

Another option is to employ multi-sensor, multi-temporal, and multi-spatial remote sensing techniques (using satellite, Unmanned Aerial Vehicle (UVA), and ground spectroradiometer) along with oceanographic data and in-field measurements for the chemical composition of the algae. This approach is usually employed to predict macroalgal blooms (Xing *et al.*, 2019) and potential global productivity (Lehahn *et al.*, 2016a). The use of these techniques for macroalgae monitoring is still in its infancy, and more data regarding the chemical composition of the algae is required.

For improving the cultivation monitoring methods and models, macroalgae agriculture can adopt the Precise Agriculture (PA) approach, which was recently employed for traditional land agriculture. PA means supplying each plant the precise quantities of water and fertilizers according to its need (suitable composition, precise quantities and timing, and suitable mechanism). The approach aims to reorganize the agriculture system towards a low-input, high-efficiency, sustainable agriculture (Zhang *et al.*, 2002a). This becomes possible by using different kinds of remote and in-field sensors. In addition to increased profitability, as a result of better management practices and the development of farm information systems, PA can provide additional benefits such as increased crop quality, improved sustainability, lower management risk, and environmental protection. Early adopters that applied some careful and efficient PA practices to land agriculture (such as yield mapping, selected site-specific practices, including crop nutrition, and precise farm information systems), have in most cases benefited financially from improved soil and input management, crop yield and quality, and marketing (Robert, 2002). Hence, adopting similar PA techniques for macroalgae could potentially offer smart and efficient monitoring solutions and help to maximize the growth of the algae, optimizing its quality, and minimizing the negative environmental effects and the cost of the process.

However, adopting PA to macroalgae requires overcoming several barriers:

(1) Lack of necessary information and technological difficulties. There is a need to acquire a better understanding and quantification of algae's temporal and spatial responses and adaptations to the environmental conditions. The data required must be generated by automatic rapid sensor systems adapted to the marine environment (Stafford, 2000). Furthermore, if the algae is the product and the best sensor of its environment, then rapid online sensing of algae composition may provide information on crop condition necessary to direct spatially variable inputs (Stafford, 2000).
(2) Precision agriculture is an information-intensive technique, and therefore cannot be realized without advances in networking and computer processing power (Stafford, 2000). This computer can also be programmed to make real-time information-based decisions, thus controlling cultural practices such as fertilizer, herbicide, and pesticide application.

Overcoming the barriers and adopting the farm management concept can meet much of the increasing market and environmental demands. Precision agriculture is seen to be the correct way ahead for crop producers shortly, as optimized inputs will lead to improved yields and crops and reduced costs and environmental impact, followed by the audit trail that consumers and legislation increasingly require (Stafford, 2000; Van Alphen and Stoorvogel, 2000).

6.10 Chemical Composition Characterization Methods

In recent years, biomass from marine macroalgae has been globally recognized as a source for valuable chemical compounds for a wide range of applications (such as human and animal nutrition, pharma, fertilizers, and bioenergy) (Stengel and Connan, 2015), owing to their rich composition of unique functional polysaccharides, pigments, lipids minerals, and proteins (Charrier *et al.*, 2018). The richness, diversity, and complexity of macroalgal compounds, along with a growing commercial interest and awareness, demand reliable, fast, and efficient extraction and measurement techniques that allow accurate determination of the chemical characteristics of the algal biomass and safe provision of the target compounds (Charrier *et al.*, 2018).

Quantitative chemical composition characterization methods adapted explicitly to macroalgae are essential for acquiring a broad database on algal biomass for both basic and applied research. Expanding this database can help address essential questions in algal physiology, developmental biology, chemical ecology, sustainable aquaculture, cultivation control, and technological processing techniques. Therefore, the characterization methods should be based on simple, affordable, analytical techniques, to be accessible to everyone interested in the information, such as phycologists, algae producers, engineers, or consumers interested in the nutritional value of the algae (Barbarino and Lourenço, 2005).

The next section describes methods to extract and measure major compounds from macroalgae such as proteins, carbohydrates, ash, and lipids.

Proteins

Proteins are polypeptides formed from amino acids through amide linkages. Macroalgae-based proteins have recently become a focus of applied research, in particular for application as human and animal nutrition, health, fertilizers, and plant growth stimulants (Angell *et al.*, 2016). These studies have generated a significant database on the protein biochemistry of seaweeds across diverse disciplines (Angell *et al.*, 2016). However, comparing the protein content of macroalgae reported in different studies can be difficult owing to methodological differences (Øverland *et al.*, 2019).

The commonly used methods to quantify protein in macroalgae are (1) Direct extraction procedures and subsequent quantification of soluble protein based on colorimetric assays (mainly the alkaline copper assay (Lowry *et al.*, 1951) and the Coomassie Brilliant Blue assay (Bradford, 1976); (2) Estimation of the total nitrogen content (mainly using either the Kjeldahl method or through combustion using CHN analyzers) followed by the use of a nitrogen-to-protein conversion factor (Angell *et al.*, 2016; Charrier *et al.*, 2018); and (3) The sum of amino acid residues.

Protein determination using direct extraction procedures employs multiple options for the extraction component and the subsequent quantification of soluble protein. Both the extraction of the protein and the quantification of the extracted soluble protein are susceptible to inaccuracies (Angell *et al.*, 2016). The extraction of protein from macroalgae is challenging due to the complex, viscous, and often charged polysaccharide cell wall and extracellular matrixes (Øverland *et al.*, 2019; Polikovsky *et al.*, 2016a). Several methods have been used to increase the extraction yields, such as osmotic shock, mechanical grinding, high shear force, ultrasonic treatment, acid, and alkaline pre-treatment, polysaccharides-aided digestion, and their combinations (Polikovsky *et al.*, 2016b). Furthermore, each method is not a standardized process and consequently varies between studies. For example, protein extraction procedures vary with the pre-treatment of the sample (raw, milled, enzymatic digestion, etc.), the volume of water, and exposure time used for the extraction of water-soluble proteins, the type and exposure time of buffer used, whether or not the protein is precipitated, centrifuge time and force, and the type of standard used (Angell *et al.*, 2016). The efficiency of these direct extraction procedures differs in their own right and is also influenced by the chemical and morphological features of the seaweeds themselves (Barbarino and Federal, 2005).

Also, the main used methods for quantifying protein in the extract are subjected to interference from several factors depending on the biochemistry of the seaweed (Angell *et al.*, 2016). The Lowry method detects protein by a reaction catalyzed by copper, a component of the folin phenol reaction. The chemical reaction detects peptide bonds and is also sensitive to some amino acids, such as tyrosine and tryptophan. In the Bradford method, the Coomassie Brilliant Blue dye is bound to protein mainly by arginine residues and to a lower degree by histidine, lysine, tyrosine, tryptophan, and phenylalanine residues. The binding between the dye and amino acids is attributed to van der Waals forces and hydrophobic interactions. As a consequence, the reactivity of both methods depends on the presence of certain amino acids (Barbarino and Federal, 2005; Charrier *et al.*, 2018). Furthermore, the extracts may contain several colored substances that may influence the measurements (Øverland *et al.*, 2019). Taken together, the number of unique combinations of the extraction procedure, colorimetric assays, and type of seaweed substrate lead to considerable variation in the quantitative determination of proteins (Angell *et al.*, 2016).

In contrast to the technical issues related to the direct extraction of protein, the determination of the total nitrogen content using nitrogen-to-protein factors has some important advantages. Total nitrogen analysis, carried out by Kjeldahl's method ("Association of Official Analytical Chemists — AOAC," 1990) or Hach techniques (Hach *et al.*, 1987), is fast and inexpensive. Total nitrogen from CHN analysis can also be converted to crude protein with a high degree of accuracy (Lourenço *et al.*, 2002). In addition, there is no need for extraction procedures. Thus, a better comparison of results among researchers is possible, since protein is estimated without a tricky preliminary extraction (Angell *et al.*, 2016; Lourenço *et al.*, 2002).

However, this method also requires multiplying the total nitrogen by a nitrogen-to-protein factor. The traditional conversion factor of 6.25, which is used as the standard factor for most food and feed ingredients, assumes that all nitrogen is in the form of protein. In reality, all plant material, including macroalgae, have significant sources of non-protein nitrogenous materials such as chlorophyll, nucleic acids, free amino acids, and inorganic nitrogen (e.g. nitrate, nitrite and ammonia). Therefore, the use of the 6.25 conversion factor can lead to an overestimation of the protein level in macroalgae (Angell *et al.*, 2016).

For these reasons, analysis of nitrogen and the use of a macroalgal-specific nitrogen-to-protein conversion factor has been recommended. N-protein conversion factors are based on the quantification of total amino acids, which is considered to be the most accurate way of determining protein and the independent determination of total nitrogen (Angell *et al.*, 2016). Angell *et al.* (2016) have analyzed the variation in this factor for 103 species across 44 studies that span three phyla, multiple geographic regions, and a range of nitrogen contents. They have found an overall median nitrogen-to-protein conversion factor of 4.97 and an overall mean nitrogen-to-protein conversion factor of 4.76. Therefore, they proposed using the overall median value of 5 as the most accurate universal seaweed nitrogen-to-protein conversion factor. Despite the established science for the calculation of N-protein conversion factors, which suggests that high concentrations of non-protein nitrogen are common for seaweeds, only a few empirical studies have calculated seaweed-specific N-protein conversion factors.

The sum of amino acid residues is considered as the most accurate procedure to measure the actual protein content. The method includes acid hydrolysis, which destroys all proteins, followed by an analysis of the amino acids in the sample by ion-exchange chromatography. The values for the total amino acid residues are calculated by summing up the amino acid masses retrieved after acid hydrolysis (total amino acid), minus the water mass incorporated into each amino acid after disruption of the peptide bonds. The total amino acid analysis involves some errors, such as the total (tryptophan) or partial (methionine and cysteine) destruction of some amino acids, as well as the impossibility of identifying the contribution of free amino acids in the samples. However, it indicates the maximum possible concentration of protein in the sample, considering that all amino acids are in proteins (Barbarino and Federal, 2005). However, the total amino acid analysis is an expensive and technical method that is seldom used to determine protein contents in seaweeds (Angell *et al.*, 2016; Barbarino and Federal, 2005), and the vast majority of laboratories determine proteins using either direct extraction procedures and the subsequent determination of soluble protein or the indirect method of protein determination using the generic N-protein conversion factor of 6.25.

Ash

Ash content represents the number of total minerals present in biomass (Liu, 2019). Today the ash fraction has a low market value compared to other biomass constituents; furthermore, the high ash content in most algae significantly affect various biomass-processing operations, notably thermal treatments such as pyrolysis, gasification, or combustion, which leads to equipment degradation and lower product quality. Therefore, research has been conducted to remove ash from algae (Liu, 2019; Arthur Robin *et al.*, 2018b); an accurate and reliable measurement of the ash content is critical in documenting the quality of biomass.

The ash content in algae and other biomass is commonly measured gravimetrically by burning samples in a muffle furnace at medium temperature (usually 400–600°C) for a specified duration (Liu, 2019; Arthur Robin *et al.*, 2018b). However, there is no standard method, and the ashing conditions vary among reports in temperatures, duration (1–70 h), and sample size (80 mg to several g), which results in large variations in ash content (Liu, 2019). Liu (2019) investigated the effects of sample size (1, 4g), ashing temperature (550, 600°C), and duration (6, 16 h) on ash measurement of 13 algae species (Liu, 2019). He recommended on ashing of 1–4 g samples at 600°C overnight as a standard method for measuring ash content in all biomass.

Carbohydrates

Carbohydrates are an important component of storage and structural materials in plants. In addition, they are a key target component in the application of macroalgae for the production of bio-based chemicals and biofuels. However, there is no generally accepted analytical protocol for an adequate determination of individual carbohydrate molecules in macroalgae. Therefore, determining the biochemical composition of macroalgae, and thereby estimating their economic potential, is difficult (Van Hal *et al.*, 2014).

Several conventional methods are generally used for the identification and quantification of the carbohydrate content in biological samples: (1) The phenol–sulfuric acid or Dubois method (Dubois *et al.*, 1956); (2) The anthrone method (Dreywood, 1946); (3) Hydrolysis and subsequent determination of resulting monosaccharides by high-performance anion-exchange chromatography with a pulsed amperometric detector (HPAEC–PAD); (4) The NREL method (Charrier *et al.*, 2018); and (5) The By Difference method (Merrill and Watt., 173AD).

The By Difference method calculates the total carbohydrate content in samples by subtracting the percent dry weight of protein, lipid, and ash (which are measured using other techniques) from total dry weight and assuming the remaining mass is a carbohydrate (Fiset *et al.*, 2017). The other assays are based on polysaccharide hydrolysis into simple sugars with concentrated sulfuric acid and heating and estimating the resultant monosaccharides.

In Anthrone and Dubois methods, the monomeric derivatives are quantified by colorimetry. The Dubois method includes phenol in the hydrolysis step to produce furan derivatives that form a yellow complex (absorbs at 490 nm), while the anthrone method includes an anthrone solution in the hydrolysis step to produce hydroxymethyl furfurals that form a green complex in the presence of anthrone (absorbs at 620 nm) (Fiset *et al.*, 2017; Stengel and Connan, 2015).

In both Anthrone and Dubois methods, it is assumed that all carbohydrate monomers present after hydrolysis produce the same colorimetric response as the reference standard used. However, both assays exhibit different colorimetric responses linked to the sugar composition (Fiset *et al.*, 2017; Stengel and Connan, 2015), and thus they are highly dependent on the sugar used for the calibration, and the carbohydrate contents of samples could thus be over- or underestimated.

The anthrone method is often preferred to assay hexoses; however, the anthrone reagent is expensive, and solutions in sulfuric acid are not stable. The Dubois method is widely used among the quantitative assays available for total carbohydrate estimation. It is simple, rapid, reliable, sensitive, and gives reproducible results. The reagent is inexpensive and stable, readily available, and a given solution requires only one standard curve for each sugar; moreover, the color produced is permanent and stable (Stengel and Connan, 2015). Furthermore, the Dubios method provides a more homogeneous response with total sugars and generally lower underestimations of each of the monosaccharides, compared to the anthrone method (Stengel and Connan, 2015).

The NREL method was developed by the U.S. National Renewable Energy Laboratory (NREL) for lignocellulosic biomass. The method is based on two-stage hydrolysis using sulfuric acid (second-stage hydrolysis is performed at 121°C) and subsequent determination of resulting monosaccharides (Charrier *et al.*, 2018).

All methods show substantial variability in the estimates of carbohydrates. The effectiveness of the polysaccharide hydrolyzation step depends on multiple parameters, as the types of polysaccharides present, the concentration of the acid, temperature, and duration of the hydrolysis. Therefore, it is recommended to determine optimum hydrolysis conditions according to the macroalgae class and species (Charrier *et al.*, 2018; Fiset *et al.*, 2017).

Lipids

Lipid and fatty acids (FA) compositions in marine algae have been increasingly gaining importance due to the realization of their beneficial applications in nutritional and pharmaceutical products, as fish oil replacement and as a feedstock for the production of biofuels, in particular, biodiesel through the transesterification of lipids (Kendel *et al.*, 2015; Kumari *et al.*, 2011). Furthermore, in biodiesel production, clean combustion properties of the fuel are influenced by FA structural features including chain length and degree of unsaturation. Thus, a precise quantification of FA is essential and can be used to predict the quality of biodiesel, which is reduced considerably with the increase in the number of saturated FAs (Kumari *et al.*, 2011). However, despite the number of biochemical studies exploring algal lipids and fatty acid biosynthesis pathways and profiles, analytical methods used by phycologists for this purpose are often diverse and incompletely described (Stengel and Connan, 2015).

Traditionally, the analytical approaches for the quantitative determination of total lipids (TLs) and fatty acids (FAs) are based on initial lipid extraction by solvent (solid–liquid extraction, SLE), followed by trans-methylation (i.e. conventional methods) where FA are sought, or methylated with one-step procedures wherein methylation reagent is added directly to the samples without previous extraction (direct transesterification methods, as Garcia method (GM) (Larrosa and Mar, 2012) and Lepage and Roy (LRC) (Lepage and Roy, 1984)), and FA determination by assessing the corresponding methyl esters via gas chromatography (GC) (Kumari *et al.*, 2011).

Several of the lipid extraction methods are by Folch *et al.* (1957), Bligh and Dyer (1959), and AOAC (1990). The Folch extraction method involves extracting homogenized samples in 2:1 (v/v) chloroform–methanol at a 20:1 ratio of solvent: sample, partitioning the crude extract with water or a weak salt solution to achieve 8:4:3 (v/v/v) chloroform–methanol–water, and then removing aqueous contaminants. This biphasic system is then filtered to remove the homogenate and the organic phase is removed as a purified lipid extract (Fiset *et al.*, 2017).

The Bligh and Dyer method is similar to the Folch method, except that the initial organic extraction is performed with 1:2 (v/v) chloroform–methanol at a 3:1 ratio of solvent:sample, which is followed by the addition of chloroform to adjust the solvent to 1:1 (v/v) chloroform–methanol, and has an aqueous partitioning at a ratio of 2:2:1.8 (v/v/v) chloroform–methanol–water. The AOAC 925.32 (Association of Official Analytical Chemists — AOAC, 1990) method involves initial acid hydrolysis of samples at $100\,^{\circ}C$. The hydrolysate is then partitioned with ether and the organic ether phase is removed as a purified lipid extract. The AOAC 923.05 (Association of Official Analytical Chemists — AOAC, 1990) method consists of an initial extraction of homogenized samples with hot aqueous alcohol and ether, followed by filtering of the extract with sand and asbestos. Other methods are similar, but include different solvents as petroleum ether, ethyl, or dichloromethane (Fiset *et al.*, 2017).

The conventional methods are time-consuming and use toxic solvents while direct transesterification methods are simplified, rapid, sensitive, minimize the use of solvents, and give better recoveries of FAs. Despite their disadvantages, the conventional methods of lipid extraction are the only means to study different lipid classes (Kumari *et al.*, 2011).

Furthermore, it is important to note that different solvents are better suited for particular lipids and, therefore, different methods can provide different extraction efficiencies. For example, triacylglycerides (storage lipids) can be extracted with relatively non-polar solvents such as chloroform, but more polar lipids associated with membranes require polar solvents such as methanol to be fully extracted (Fiset *et al.*, 2017).

Nowadays, there are also expensive techniques and instruments such as thin-layer chromatography coupled with flame ionization detector (TLC-FID), thin-layer chromatography matrix-assisted laser desorption and ionization-mass spectrometry (TLC-MALDI-MS), or high-performance liquid chromatography coupled with atmospheric pressure chemical ionization-mass spectrometry (HPLC-ACPI-MS), which are available and are used to separate and quantify the lipids. The total lipids are extracted from algal biomass by SLE, the major lipid classes are fractionated using silica cartridges and TLC, and the individual lipid classes are scraped from the TLC plates and converted to fatty acyl methyl esters by direct or non-direct transmethylation, before analysis by GC-FID (Stengel and Connan, 2015).

Chapter 7

Downstream Processing

7.1 Sample Handling, Post-harvest Treatment

We previously went through the different cultivation approaches for seaweeds, and the following chapters are mainly focused on how to process seaweeds, and how they can be used. However, we need first to cover the post-harvest biomass handling. This section will, therefore, talk about what to do with the seaweed biomass after harvest and before processing it and why it is an important step.

Indeed, if the cultivation technique, the harvesting time, and location can have a dramatic effect on the seaweed biomass constituents (see Chapter 6), the post-harvest handling plays a critical role in preserving the quality of biomass constituents (Neveux *et al.*, 2014; Robic *et al.*, 2008; Uribe *et al.*, 2019). As seaweed are marine organisms living in a non-sterile dynamic environment, they have considerable water content and are home to a plethora of organisms including small animals, microorganisms, other seaweeds (Ingle *et al.*, 2018b), but also inorganic materials such as sand and sea salt. Post-harvest treatments are therefore essential to remove contaminants, preserve seaweeds from biotic and abiotic degradation, and ensure the efficiency of their transport and storage before processing. Most of those treatments can be separated into three sections: rinsing, drying, and storage, which are covered separately in what follows.

7.2 Abiotic Spoilage

Following harvesting, the fresh seaweed is subject, out of its natural environment, to abiotic factors (i.e. "that are not derived from living organisms") that can lead to spoilage, Fig. 7.1, which is usually linked with the loss of water, mechanical damage, light (mostly sunlight), and oxidation. Mechanical damage will usually not directly affect the biomass constituents; however, it will drastically accelerate abiotic and biotic spoilage by increasing the surface-to-volume ratio and damaging cell wall and cell membrane which protect the tissues from spoilage. Light can alter seaweed biomass by various phenomena but mostly by the photobleaching of the pigments and photolysis of light-sensitive elements. Moreover, light, and notably UV light, can create reactive oxidative species (ROS) and damage protein, genetic materials, or polysaccharides (Sappati *et al.*, 2019; Vairappan *et al.*, 2014). Finally, as water usually has a low oxygen content due to its poor solubility in it, the oxygen in the air will increase

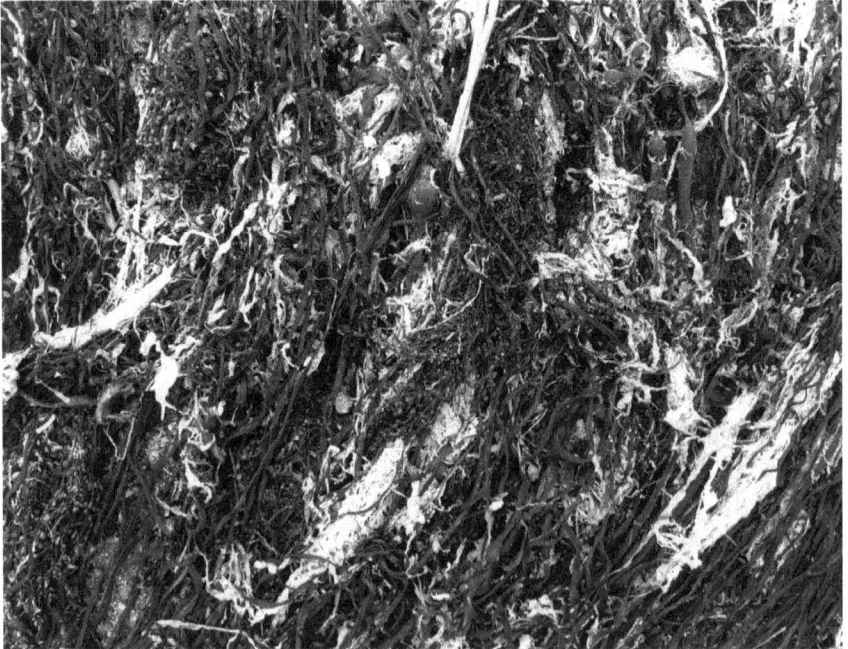

Fig. 7.1   Seaweeds on shore getting spoiled by abiotic and biotic agents.
Source: https://commons.wikimedia.org/wiki/File:Dried_seaweed.jpg, credit: public domain.

the overall oxygen content in the biomass, leading also to various damages by oxidation starting from the most oxidative-sensitive constituents such as antioxidants. Moreover, the loss of water by dehydration can lead to structural changes in the tissue to the decrease of hydrophilic interaction: those changes can sometimes be irreversible and detrimental, affecting, for example, the solubility and thus the extractability of molecules of interest such as proteins or polysaccharides (Fleurence *et al.*, 1995).

7.3 Biotic Spoilage

The biotic agents are the main sources of spoilage of seaweed biomass. First, a lot of microorganisms are living at the surface of the seaweed (bacteria, fungus, microalgae, archaea, etc.) and among them, many have the capability of spoiling seaweed tissues (Ingle *et al.*, 2018b). For example, those microorganisms, as part of the epiphytes, are studied today as a source of new enzymes, notably polysaccharides, that are capable of hydrolyzing the unique polysaccharides found in seaweeds (Coste *et al.*, 2015; Kang *et al.*, 2016). If the communities of microorganisms found an equilibrium during the life of the seaweed as a seaweed can regulate its epiphytes (Ingle *et al.*, 2018b), once harvested, this equilibrium is broken, and the unregulated epiphytes can lead to the production of enzymes capable of degrading seaweed constituents and thus causing spoilage. Such spoilage is a complex phenomenon, involving dynamic and diverse microorganisms' communities that compete or work synergistically. It is in a way similar to other land biomass, including crop and food, and thus followed the same and most accepted rule for preservation: hindering all microorganism activity to prevent spoilage.

The second biotic source of spoilage is pluricellular organisms that directly feed on seaweed biomass. Among them, most are from the marine environment: zooplanktons, small fish, crustaceans, larvae from various animals, etc. (Ingle *et al.*, 2018b) but some can come from land, such as insects. Those organisms consume seaweed biomass, thus spoiling it, and often reproduce themselves, thus being able to increase spoilage over time as microorganisms do if the conditions allow it. The third and last biotic source of spoilage is the seaweed itself. Indeed, it is important to remember that seaweed is a living organism, that its harvesting will apply stress to it and that living organisms have stress-response mechanisms to cope with stress. Those mechanisms can prevent spoilage but can also accelerate it. For example, due to the metabolism shift during harvest, seaweed tissues might rely on storage to find energy and resources and thus degrading storage constituents of interest (storage carbohydrates, protein, etc.). Other such mechanisms are the production of stress-related products that can be detrimental to future uses because of their toxicity, alteration of flavor, inhibiting properties, etc. Finally, it is important to underline the fact that the three biotic spoilage agents mentioned above usually use complex chemical processes to spoil seaweed, notably by the excretion of enzymes. Thus, preventing biotic spoilage require working on two aspects: the source of the biotic spoilage (the organisms), and their spoiling arsenal (enzymes, etc.).

7.4 Rinsing and Deashing

Rinsing with water with the help of mechanical energy (pressure, mixing, screening, etc.) is an effective way to remove macro and certain micro contaminants. However, rinsing does not alter the water content of seaweed tissues and thus is very limiting in reducing biotic spoilage. Note that the use of freshwater, deionized water, and other low-salt content water might cause osmotic shock for seaweed cells, which can increase spoilage by altering the protective role of the membrane, and the leaching of intracellular components. The removal of ash content, which is detrimental to many bioprocesses and has low value, is an effective way of improving the quality of seaweed feedstock. For these purposes, several chemical or physical processes have been proposed with various advantages and disadvantages (Gallagher *et al.*, 2018; Magnusson *et al.*, 2016a; Neveux *et al.*, 2014; Robin *et al.*, 2018c).

7.5 Drying

Cultivation and subsequent harvesting of seaweeds are only the first steps of "resource to product" processes, supplying the feedstock for the biorefinery. Next, post-harvest treatments such as cleaning, washing, size reduction, preservation, drying, storage, and energy extraction are applied (Milledge *et al.*, 2014b), Fig. 7.2. Biorefinery conversion technologies can be divided into those that require a drying step, and those that do not. Conversion methods such as direct combustion, pyrolysis, gasification, and transesterification to biodiesel require dry input (Milledge *et al.*, 2014b).

Fig. 7.2   Drying green seaweed in Vientiane, Laos (by Paul Arps, CC-BY-2.0).

Drying or dewatering (mechanical removal of water) are key process-ing stages in obtaining algal biomass as a product. Algal biomass needs to be dried to increase its "shelf-life" and to be more suitable for oil or metabolite extraction (Milledge *et al.*, 2014b; Shahid *et al.*, 2017). Harvested macroalgae have high initial water content (80–90%) (Milledge and Harvey, 2016b). Wet biomass is subjected to rapid deterioration in its quality and is heavy to transport (Gallagher *et al.*, 2017). Reducing the water content (<30%) prevents the growth of spoilage-causing microor-ganisms, slows down detrimental enzymatic reactions, allows stable stor-age, and reduces transportation costs for its customers (Chan *et al.*, 1997; Milledge and Harvey, 2016b; Shahid *et al.*, 2017).

At the same time, the drying methods are energy-intensive processes. So special care has to be spent on choosing the drying technology. However, drying can also be detrimental by affecting seaweed constitu-ents (Chan *et al.*, 1997; Uribe *et al.*, 2019; Wong and Chikeung Cheung, 2001). As it often involves high temperature, sunlight, and/or increase of surface-to-volume ratio to improve the contact with air. This will promote the degradation of temperature and oxidation-sensitive constituents by abiotic factors or can enhance degradation by biotic agents (notably enzymes). Another consequence is structural as the cell walls of most seaweeds are made of charged polysaccharides, which absorb a lot of water, provoking the swelling of the tissues and contributing to the mechanical strength of the tissues. Removing water changes the chemical bonds between cell wall components, their swelling, and thus altering some of its properties, sometimes irreversibly. For example, pigment removal in seaweed has been reported to be more efficient on fresh undried seaweeds. Finding a proper drying method is, therefore, finding the balance between reducing spoilage, saving energy, and preserving sensitive materials (Robic *et al.*, 2008; Uribe *et al.*, 2019). It will depend on the purpose of the harvester, the seaweed species, and the subsequent processing chains or usage.

Water evaporation requires a lot of energy (at least 800 Kcal for evaporating 1 kg of water) and has generally been recognized as the economic bottleneck of algae processing. It can account for up to 70% of the total cost (Walker *et al.*, 2016). On the other hand, mechanical removal of water (dewatering) generally uses less energy than evaporation. Therefore, it can be preferable to minimize the water content of the harvested algae before drying (Milledge and Harvey, 2016b). Dewatering to 20–30% of water content is usually a good objective. It stabilizes the biomass, allows transportation without too much water, and reduces the energy consumption if further drying steps are required (Lyons, 2009).

Unfortunately, while dewatering techniques for microalgae have been widely studied, research with macroalgae is comparatively much less advanced (Gallagher *et al.*, 2018). Currently, established methodologies (as pressing or centrifugation methods) are not always able to deliver sufficiently low water content (Mahmoud, 2010). The small number of studies that have been published showed that dewatering is highly dependent on algal species and season of the collection (Gallagher *et al.*, 2018) and that several pre-treatments can improve dewatering results (Mahmoud, 2010; Raghavan, 2015). For example, kelps were found to be challenging to dewater because their alginate content gives a slippery gel-like nature which strongly binds water. Only <1% of the water was removed by screw-pressing (Gallagher *et al.*, 2018; Raghavan, 2015). These works have found that mineral acid pre-treatments followed by pressing are suitable and effective for dewatering (Gallagher *et al.*, 2018;

Raghavan, 2015). Another useful technique was the combination of mechanical pressing with electro-osmosis (Mahmoud, 2010). Further studies need to be done on macroalgae dewatering techniques. Establishing an industrial dewatering method to achieve sufficient low water content will decrease the need for further thermal drying and the overall cost of algal-based products.

When further drying is required, several technologies are available. In choosing the most appropriate drying method, one needs to consider the environmental and economic factors, as well as the purpose the macroalgae will eventually be used for. The drying procedure might harm the chemical composition of the algae. Therefore, whether the macroalgae will be used as food, medicine, animal feed, fertilizer, or a source of specific nutrients will determine how it should be dried. Among the parameters that can affect the quality of the product, energy consumption, or efficiency of the process are algae size and thickness, its initial and final moisture content, temperature and relative humidity of drying gases, their velocity, and exposure time. The next sections will describe conventional drying methods.

Sun-drying: Solar methods are the least expensive drying option, but they are weather and volume-dependent (Milledge and Harvey, 2016b), require more extended time frames, and are more labor-intensive than the other methods (Chan *et al.*, 1997), Fig. 7.2. For example, sun-drying in tropical locations can take 2–3 days in sunny weather and up to 7 days in rainy seasons. Also, large areas are required as only around 100 g of dry matter can be produced from each square meter of sun-dried surface (Milledge and Harvey, 2016b). Furthermore, intense solar radiation can dehydrate algal chlorophyll and affect the texture and color of the final product (Shahid *et al.*, 2017). Depending on temperature and time of exposure, nutritional value can also be damaged (Chan *et al.*, 1997).

Freeze-drying (lyophilization): This refers to the sublimation or removal of water content from the frozen product under vacuum conditions (Lombrana, 2009). The process can be described as a sequence of three operations:

A. Freezing stage. The material is cooled below its triple point (the lowest temperature at which the solid, liquid, and gas phases of the material can coexist). This stage should be done slowly to induce a desired large ice crystal structure within the sample, which facilitates drying.

B. Primary drying phase. Drying of ice by sublimation at a pressure below that corresponding to the triple point (a range of few millibars). This stage removes 70–95% of the total sample moisture.

C. Secondary drying phase. Desorption of unfrozen water resident in the dried cake. Removes 5–10% of the total sample moisture (Adams *et al.*, 2015; Lombrana, 2009). After the freeze-drying process is complete, the vacuum is usually broken under an inert environment, and the material is sealed under controlled conditions (humidity, etc.). At the end of the operation, the final residual water content in the product can be extremely low (< 4%).

Freeze-drying provided seaweeds with the best nutritional quantity and preservation of the bioactivity of its compounds. However, the equipment and operating costs for freeze-drying are higher, and its drying capacity is smaller than other drying methods (Chan *et al.*, 1997). Since this method is expensive, it is used mainly to produce high-value products (for nutritional supplements or pharmaceutical use).

Rotary drying: Rotary dryer is an industrial dryer employed to reduce the liquid moisture content of the algal biomass by direct contact with a heated gas. The dryer is designed to have a slope rotating cylinder; hence, it can move the macroalgae inside the dryer from the feeding inlet to the discharging outlet by gravitational force. Wet biomass is fed continuously at a steady rate. During its progression to the discharge, the biomass is lifted with the dryer rotations and a series of internal fins lining the inner wall of the dryer. When the biomass gets high enough to roll back off the fins, it falls back down to the bottom of the dryer, passing through the hot gas stream. The gas flow can be set up in countercurrent or co-current modes. The countercurrent mode is preferable to maintain constant and uniform heat-exchange along the process (Aziz *et al.*, 2014; Shahid *et al.*, 2017). This method is suitable for industrial applications due to the large heat transfer area leading to high thermal efficiency, excellent drying control, comfortable handling capability, easy implementation for continuous operation (Aziz *et al.*, 2014), and relatively small variability in the quality of the product. Algae moisture content, nutritional value, and thermal efficiency of the process can be determined by gas temperature and velocity, rotation velocity of the drum and contact surface between biomass and gas (size of algae, their feeding rate).

Spray drying: This is a typical industrial drying process in the production of powder and granules from liquid materials. Atomized droplets of media are sprayed vertically through hot gas, which helps to dry the biomass within a short time (seconds to few minutes) (Shahid *et al.*, 2017). Typically, the hot gas is air, but for sensitive materials such as pharmaceutical products, nitrogen gas can be used instead. The liquid feed can be in the form of a solution, suspension, dispersion, emulsion, while the dried product can be obtained in the form of powders, granules, or agglomerates depending upon some process variables, such as physical and chemical properties of the liquid feed (particle size, flow rate, viscosity, solubility) and of the drying air (inlet and outlet temperature, pressure) as well as the dryer design (a type of atomizer) (Bono *et al.*, 2011). The advantages of the method are high ability to control the product particle size, and suitability for thermosensitive products as the biomass is exposed to heat for a very short time, which permits drying with minimal thermal degradation (Bono *et al.*, 2011; Shahid *et al.*, 2017). The experimental result has shown that there is a connection between powder size, antioxidant activity, moisture content and process operation conditions (Bono *et al.*, 2011; Tun *et al.*, 2016). However, the method is quite energy-intensive in many cases and requires pre-treatment for biomass as washing, chopping, liquidizing, and filtration. Also, the high pressure inside the tank can rupture the intact cells, which may deteriorate the product quality (Shahid *et al.*, 2017).

Vacuum-shelf drying: Biomass is placed inside the dryer where the water molecules are allowed to escape by heat and pressure. The use of pressure makes the drying faster and cheaper, but the entire process requires the higher installation and operational costs (Shahid *et al.*, 2017). Also, the method can be done only on a batch mode, and it is not very efficient; the biomass near the source of the dried air is over-dried, while that away from the source is left under-dried. As a result, there could be a moisture gradient in the product.

7.6 Storage

Storing seaweed biomass in proper conditions can help to preserve its constituents and limit spoilage (Robic *et al.*, 2008). For example, storing seaweeds at low temperature or below the freezing point of water will prevent most spoilage. However, they are not viable options for pilot or industrial quantities. Ensilage is also a viable method, using lactic bacteria to lower the biomass pH and preserve it against spoilage, however, seaweed ensilage is not a straightforward technic for all seaweed species and needs to be tuned and monitored properly (Cabrita *et al.*, 2017). As storage has a cost in terms of infrastructure, surface usage, or energy expenditures, it should be considered for the lowest amount of time, under ambient conditions. Protecting the seaweed from light, at ambient temperature, and under local weather conditions is relatively easy and cheap but requires an important drying step to prevent spoilage after few hours. There is an interdependency between the drying process and the subsequent storage step, but the cost of both can be avoided by directly processing the biomass close to the harvesting site, which should be the preferred approach, notably for energy considerations (Milledge and Harvey, 2016b). Rinsing, dewatering, deashing, drying, and storage are post-harvest treatments that can be critical to improving and safeguarding the seaweed constituents and their value for biorefineries. However, such processes have advantages and disadvantages in terms of efficiency, cost, and feasibility, and whose choices depend on the seaweed species and the subsequent usages. Usually, the lower the number of processing steps, the better it is in terms of cost, energy expenditure, and effect on the biomass.

7.7 Milling

Milling, grinding, or chopping are mechanical pre-treatments commonly used to reduce algae particle size and increase their accessible surface area. Smaller particle size improves enzymatic reaction efficiency (e.g. during anaerobic digestion for biogas, fermentation for alcohols, and hydrothermal liquefaction for bio-oils) than un-milled algae (Ghadiryanfar *et al.*, 2016; Maneein *et al.*, 2018; Montingelli, 2015; Roesijadi *et al.*, 2010a).

Mechanical pretreatments have been mainly applied to lignocellulosic feedstock (Montingelli, 2015). The inherent properties and the composite structure of lignocellulosic materials make them resistant to enzymatic attack. The crystallinity of cellulose, its accessible surface area, protection by lignin and hemicellulose, degree of cellulose polymerization, and degree of acetylation of hemicelluloses are the main factors affecting the rate of digestibility of lignocelluloses. Hence, the treatment of biomass is needed to increase cellulose and hemicellulose accessibility and enzymatic reaction effectivity (Kratky and Jirout, 2011).

Biomass can be comminuted by a combination of chopping, milling, and grinding. Usually, after chopping a particle size of 10–30 mm can be achieved, while grinding or milling allow finer sizes in the range of 0.2–2 mm (Montingelli, 2015). Several milling machines are commercially available. The choice of suitable grinding or milling machine depends on the moisture content of the biomass and energy consumption. Colloid mills and extruders are suitable wet materials with moisture contents of more than 15–20% (wet basis), whereas two-roll, attrition, hammer, or knife mills are suitable for dry biomass with moisture contents of up to 10–15% (wet basis). Ball mills can be used for either dry or wet materials (Kratky and Jirout, 2011). Generally, the energy requirements of the process depend on the type of the machine, initial and final particle sizes, and biomass characteristics (i.e. processing amount, composition, and moisture content) (Kratky and Jirout, 2011; Montingelli, 2015).

Unfortunately, the same milling techniques used for lignocellulosic terrestrial plants may not produce the expected increase in surface area for seaweeds (Adams *et al.*, 2017). For instance, milling of *L.digitata* did not significantly increase its surface area, and mechanical wet milling of *L. digitata* using cutting discs did not enhance glucose release (Manns *et al.*, 2016). In addition, different species of macroalgae may exhibit different beneficial values from mechanical treatment. The differences can be due to diverse cell wall structure, where macroalgae with more fibrous cell walls would benefit more from size reduction (Paul *et al.*, 2016). For example, Amamou *et al.* (2018) showed that neither vibro-ball milling nor centrifugal milling of *Ulva Lactuca* affected its sugar release (Amamou *et al.*, 2018). Centrifugal milling of *U. Lactuca* showed a 4.4% reduction in ethanol yield. However, the same treatment on *Gelidium sesquipedale* showed up to a 129% increase in sugars released, increasing the ethanol yield by 80% compared to only cutting milled seaweed. Ball milling of *Chaetomorha Linum* also enhanced bioethanol production by 63.6% compared to non-milled biomass (Maneein *et al.*, 2018; Schultz-Jensen *et al.*, 2013).

The size reduction step is a very expensive operation that can consume up to 33% of the total electrical demand of the process (Kratky and Jirout, 2011; Montingelli, 2015). Therefore, a thorough and careful evaluation of the machines employed as well as of the macroalgae characteristics and beneficial value from mechanical treatment is required to improve the economics of the whole process (Montingelli, 2015).

7.8 Bioprocessing of Seaweed Biomass: Some Classic and Emerging Technologies

Although drying (see Section 7.5) the algae can extend storage time and decrease feedstock transport cost (Lehahn *et al.*, 2016b), it requires high energy inputs (Milledge *et al.*, 2014b), which is problematic if the energy source is expensive or non-renewable. In addition, these methods are less suitable for seaweeds due to their high ash and alkali contents (Na and K) (Bruhn *et al.*, 2011a). For this reason, this part of the chapter focuses on conversion methods that do not require drying and can utilize wet algae, such as hydrothermal treatments, fermentation, and anaerobic digestion (Milledge *et al.*, 2014b). We then choose to cover the recent trends in emerging sustainable processes as classical processes are becoming or will become obsolete, even though there is no proper commercial seaweed biorefinery to relate at the time of the writing.

7.8.1 *Hydrothermal treatments*

Hydrothermal subcritical water technologies utilize liquid state high-pressure-high-temperature (100–374°C) water to process the biomass into a variety of products (Cocero *et al.*, 2018). These can be used as, or further processed too, different types of biofuels and other products such as artificial soil, fertilizers, activated carbon, and more (Libra *et al.*, 2011). Hydrothermal hydrolysis, which occurs in water heated to 100 to 240°C is the break-down of polymers into monomers such as simple sugars, which can be fermented into organic chemicals such as ethanol, butanol, and acetone (Roesijadi *et al.*, 2008). In hydrothermal carbonization (HTC), which occurs in water heated to 180 to 250°C, the carbon fraction in the solid residue (hydrochar) (Kambo and Dutta 2015) is enhanced, thus providing carbon-based products (Libra *et al.*, 2011) and increasing the residue's calorific value. In hydrothermal liquefication (HTL), which occurs at water temperatures above 280°C, biomass is liquified to a high-energy liquid bio-oil (Toor *et al.*, 2011). This bio-oil can be upgraded using different techniques (solvent addition, emulsification, esterification, hydro or zeolite cracking, and others) and refined for different fuel applications (Saber *et al.*, 2016).

Results from hydrothermal treatments performed on green seaweed biomass in recent years are presented in Table 7.1. Daneshvar *et al.* (2012), for example, reported that soluble sugar production in *C. fragile* began at 170°C and reached a maximum at 210°C where more than 50% of the dry algae biomass was converted to soluble carbohydrates. The higher heating value (HHV), calculated by the Boie Eq. 7.1 (Mason and Gandhi, 1980), increased constantly between 140°C to 230°C, till a maximum of 22.6 MJ kg⁻¹, an increase of 6 MJ kg⁻¹ compared to the algal initial HHV.

$$Q = 151.2 \ C + 499.77 \ H + 45.0 \ S - 47.7(O) + 27.0 \ N \qquad (7.1)$$

where Q is the gross heating value in Btu/lb on the dry basis and C, H, S, (O), and N are the respective contents of carbon, hydrogen, sulfur, oxygen, and nitrogen in DW percent. Neveux *et al.* (2014b) conducted HTL on six green macroalgae (Table 7.1) and measured HHV of 4–20.5 MJ kg⁻¹ for the hydrochar, and 32.5–33.8 MJ kg⁻¹ for the bio-oil. The highest hydrochar HHV was measured for the *Chaetomorpha* genus, while bio-oil HHV had no significant differences between species. Zhou *et al.* (2010) liquified *U. prolifera* and found that the bio-oil yield, calculated with the Dulong Eq. (7.2) (Mason and Gandhi, 1980), increased with temperature up to a maximum of 20.4% (w/w) and HHV of 28.7 MJ kg⁻¹ at 300°C.

$$Q = 145.44 \ C + 620.28 \ H + 40.5 \ S - 77.54(O) \qquad (7.2)$$

where Q is the gross heating value in Btu/lb on the dry basis and C, H, S, and (O) are the respective contents of carbon, hydrogen, sulfur, and oxygen in DW percent.

Hydrothermal hydrolysis of green seaweeds, which does not involve any hazardous chemicals, can potentially serve as a green preliminary fermentation step. However, this technology is not yet efficient enough and faces optimization challenges due to temperature overlaps with monosaccharides' deconstruction (Toor *et al.*, 2011) and carbonization processes. In this instance, HTC and HTL have the advantage of achieving possible full conversions into hydrochar or bio-oil. However, beyond the high energy consumption disadvantage, large-scale implementation of HTC and HTL on seaweeds is challenging because of its high ash content, which causes fouling issues in large-scale continuous flow reactors due to the presence of alkali metals, earth alkaline metals, or halides

Table 7.1 Hydrothermal treatments for the processing of seaweed biomass.

Species	Treatment	Temperature/ Pressure range	Reaction time	Maximal yield	Optimal conditions	Reference
Ulva pertusa Kjellman	Hydrolysis	100–200°C	2–12 minutes	8.5% glucose (w/w)	180°C, 10.48 bar and 8 minutes	Choi *et al.* (2013)
C. fragile	Hydrolysis	100–240°C	10 minutes	>50% soluble carbohydrates (w/w), HHV of 22.6 MJ kg^{-1}	210°C	Daneshvar *et al.* (2012)
Ulva, Derbesia, Chaetomorpha, Cladophora and *Oedogonium* Link ex Hirn	HTC, HTL	330–341°C/ 140–170 bar	5 minutes	20.5 MJ kg^{-1} for hydrochar and 33.8 MJ kg^{-1} for bio-oil		Neveux *et al.* (2014b)
U. prolifera	HTL	220–320°C		20.4% bio-oil (w/w) and HHV of 28.7 MJ kg^{-1}	300°C	Zhou *et al.* (2010)

Source: Adapted with permission from Zollmann *et al.* (2019a).

(Neveux *et al.*, 2014a). This problem may be solved by pre-treatment ash removal, for example, by PEF (Robin *et al.*, 2018b) or rinsing (Neveux *et al.*, 2014a).

7.8.2 *Fermentation for ethanol production*

Ethanol is a common fermentation product and is commonly blended into transportation fuels (Mussatto *et al.*, 2010; Wang *et al.*, 2011). Ethanol production from seaweeds was examined by the American government as an economic bio-alternative to fossil fuels in the 1980s (Mcintosh, 1985; Wagener, 1981).

Before fermentation, polysaccharides in seaweeds must be hydro-lyzed into monosaccharides. Macroalgae contain several distinctive monosaccharides, including rhamnose, xylose, gluconic acid, galact-uronic acid, fucose, and more (Kim *et al.*, 2011; Robin *et al.*, 2017). Acid hydrolysis is widely used in biomass degradation processes. However, detailed parameter optimization of the thermochemical hydrolysis of *Ulva* was performed only recently (Jiang *et al.*, 2016b). Although common, acid hydrolysis produces non-sugar by-products, causing environmental hazards and slowing subsequent fermentation (Palmqvist *et al.*, 1999). These by-products include formic acid, levulinic acid, acetic acid, 5-hydroxymethylfurfural (HMF), phenols, and heavy metals (Trivedi *et al.*, 2015; Wu *et al.*, 2014). An alternative that is considered environ-mentally friendlier, but slower (more than a day compared to less than an hour) and more expensive, is enzymatic hydrolysis (Trivedi *et al.*, 2015). An effective enzymatic hydrolysis process requires a pre-treatment which enables the enzymes better accessibility to the cellulose by increasing available surface area and removing structural interferences (for example, hemicellulose) (Alvira *et al.*, 2010). This energy and/or chemical demand-ing pre-treatment can be eliminated if hydrolysis is performed after prior extraction of other biomass fractions, following a biorefinery cascading extraction process. An optimization study by Trivedi *et al.* (2013) for *Ulva fasciata* Delile achieved a maximal sugar yield of 206.82 ± 14.96 mg/g after a pre-heat treatment in aqueous medium at 120°C for 1 h, followed by incubation in 2% (v/v) of the enzyme cellulase 22,119 for 36 h at 45°C. This study also confirmed that enzymes can be used twice without com-promising saccharification efficiency, which is an industrial and environ-mental advantage. Enzymes that can be used for this purpose include cellulase for cellulose (El-Dalatony *et al.*, 2016; Trivedi *et al.*, 2015), amyloglucosidase and α-amylase for starch (Korzen *et al.*, 2015a), and advanced enzymatic complexes such as Viscozyme, combining cellulase, *β*-glucanase, hemicellulase, and xylanase (Kim *et al.*, 2014). Another hydrolysis method is hydrothermal hydrolysis for which applications on macroalgae are discussed in what follows.

The most effective fermenting microorganism known today is *Saccharomyces cerevisiae*. This yeast has high fermentation rates for glucose, fructose, and mannose (van Maris *et al.*, 2006), and in anaerobic conditions achieves high yields of ethanol (Wang *et al.*, 2004). However, saline environments are toxic to *S. cerevisiae* (Tekarslan-Sahin *et al.*, 2018), and therefore algal pre-washing is required (Roesijadi *et al.*, 2010). Pre-wash may be avoided in the future by utilizing the newly developed salt-resistant strains (Tekarslan-Sahin *et al.*, 2018). In addition, this yeast has limited fermentation of non-glucose sugars. Therefore, *Escherichia coli* was genetically engineered to produce ethanol from pentose and hexose sugars, offering a seaweed fermentation alternative (Asghari *et al.*, 1996; Golberg *et al.*, 2014; Vitkin *et al.*, 2020). Thus, *E. coli* can be used to ferment monosaccharide sugars such as rhamnose and glucuronic acid, which occur in large quantities in green macroalgae (Kim *et al.*, 2011), but with lower ethanol yields (Kim *et al.*, 2011; Saha *et al.*, 2003). Other alternative microorganisms are *Clostridium* species that can produce acetone, butanol, and ethanol by anaerobic fermentation from a wide variety of sugars (van der Wal *et al.*, 2013).

Fermentation can be performed after the hydrolysis process, in the Separate Hydrolysis Fermentation (SHF) method, or simultaneously with hydrolysis, in the Simultaneous Saccharification and Fermentation (SSF) method. In SHF, both hydrolysis and fermentation steps can be optimized separately, thus enabling higher sugar and ethanol yields. However, this comes at a cost of time and capital (Olofsson *et al.*, 2008). In comparison, the SSF saves time and reduces process steps and capital costs. However, the prominent downside of the SSF is operating the enzymes in sub-optimal temperature conditions, leading to lower yields of sugar and ethanol (Olofsson *et al.*, 2008). For the sake of future optimization, Vitkin *et al.* (2015) have built BIO-LEGO, a web-based application for biorefinery design and evaluation of serial biomass fermentation. Table 7.2 presents results from ethanol production experiments performed in recent years on marine macroalgae, *Ulva* sp.

Fermentation is a long known and well-industrialized process. However, the fermentation processes of marine biomass still need to be

Table 7.2 Comparison of reported ethanol production from *Ulva* sp., using different microorganisms.

Species	Hydrolysis method	Total sugar (mg g^{-1} DW)	Microorganism	Ethanol yield (g/g sugar)	Reference
U. lactuca	Acid + enzyme	343	*Clostridium beijerinckii*	0.4	Bikker *et al.* (2016)
Ulva sp.	Enzyme	200	*E. coli* Ko11[a]	0.4	Kim *et al.* (2011)
U. fasciata	Hot buffer + enzyme	±112	*S. cerevisiae* MTCC No. 180	0.47	Trivedi *et al.* (2015)
U. Lactuca	Acid	113	*S. cerevisiae*	0.55	El-Sayed *et al.* (2016)

Note: [a]*E. coli Ko11* was modified to allow more efficient sugar utilization.
Source: Adapted with permission from Zollmann *et al.* (2019a).

optimized. Carbohydrates from green seaweed include very low lignin ratios compared to terrestrial crops (Dave *et al.*, 2013), which is an advantage due to lignin's structural interference with cellulose extraction (Cheng, 2017). However, a major challenge is maximizing the conversion ratios of monosaccharides to ethanol, which is challenging due to the limitations of the fermenting organisms to ferment non-glucose sugars. In addition, this partial conversion leads to high organic matter content in the process effluent, which requires additional treatment before discharge (Pimentel, 2003). Other challenges relate to the hydrolysis step that is subject to toxicity or efficiency problems, depending on the method used. Finally, despite its potential, sustainable ethanol production from algae is not mature yet and requires more research to increase yields before future implementation.

7.8.3 *Anaerobic digestion*

Macroalgae are a potential source of anaerobic digestion (AD). The AD product is biogas, a gaseous mixture containing about 60–70% methane and 30–40% CO_2 (Hosseini and Wahid, 2014), and variable trace amounts of CO, N_2, O_2, H_2, and the undesirable H_2S. H_2S must be removed before downstream conversion (Roesijadi *et al.*, 2010). Biogas can be exploited directly as a fuel or used as a raw material for the production of synthetic gas or hydrogen (Hosseini and Wahid, 2014; Song *et al.*, 2015).

Early attempts to cultivate macroalgae as AD feedstock were made in the United States during the 1970s and 1980s (Roesijadi *et al.*, 2010). Later, green macroalgal genera such as *Ulva* were examined, too (Habig *et al.*, 1984). A major advantage of green macroalgae is the very low lignin content, which is beneficial for methane generation (Dave *et al.*, 2013). For instance, *U. lactuca* contains 1.56 ± 0.08 (% w/w on a dry basis) lignin (Yaich *et al.*, 2011). In addition, for green macroalgae such as *Ulva* and *Derbesia tenuissima* were demonstrated high production potentials of 45–56 and 138 tons DW per hectare per year, respectively (Bruhn *et al.*, 2011; Mata *et al.*, 2016), 58% of DW (Rasyid, 2017). This combination of rapid growth, high carbohydrate, and low lignin levels make *Ulva* appropriate biomass for biogas production (Dave *et al.*, 2013) (Table 7.3). However, pre-processing such as washing is necessary (Roesijadi *et al.*, 2010) to prevent salt inhibition. Inhibition can be caused also by increased H_2S content. This is because sulfur, which appears in green macroalgae mainly as SPs, dimethylsulfoniopropionate (DMSP), and sulfur-containing, amino acids (Stefels 2000; Wang *et al.*, 2014) is hardly removed during the wash (Bruhn *et al.*, 2011). Additional pretreatment procedures such as mechanical maceration, drying, thermal treatment, or solid/liquid separation can further increase methane yields, but again, these come with an energetic cost (Nikolaisen *et al.*, 2011). Methane production is also affected by feedstock C:N ratio, as low nitrogen may be harmful to methanogenic microorganisms. Compared to the optimal range of 15.5–19 C: N (Sievers and Brune, 1978), *Ulva* can be

Table 7.3   Potential methane yield for green macroalgae.

Species	Reported potential methane yield	References
Ulva sp.	0.22–0.33 m^3 kg^{-1} volatile solids (VS)	U.S. Department of Energy Roesijadi *et al.* (2010)
U. lactuca	4000–7000 m^3 CH_4 $hectare^{-1}$	Bruhn *et al.* (2011)
Cladophora and *Chaetomorpha*	0.48 m^3 kg^{-1} VS	Gunaseelan (1997)

Source: Adapted with permission from Zollmann *et al.* (2019a).

found in a wide C: N range of 7.9 up to 24.4 (Bruhn *et al.*, 2011), depending on cultivation conditions.

Finally, although green macroalgae present a high potential for methane production, it is still not widely implemented as an AD source. Some reasons are the high cultivation and pre-treatment costs and the current difficulty of ensuring a reliable and constant supply of suitable feedstock. However, a major opportunity may be the co-utilization of seaweeds for bioremediation and bio-energy needs. Another issue is high biomass water content, which, until cost-efficient concentrating methods are developed, lead to AD digesters too big to be economically feasible (Bruhn *et al.*, 2011).

7.8.4 *Green solvents for clean processing*

Most biomass processes take place in liquid media. Thus, solvents become a major process input. Their cost, availability, and recyclability are crucial. Moreover, with the strengthening of environmental and safety regulations, solvents that are hazardous, polluting, and/or non-renewable (such as petroleum-based solvents) are becoming less relevant (Chemat *et al.*, 2012). The quest for green solvents has therefore been a key step toward green processes and green chemistry. The following section presents some of the major green solvents, i.e. water, bio-based solvents, supercritical fluids, deep eutectic solvents, aqueous two-phase system and ionic liquids, and their applications for macroalgal processing.

Water is by far the most used green solvent and is the solvent of choice for any green process, notably because it is also a major component of fresh biomass (Chemat *et al.*, 2012; Herrero and Ibáñez, 2015; Rombaut *et al.*, 2014). As "classic" uses of water, such as in acidic, enzymatic, or hydrothermal treatments are discussed above, here we present two promising water-based solvent extraction systems: Pressurized Liquid Extraction (PLE) and Aqueous Two-Phase System (ATPS).

PLE is the reduction of the high water polarity by changing temperature or pressure, leading to the extraction of low-polarity components that are usually poorly soluble in water without the need of adding any chemicals or catalysts (Herrero *et al.*, 2006). Classic solid-liquid extraction methods are sometimes enough to extract compounds with low polarity from seaweed (Tierney *et al.*, 2013). However, PLE potential was demonstrated by Fayad *et al.* (2017) on the brown alga *Padina pavonia (Linnaeus) They*. In this study, PLE (two 60 second cycles at 60°C and 150 bar) was compared to other green extraction technologies (i.e. supercritical fluid extraction, electroporation extraction, and microwave-assisted extraction). Combined with microwave-assisted extraction, PLE was the most efficient extraction of the cosmetically valuable anti-hyaluronidase (Fayad *et al.*, 2017).

Aqueous Two-Phase System (ATPS) are solvent systems made of two immiscible phases, each of them being water-based. This is rendered possible due to the different types of interactions of the water molecules between themselves and the different substances in solution, with a different additive (polymers or salt) in each phase. One of these substances (defined as chaotropic) is weakening the hydrogen bonds between water molecules and reducing the hydrophobic effect in solution. While the second substance (kosmotropic) reinforces the hydrophobic effect by strengthening the hydrogen bonding between water molecules (Diamond and Hsu, 1992). To phrase it simply, we can caricature that one phase contains a chaotropic agent that makes water molecules interact "more" with each other and other substances, while the second phase contains a kosmotropic agent that makes water molecules interact more with themselves and thus less with other substances. The density difference of each phase will then result in an upper and a lower phase. Thus, ATPS also called ABPS (Aqueous biphasic system) are characterized by a couple of substances that would induce the separation of the solution in two phases after mixture. They are unique solvent systems that are particularly used to extract and purify sensitive proteins, notably enzymes, preserving their functionalities. Such systems are used to extract and purify R-phycoerythrin from the red seaweed *Gilidium pusillum* and *Grateloupia torture* using polyethylene glycol (PEG)/potassium phosphate and PEG/potassium carbonate ATPS, respectively (Denis *et al.*, 2009; Mittal *et al.*, 2019). It was also used successfully for the extraction and isolation of bromoperoxidase from the brown seaweed *Laminaria digitate* and can be used for similar enzyme extraction from seaweed (Jordan and Vilter, 1991).

Bio-based solvents are renewable and usually less toxic and more environmentally friendly than their petroleum counterparts. These include ethanol, acetone, butanol, glycerol, methanol, and limonene, which can be used to extract low polarity components from various biomass, including seaweeds. Simple solid/liquid extraction or enhanced processes can be used (for example, Ultrasound Technologies (UT), Microwave

Technologies (MWT), or Pulsed Electric Field (PEF) (Herrero and Ibáñez, 2015). These solvents can be used alone, blended with other solvents (usually water), or as a biphasic solvent system. They are effective in extracting phenolic compounds, lipids, or polysaccharides from seaweed (Cho *et al.*, 2010; Ray and Lahaye, 1995; Wang *et al.*, 2009).

Green Supercritical Fluids (SF) are green solvents in temperature and pressure conditions above their critical point, which gives them a unique set of properties of liquid and gas (Herrero *et al.*, 2006; Turner, 2015). Those properties can be used to enhance performance during the extraction of reaction catalysis, but they can also be tuned according to process temperature and pressure, making it a versatile and robust method (Herrero and Ibáñez, 2015; Turner, 2015). The most used SF is CO_2, due to its moderate critical state condition, non-toxicity, and high recyclability. However, sCO_2 (supercritical CO_2) applications are limited due to its very low polarity unless mixed with a co-solvent (such as bio-based green solvent or water), thus improving its efficiency for extraction of polar compounds (Herrero *et al.*, 2006; Turner, 2015). Today, applications on seaweeds focus on the extraction of lipids, phenolic compounds, pigments, fibers, and other bioactive compounds (Fabrowska *et al.*, 2017, 2016; Herrero *et al.*, 2006; Messyasz *et al.*, 2017).

Ionic liquids (ILs) are a relatively new type of green solvents, for example, 1-Butyl-3-methylimidazolium hexafluorophosphate (Fig. 7.3). ILs are made of ionic compounds that are liquid and stable at relatively low-temperature conditions (Isik *et al.*, 2014). They are unique in the fact they are made of a mixture of anion and cation, which by their interaction remain liquid at temperature range much below their melting point. Their properties as solvent are flexible and tunable, they are stable even at high

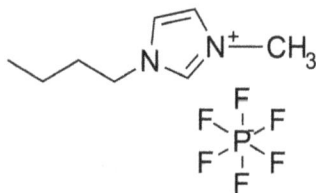

Fig. 7.3 Chemical structure of 1-Butyl-3-methylimidazolium hexafluorophosphate a commonly used ionic liquid.

Source: https://commons.wikimedia.org/wiki/File:Bmim.svg, credit: own work by grhowes CC-ASA-3.0, public domain.

temperatures and non-flammable. Although there is some controversy regarding the environmental toxicity of some of the ILs (Cvjetko Bubalo *et al.*, 2014), they have proven to be excellent at dissolving what other green solvents could not, such as cellulose (Isik *et al.*, 2014). Therefore, ILs could play a role in the extraction, recovery, or hydrolysis of cellulose (Pezoa-Conte *et al.*, 2015). However, the efficiency of the process needs to be compared to other processes (Jmel *et al.*, 2017). Additionally, it was found that ILs can help in hydrolyzing various seaweeds as well as extracting seaweed polysaccharides, phenolic compounds, or proteins (Gereniu *et al.*, 2018; Malihan *et al.*, 2014; Martins *et al.*, 2016; Uju *et al.*, 2015; Vo Dinh *et al.*, 2018; Weldemhret *et al.*, 2020).

With some similarities to the ionic liquid, deep eutectic solvents (DES) are a mixture of quaternary ammonium salt with a Lewis or Bronsted acid. They are liquid at a temperature much lower than their constituents' melting point, usually biodegradable, non-toxic, non-flammable, and capable of dissolving a wide variety of biomolecules. They are usually cheaper and more environmentally friendly than ILs (Chen and Mu, 2019; Lee and Row, 2016). They have been used to extract various molecules from seaweeds, mainly polysaccharides from *Kappaphycus alvarezii*, but also brown seaweeds (Das *et al.*, 2016; Saravana *et al.*, 2018).

Finally, the solvent choice is important for the solubilization of the target compound(s). Achieving enhanced efficiency will usually depend on increased mass transfer, solvent diffusion, and tissue damage that can be promoted by stirring or combining other extraction technologies such as UT, MWT, or PEF. This was partially reviewed recently by Cikoš *et al.* (2018), who focused on the extraction of bioactive molecules from seaweed utilizing technologies such as UT, MWT, PLE, and SF (Cikoš *et al.*, 2018).

7.8.5 *Emerging technologies*

Increased interest in biomass processing has pushed towards investigating new smart technologies for biomass processing. Here, we refer to smart processing technology as a technology with some of the following attributes: energy-efficient, quick, non-chemical (does not require the addition of chemical substances other than water), zero-waste, non-hazardous, environmentally friendly, scalable, cheap, applicable to untreated biomass, versatile, and combinable with other green processes. In addition, it provides a unique effect leading to enhanced efficiency. PEF, MWT, and UT are examples of such technologies, and their application to seaweed processing are as follows.

Pulsed electric field: Pulsed electric field (PEF) is a versatile technology that combines an environmentally friendly process with a unique output: increased permeability of cell membranes. This phenomenon, called electro-permeabilization or electroporation, is created by applying pulses of the electric field. PEF has been used to introduce molecules (i.e. drugs, DNA, and dyes) into cells, to extract intracellular components (i.e. water, ions, sugars, proteins, and secondary metabolites), or to induce lethal or non-lethal stress. Besides its versatility of applications and targets (bacteria, yeast, microalgae, mammalian cells, plant tissues, and seaweeds), PEF treatment does not require chemical additions and has low energy (from few 100 kJ kg^{-1} down to 1 kJ kg^{-1} or even lower in certain applications) and water consumption compared to other cell permeabilization processes (Golberg *et al.*, 2016). PEF also has several industrially relevant advantages as it is quick (few seconds to minutes), applicable to fresh biomass, food-grade, scalable, and mild-thermal (thus preserving most thermosensitive components) (Golberg *et al.*, 2016).

Although PEF has been applied in many different tissues, applications of PEF on seaweeds are scarce. However, potential PEF applications for integrated macroalgal biorefinery were suggested by Robin and Golberg (2016) and listed as "feedstock improvement by genetic engineering, dehydration, valuable chemical extraction, pre-treatment to enhance hydrolysis or biochemical reactions, and process waste treatments". Nevertheless, Polikovsky *et al.* (2016) showed that PEF could be used for specific protein extraction from *Ulva* sp. Postma *et al.* (2017) also showed that PEF improved protein extraction from *Ulva* sp. and was more energy-efficient (6.6 kW kg^{-1} protein) compared to protein extraction by high shear homogenization (>300 kW kg^{-1} protein). Robin *et al.* (2018a) found that PEF could also improve protein extraction, producing an extract with high antioxidant activities. However, those studies reported a rather low protein yield (<15% of initial protein content), which is lower than homogenization (39%), enzymatic-assisted extraction (26.1%), or 24 h osmotic shock extractions (19.5%) (Postma *et al.*, 2017), and similar to values (10–11%) obtained by aqueous or chemical extractions with or without ultrasonic treatment (Kazir *et al.*, 2019). Robin *et al.* (2018a, 2018b) also reported that the protein content in the residue increased due to the removal of a significant fraction of the salt, and suggested the application of PEF for the de-ashing of green seaweed biomass. Using PEF treatments (20–50 kV cm^{-1}, 20–50 pulses of 5 μs) coupled with hydraulic pressing, they obtained up to 45% removal of initial ash content from fresh biomass of *Ulva* sp., compared to 18% removal by pressing (Robin *et al.*, 2018b), 7–83% by washing (Magnusson *et al.*, 2016; Neveux *et al.*, 2014a), and above 80% by extensive acid wash (Hu *et al.*, 2017). Compared to ash removal from seaweed by water rinsing (Magnusson *et al.*, 2016; Neveux *et al.*, 2014a) or extensive acid washing (Hu *et al.*, 2017), PEF treatment was among the quickest (few minutes compared to hours), the least water-intensive (i.e. 100 ml for 140 g of fresh biomass compared to 1 liter per 100 g in a washing treatment), and the simplest (no pre-treatment, 2 steps), while using no chemicals (Robin *et al.*, 2018b). Process conditions of recent PEF studies on green seaweeds are detailed in Table 7.4.

Finally, although current signs are promising, it should be noted that the study of the effect and applications of PEF on seaweed tissues is still in its infancy. Therefore, more research is needed to support the use of PEF for macroalgae processing.

Table 7.4   Pulsed electric field treatment for the processing of *Ulva* sp.

Extracted product	Solvent	PEF treatment conditions*	Other treatment conditions	References
Protein	H_2O	75 pulses/5.7 μs/ 2.96 kV/cm/ 0.5 Hz/30.81 kJ/ kg fresh algae	Coupled with hydraulic pressing (5 min, 45 daN/cm²)	Polikovsky *et al.* (2016)
Protein and carbohydrate	H_2O	2 pulses/0.05–5 μs/ 3–7.5 kV/cm	1 h diffusion time	Postma *et al.* (2017)
Ash	H_2O	10–50 pulses/ 4–6 μs/2–6 kV/ cm/0.5 Hz	Coupled with hydraulic pressing (5 min, 45 daN/cm²)	Robin *et al.* (2018b)
Protein	H_2O	10–50 pulses/4–6 μs/2–6 kV/ cm/0.5 Hz	Coupled with hydraulic pressing (5 min, 45 daN/cm²)	Robin *et al.* (2018a)

Note: *PEF conditions: number of pulses/pulse duration/field strength/pulse frequency/specific energy consumption.
Source: Adapted with permission from (Zollmann *et al.*, 2019a).

Microwave and ultrasound technologies: Microwave and ultrasound are promising examples of "smart" and "green" technologies (Li *et al.*, 2016; Mason *et al.*, 2011; Chatel *et al.*, 2014; Tiwari, 2015). Microwave technologies (MWT) are based on non-ionizing electromagnetic waves in the frequency band of 300 MHz to 300 GHz (Routray and Orsat, 2012). MWT improves mass transfer, solvent diffusion, and tissue disruption, thus improving common bioprocesses such as extraction of molecules from tissues, chemical or enzymatic hydrolysis, and chemical reactions, in a quick (few minutes). It is energy-efficient (fast uniform volumetric heating; energy consumption of 1800 kJ kg^{-1} of fresh materials at the pilot scale) (Périno *et al.*, 2016). In addition, it is a non-chemical treatment that can be applied directly to fresh biomass and is available on an industrial scale (Mason *et al.*, 2011; Leonelli and Mason, 2010; Li *et al.*, 2016).

MWT has been successfully applied on seaweeds to enhance SPs extraction from *Ulva* sp. (10.79% yield) (Wang *et al.*, 2011), *Ulva meridionalis* (Horimoto and Shimada) (40.4% yield), *Ulva ohnoi* (Hiraoka and Shimada) (36.5% yield) and *Monostroma latissimum* (Wittrock) (53.1% yield) (Tsubaki *et al.*, 2016), surpassing the yields of hot water extraction, hot water reflux extraction, and ultrasound-assisted extraction (Wang *et al.*, 2011; Tsubaki *et al.*, 2014, 2016). The combined use of MWT and poly-oxometalate (POM) for hydrolysis of polysaccharides from *Ulva* sp. achieved a saccharification yield of 35–44%, which is 1.7–6.3 times higher than combined MWT and acid hydrolysis, and 5–33% higher than combined POM and conduction heating hydrolysis (Tsubaki *et al.*, 2014). MWT also improved the extraction of essential oil from *Ulva* sp. (Patra *et al.*, 2015, as *Enteromorpha*), the extraction of plant biostimulants (Michalak *et al.*, 2015), and the extraction of pigments from various algae species. This achieved higher yields than Soxhlet apparatus extraction and supercritical CO_2 extraction, and higher or similar yields compared to ultrasound extraction (Fabrowska *et al.*, 2017). MWT has also been used for pyrolysis of seaweed biomass (Budarin *et al.*, 2011). Yuan and Macquarrie (2015) demonstrated the full potential of MWT using a step-by-step microwave treatment to obtain various products (fucoidan, alginate, and biochar, subsequently) at different microwave-assisted extraction conditions in a biorefinery approach.

Ultrasound technologies (UT) are based on the application of ultrasonic waves in the range of 20 kHz to 1 MHz. The wave creates cavitation microbubbles that burst and deliver high-energy mechanical shockwaves as well as heat, which enhance mass-transfer and disrupt cell walls (Mason *et al.*, 2011). UT treatments are therefore similar to MWT in terms of output (enhanced mass transfer, solvent diffusion, and cell wall disruption), but apply different mechanisms and may achieve different results by causing less biomass heating. The advantages of UT include speed (few minutes) and low-energy consumption (around 50 kJ L^{-1} for a pilot continuous system) (Alexandru *et al.*, 2013). Furthermore, it is an environmental-friendly, non-chemical, mild-thermal, versatile method, applicable to fresh biomass and available on an industrial scale (Mason *et al.*, 2011; Leonelli and Mason, 2010; Tiwari, 2015). UT was successfully applied to seaweed processing to enhance saccharification from *Codium tomentosum* Stackhouse, but achieved similar or lower results than those obtained with hot water extraction (24 h, 50°C) or enzymatic treatment (Rodrigues *et al.*, 2015b). This apparent inferiority was confirmed in another study on *Ulva rigida C. Agardh* (Karray *et al.*, 2015) where a UT treatment improved the biodegradability of seaweed (57.1%) only slightly compared to untreated biomass (53.5%). At the same time, hot acid and hot alkali treatments proved less effective (16.12% and 35.24%, respectively), and enzymatic treatments achieved better results (62.7% and 57.7%). Nevertheless, using *Ulva* sp. Korzen *et al.* (2015b) combined UT, enzymatic treatment and ethanol fermentation in one step, obtaining almost 20% (w/w) glucose, and glucose-to-ethanol ratio of 0.33. UT was also listed as one of the most promising green technologies to enhance polysaccharide extraction from seaweeds (Wu, 2017). Process conditions of recent MWT and UT studies on green seaweeds are detailed in Table 7.5.

7.8.6 *Halobiorefinery: Concept, potentials, and benefits*

The development of biorefinery is expected to play a key role in the future bioeconomy. The production of biofuel, energy, food, animal feed, and chemicals from sustainable feedstock such as lignocellulosic biomass or agroindustrial waste has proven to be a promising solution for a decarbonized and sustainable economy. However, doubts and critics have also been raised about their economic reliability as well as their environmental impacts. For example, first-generation bioethanol production is known to use a considerable amount of water from the production of the feedstock (corn, etc.) to the distillation of ethanol (Fingerman *et al.*, 2010; Pfromm, 2008; Wu *et al.*, 2009). If second or even third generation is expected to

Table 7.5 Microwave and ultrasound treatments for green seaweed processing.

Species	Product	Solvent	Treatment conditions[*]	References
Ulva sp. (previously *Enteromorpha* sp.)	Sulfated polysaccharide	H_2O	MWT/700 W/70°C/25min/1:40	Wang *et al.* (2011)
U. meridionalis, *U. ohnoi* and *M. Latissimum*	Sulfated polysaccharide	H_2O	1kW/2.45GHz/100–180°C/14 min/1:20	Tsubaki *et al.* (2016)
Ulva sp.	Hydrolysate	H_2O, various acids and 2–50 mM POM	MWT/1kW/2.45GHz/140°C/14 min/1:20	Tsubaki *et al.* (2014)
Ulva sp. (previously *Enteromorpha* sp.)	Essential oil	H_2O	MWT/40W/15GHz/240 min/1:10	Patra *et al.* (2015)
Ulva sp., *Cladophora* sp., red seaweed	Plant biostimulants	H_2O	MWT/1000W/25–60°C/30 min/1:3	Michalak *et al.* (2015)
Cladophora glomerate (*Linnaeus*) Kützing, *Cladophora rivularis* (*Linnaeus*) Kuntze and *Ulva flexuosa* (Collins and Hervey) Wynne	Chlorophyll and carotenoids	Ethanol water 7:3	MWT/800W/40°C/60 min/1:25	Fabrowska *et al.* (2017)
C. tomentosum	Saccharification	H_2O	UT/400 W/50–60 kHz/50°C/60 min/1:25	Rodrigues *et al.* (2015b)
U. rigida	Pre-treatment for anaerobic digestion	H_2O	UT/120 W/40kHz/5 min	Karray *et al.* (2015)
U. rigida	Saccharification and ethanol production	H_2O	UT/120 W/40kHz /180 min/37°C/1:48	Korzen *et al.* (2015b)
C. glomerate, *C. rivularis* and *U. flexuosa*	Chlorophyll and carotenoids	Ethanol-water 7:3	UT/60 min/40°C/1:25	Fabrowska *et al.* (2017)

Note: [*]Treatment conditions: Technology/power/frequency/temperature/duration/biomass-to-solvent ratio.
Source: Adapted with permission from Zollmann *et al.* (2019a).

be less water-consuming, the pre-treatment, fermentation and distillation steps needed to produce ethanol still require a minimum of 3L of water per L of ethanol produced (Pfromm, 2008). Therefore, the production of bio-based products not only puts increased pressure on the water resource of the region where it is implemented but also limits the geographical area where biorefinery could be built in the future as fresh or drinking water shortages are experienced in an increasing number of regions throughout the globe.

In parallel, the so-called third-generation biomass feedstocks for biorefinery are getting increasing attention, notably micro- and macroalgae. Both can be grown in sea or brine water and thus present considerable advantages toward first- and second-generation biomass in terms of water input. Indeed, the use of seawater or brine groundwater for cultivation of algae or halophyte plants has led to the development of biomass production without the requirement of freshwater resources and arable land (Abideen *et al.*, 2011; Baghel *et al.*, 2015; Fernand *et al.*, 2016; Rui Jiang *et al.*, 2016; Konda *et al.*, 2015; Lehahn *et al.*, 2016a; Wei *et al.*, 2013). This is particularly relevant for coastal regions and arid or semi-arid locations. If the feedstock production can now be done without freshwater, one could ask about its processing.

Indeed, biomass with high salt content is considered, as of today, low-quality feedstock for biorefinery. The main reason for such consideration is mainly historical: bioprocess in biorefinery was developed and designed for processing land biomass with low salt content. As a consequence, salt is interfering in various ways with traditional biomass processing, mainly by damaging equipment (corrosion, etc.), affecting process yield and product quality, or severely inhibiting certain processes (fermentation, etc.) (Asadullah, 2014; Chen *et al.*, 2014; Das *et al.*, 2004; He *et al.*, 2014; Huang *et al.*, 2016; Maiorella *et al.*, 1984; Milledge and Harvey, 2016b; Pattiya *et al.*, 2013; Stefanidis *et al.*, 2015; Wang *et al.*, 2015). This is true for upstream and downstream processes, regardless of whether they are mechanical, thermochemical, biochemical, or microbiological. In response, it was suggested to remove the problem from the biomass, by removing the salt, but this additional step increases the complexity, cost, and environmental impacts of the biomass processing while affecting the biomass quality (He *et al.*, 2014; Huang *et al.*, 2016; Arthur Robin *et al.*, 2018e). On the other hand, salt is starting to be seen as not-so-great an enemy as certain processes are made halotolerant and even sometimes what we proposed to call "halophiles". Halophile is a term from Greek origin that means "salt-loving", it is usually used for organisms that thrive in a salty environment. We proposed to widen its uses to biomass processes — and biorefinery in general — that are designed to use salt as a technological, economic, and ecological ally. All hail the Halobiorefinery.

Halobiorefineries would require that almost every step of the biomass processing that traditionally requires freshwater be undertaken with salty water instead. This is a completely unexplored field, as — to the best of our knowledge — only works on specific processing steps using salt or seawater have been published but none on salt-based biorefinery system. The development of halotolerant and halophile processes requires either the need for new technology or the adaptation of existing ones to this new condition. The use of corrosive-resistant equipment will require additional capital cost, but this would be covered by the advantages of halophile processing in terms of operational costs. Nevertheless, some advantages proposed as follows support the relevance of such a concept.

The first advantage of halobiorefineries would be the use of abundant and cheaper utility costs due to the use of seawater or salty water instead of freshwater (Fig. 7.4). Utilities are usually a minor cost in biorefinery

Where is Earth's Water?

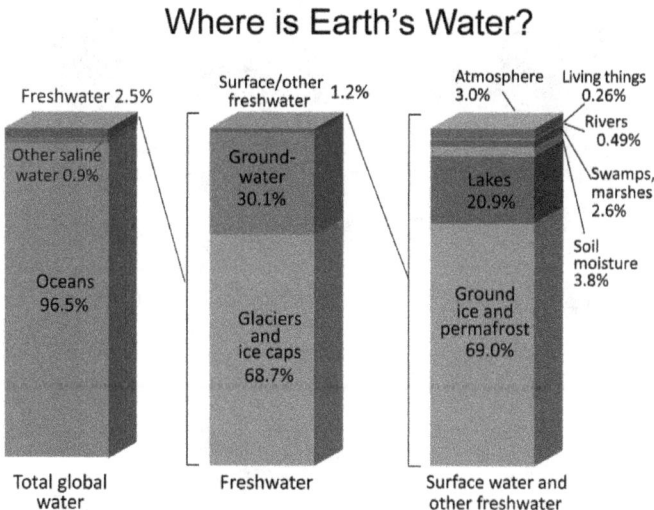

Fig. 7.4   Distribution of Earth's Water.

Source: USGS, Credit: Public Domain, https://water.usgs.gov/edu/gallery/watercyclekids/earth-water-distribution.html).

economics, yet, this cost is expected to increase due to the increasing pressure on water resource and the instability of its availability due to climate change. Moreover, stricter rules on wastewater treatment are expected to be set in place to allow for the recycling of the resource in human society. When the use of biorefinery for water can be in the order of 10–100 times per unit of mass of feedstock, this can be critical in the region where water is already a problem, in arid, semi-arid, highly urbanized, or desertic regions. Independency to freshwater is also useful in a region where the problem of water scarcity is seasonal: there were recently water shortages in countries from all continents such as in South Africa, Australia, the State of California, or even Mediterranean countries during prolonged months. In addition to seawater or brine, high-salt effluents such as desalination plant effluent could be used as inputs, thus integrating halobiorefineries in existing industrial systems, in particular in the coastal areas (Fig. 7.5).

Fig. 7.5   Salt Marshes in Guérande, France.

Source: LucasD — Own work, CC BY-SA 4.0, https://commons.wikimedia.org/w/index.php? curid=69056299.

The second advantage is linked to potential gain in energy or other natural resource uses due to some benefits obtained from salt processing. This advantage is, of course, dependent on the process itself. Many examples are presented in what follows where energy, efficiency, and sustainability are improved in haloprocesses. However, this advantage can also impact the whole plant design. For example, the use of extreme environment throughout a cascading biorefinery can prevent the contamination of raw materials, hydrolysate, fermentation broth, filtrate, and other fluxes due to their high salt content. This can drastically reduce the cost of plant design and energy expenditure by working in less drastic conditions and removing the need for sterilization and other expensive contamination management procedures. For instance, this approach is used for the production of the bioplastic polyhydroxyalkanoates (PHA) by halophilic archaea in the company BluePha (Fig. 7.6), a spin-off of Tsinghua University in China (Ye *et al.*, 2018) or by halophyte archaea grown on seaweeds (Ghosh *et al.*, 2019).

A third advantage concerns the feedstock: halobiorefineries would be able to process any biomass, regardless of its salt-content, thus having

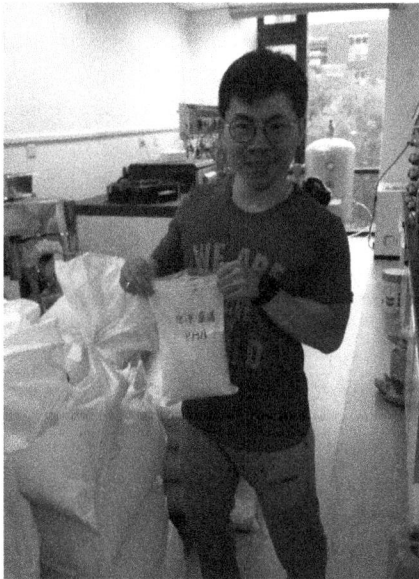

Fig. 7.6 BluePha worker holding a bag of white polyhydroxyalkanoates produced by halophilic microorganisms cultivated in a high salt environment in Beijing, China.

Source: Credit photo: Arthur Robin (2017).

Fig. 7.7 Salicornia trials at International Center for Biosaline Agriculture in Dubai (by Mohammed Shahid, own work, from International Center for Biosaline Agriculture.

Source: Credit: CC-BY-SA-4.0, https://commons.wikimedia.org/wiki/File:Salicornia_trials_at_ International_Center_for_Biosaline_Agriculture.jpg#/media/File:Salicornia_trials_at_International_ Center_for_Biosaline_Agriculture.jpg).

greater feedstock flexibility than most current biorefineries. This is crucial as feedstock availability and cost is the cornerstone of any biorefinery (Ghatak, 2011; Shastri and Ting, 2013). Notably marine biomass such as fishery waste, marine microalgae, marine macroalgae, halophytes, and others (Fig. 7.7) are all untapped resources for biorefineries as of today.

However, the potential benefit of halobiorefinery over its un-salty current counterpart needs to be assessed. For the moment, the vast majority of work on biorefineries and biomass processing has been done using freshwater, however, some old and recent examples showed that this "mind game" is all but theoretical.

For example, enzymes and microorganisms are two major biocatalysts in biorefinery, and both are known to be highly sensitive to salt content. However, the potential of marine and salty environments as a source of salt-tolerant microorganisms (and thus enzymes) starts to raise interest as an untapped source of new products, enzymes, or microorganisms. Indeed, marine yeast and bacteria have already been proven to undertake fermentation under high-salinity conditions for the production of various products (ethanol, lactic acid, etc) from salt-containing biomass (including seaweed) (Di Donato *et al.*, 2018; Khambhaty *et al.*, 2013; Margesin and Schinner, 2001; Oren, 2010; Qiao *et al.*, 2020; Uchida and Murata, 2004; Zaky *et al.*, 2014). Many new halophilic enzymes have been isolated from marine sources (including from seaweed or seaweed epiphytes) (Di Donato *et al.*, 2018; Gutiérrez-Arnillas *et al.*, 2016; Karbalaei-Heidari *et al.*, 2013; Margesin and Schinner, 2001; Moshfegh *et al.*, 2013; Oren, 2010; Qiao *et al.*, 2020; Zhang *et al.*, 2011). Thus, not only can halophilic enzymes and microorganisms help to process high-salt biomass, they can also provide new tools to process "common" land biomass and be used for innovative applications.

Speaking about innovative applications, writing this chapter allows presenting one example of how things sometimes nest themselves in each other in a "sexy" manner. On the one hand, halophilic enzymes with their capacity to keep their activity in high-salt conditions open a door for enzymatic halo treatment. On the other hand, a field of research is booming on the use of ionic liquid — peculiar mixtures of ions with a low melting point — for biomass processing (see Section 7.8.4) (Amarasekara, 2019). Interestingly, it was found that halophilic enzymes make excellent candidates for catalyzing reactions in certain ionic liquids, including one of the most important in biomass processing: polysaccharide hydrolysis! (Gunny *et al.*, 2014; Karbalaei-Heidari *et al.*, 2013; Zhang *et al.*, 2011). This is an exciting example where ionic liquids could become major solvent systems in halobiorefineries, taking advantage of the halophilic properties of the catalysts. Moreover, the presence of salt can even help the process of recycling the ionic liquid, adding to the synergy between ionic liquid and salt processing (Gao *et al.*, 2019). Other solvent systems based on salt are highly relevant for halobiorefineries, this is the case for deep eutectic solvent (DES, a mixture of chaotropic and kosmotropic salts with peculiar characteristics) and aqueous two-phase systems (ATPS) which also have promising applications in biomass processing (Andersson and Hahn-Hägerdal, 1990; Chen and Mu, 2019). Similarly, halophilic enzymes are compatible with DES (Gunny *et al.*, 2019). Aqueous two-phase systems can be used for efficient fractionation of biomass, including in saline environment and from salty biomass (Suarez Ruiz *et al.*, 2020). Thus, those three systems, ILs, DESs, and ATPSs, can offer a wide range of solvent solutions for any applications in biorefineries, but their chemical natures, compatibility, and efficiency in a saline environment make them even more relevant in halobiorefineries.

Also, some authors have previously investigated the benefit of using salt for improved separation of organic compounds from water, including organic solvents and platform chemicals. This phenomenon used the salting-out effect where a high concentration of salt leads to salt ions winning the competition of interacting with water molecules over other water-soluble organic compounds, improving their separation from the water phase (Card and Farrell, 1982; Fu *et al.*, 2017, 2015; I. D. Gil *et al.*, 2008; Lei *et al.*, 2003, 2002; Malinowski and Daugulis, 1994; Sun *et al.*, 2014; Wen *et al.*, 2018). Thus, salt can be an ally in downstream processes, this is uniquely important as downstream processes are usually of the highest cost in a biorefinery. The opportunity of having an efficient separation of bio-based solvents, organic acids, and other platform chemicals using salt will be even more relevant for halobiorefineries as the salt will already be present in the fermentation broth. Interestingly, those molecules, including ethanol, butanol, acetone, and platform chemicals are the main outputs of biorefineries today and the center of attention in bio-industries (Euroview, 2010; Parisi and Ronzon, 2016). Salting-out is also used as a processing aid in distillation, improving the yield, reducing the required time and the overall energy input of the process, it could, therefore, be considered that distillation of fermentation broth containing high salt would be easier, if not similar (Gil *et al.*, 2008a; Lei *et al.*, 2002, 2003).

Distillation is not the only process with halophile or halotolerant perspectives. Indeed, if thermal processes such as carbonization, pyrolysis, or gasification are considered sensitive to ash content (Hupa, 2012; Stefanidis *et al.*, 2015; Wang *et al.*, 2015), certain technologies or reactor designs have proven to get rid of that inconvenience, even turning it beneficial. For example, the use of molten salt in the thermal processing of solid waste and biomass has shown considerable advantages in terms of energy efficiency and environmental aspects (Nygård and Olsen, 2012). Other thermal processes have shown to have a better yield in seawater compared to freshwater (Greiserman *et al.*, 2019). This is notably due to the catalyst activity of certain alkali metal during the process (Kruse and Faquir, 2007). And

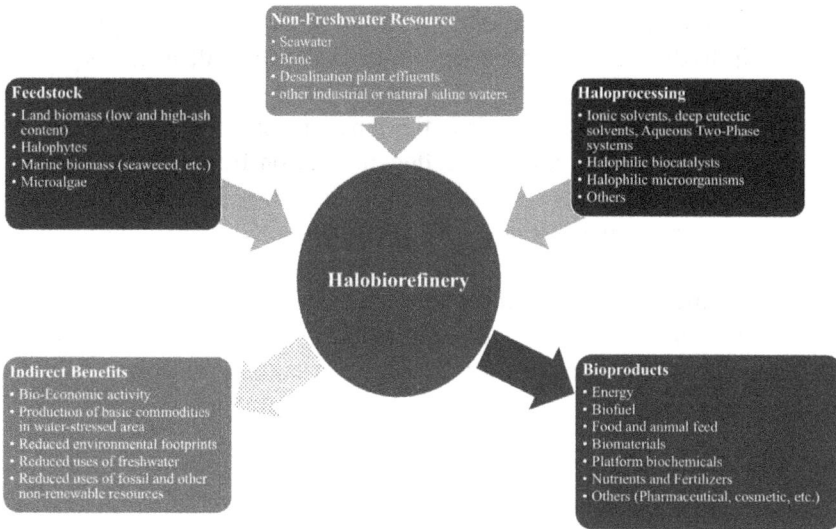

Fig. 7.8   Concept of Biorefinery: Inputs and outputs.

Fig. 7.9   Desalination plant in Hadera, Israel. Desalination plant produces huge quantities of effluents with high salt content that are today discarded in the marine environment.

Source: (PikiWiki Israel 6297 Environment of Israel.jpg.) Wikimedia Commons (credit: public domain).

solutions are existing to alleviate the effect of salt in thermal processes notably in hydrothermal processes by trapping the salt in the aqueous phase (Kruse *et al.*, 2010).

To conclude, halobiorefineries offer a new perspective to the future of biomass processing by reducing its dependence on freshwater resources, extending its feedstock compatibility to high-ash biomass including marine- and saline-based ones, and expanding its range of processing tools (Fig. 7.8). A tremendous amount of work remains before the first industrial-scale halobiorefinery opens its first valve, yet, this chapter offers a conceptual milestone, to begin with, offering a new option for the development of sustainable, integrated, for example, with desalination facilities (Fig. 7.9), and resource-efficient bio-economies.

Chapter 8

Examples of Applications

8.1 Seaweed Aquaculture

Out of ca. 25,000 macroalgae species taxonomically accepted so far (algaebase.com), only about 1% are currently cultivated and, of these, roughly ten species are intensively used for large biomass production. Among them, the kelps *Saccharina japonica* and *Undaria pinnatifida* and the red seaweeds *Porphyra/Pyropia, Kappaphycus alvarezii, Eucheuma striatum* and several strains of *Gracilaria* and *Gracilariopsis* are the most common (Fig. 8.1).

There are two typical approaches for seaweed aquaculture. One is the so-called extensive farming method that takes place in open waters, namely offshore or near shallow coastal waters. The second approach involves semi-controlled, intensive cultivation that uses land-based settings, commonly tanks and ponds. In the extensive cultivation approach, seaweeds are grown in bottom-mounted racks and nets, or along suspended long lines and rods. In the intensive approach, seaweeds are stocked in flow-through tumbled tanks to create water motion and aeration. Water motion through strong aeration is critical for uniform exposure to light and essential nutrients (N and P) while keeping a balanced gas-exchange (O_2 and CO_2) between air and seawater to avoid growth inhibition. Hence, tank cultivation requires the addition of nutrients, which are usually added externally by dripping in highly concentrated solutions or pulse-chasing the seaweed cultivars for several hours. Generally, the intensive seaweed aquaculture approach yields much higher biomass on a surface area basis; nonetheless, these methodologies are far more

Fig. 8.1 *Kappaphycus alvarezii* cultivation in India (a) cluster of three rafts anchored at study site — Umayalpuram, India (b) a 60-day old raft ready for harvest, (c) a single-seeded net bag tied to the long line ready for floating, (d) a 60-day old net bag ready for harvest, and (e) a set of nine long line net bag lines floated at study site — Umayalpuram. (f) *Saccharina japonica* farming in Alaska. Image credit: *Anchorage Daily News*.

Source: Adapted with permission from Selvavinayagam and Dharmar (2017).

expensive as they require workforce, electricity, and the above resources. Tank cultivation is more secure since tank facilities are readily accessible and enable controlling critical parameters. Hence, excessive light, potentially inducing photo-inhibition, and optimal nutrient dosing for maximal photosynthesis can be regulated and optimized for maximal productivity. Also, epiphyte damage, which may occur in any cultivation setting, can be more efficiently controlled in tank culture, resulting in less frequent disease outbreaks using this approach. Yet, as costs of tank seaweed cultivation are high, in many cases, this farm setting would prove profitable and economically viable only when producing specific products of high commercial value.

The growing market of seaweed biomass and the implementation of efficient cultivation techniques, both in sea and land-based, have developed significantly over the last several decades. These improvements in the global seaweed aquaculture will need to be backed up by the development of strains through advanced breeding tools. Acknowledging the expected challenges posed by global climate change, improved seaweed strains will need to be characterized by thermo tolerance, resistance to diseases, fast growth, and high levels of commercial products. Land-based culture also needs to focus on reducing production costs, while offshore activities of seaweed aquaculture require the development of cost-efficient technologies, particularly in highly exposed marine environments.

8.2 Food

Seaweed is traditionally used for human consumption and it is probably the oldest application of seaweed biomass by human beings (Wells *et al.*, 2017). Today, seaweed remains a common food ingredient in all the coastal regions and islands of Eastern Asia and Oceania. Fresh, dried, or cooked, seaweeds from harvested or cultivated biomass are simply processed, and then used as a raw ingredient for various dishes such as salads, soup, sushi, and many other delicacies (Figs. 8.2–8.5). Since this book is not about cooking seaweed (well, actually, it is, from a certain point of view!) we will focus instead on their nutritional properties, innovative uses, safety, and recent trends in food applications.

Benefits of seaweed for human consumption are numerous and they are often considered as « superfood » notably in Western countries where they are unusual ingredients (Barrow and Shahidi, 2008; Holdt and Kraan, 2011; Mohamed *et al.*, 2012; Wells *et al.*, 2017).

Mainly, seaweed is a good source of protein, fiber, mineral, and various other molecules of interest (antioxidants, vitamins, etc.) (Burtin, 2003;

Fig. 8.2 Plates of sushi rolls, Japanese food delicacies rolled in seaweed (nori, of the genus Pyropia) that conquered the world.

Source: Marco Verch, credit: CC-BY-2.0, https://flic.kr/p/213XM8e.

Fig. 8.3   Asian salad of fresh wakame (*Undaria pinnatifida*).

Source: Vegan Feast Catering — Seaweed Salad, CC BY 2.0, https://commons.wikimedia.org/w/index.php?curid=35619473.

Dawczynski *et al.*, 2007; Mohamed *et al.*, 2012; Sánchez-Machado *et al.*, 2004). Moreover, they bring unique texture and flavors (notably the « umami » taste) to the meal. Their (mostly red and green seaweeds) protein content can reach more than 20% of the dry weight with an attractive amino acids profile, thus making them a good dietary protein source, while being « vegan » (Bleakley and Hayes, 2017; Fleurence, 1999; Kazir *et al.*, 2019; Wong and Cheung, 2001). Their calorific value is quite low due to their high content in mineral and fibers, which make them increasingly attractive in a Western diet lacking mineral and fiber, and high in calories (Burtin, 2003; Dawczynski *et al.*, 2007; Mohamed *et al.*, 2012; Wong and Cheung, 2000). In addition, they have a low content of high-quality lipids (Burtin, 2003; Dawczynski *et al.*, 2007; Mohamed *et al.*, 2012; Rohani-Ghadikolaei *et al.*, 2012; Sánchez-Machado *et al.*, 2004).

Fig. 8.4 Jelly dessert made of agar and pomegranate juice.

Source: Sahua, Jelly_1, CC-BY-2.0, https://flic.kr/p/8eidJJ.

Fig. 8.5 Tahini miso soup with nori (*Undaria pinnatifida*).

Source: Dave Miller, Tahini Miso Soup, CC-BY-2.0, https://flic.kr/p/kmNzNR.

Their fibers such as fucoidan, ulvan, carrageenan, alginate, agar are unique and have many health benefits (Alba, 2018; Barrow and Shahidi, 2008; Mohamed *et al.*, 2012; Tanna and Mishra, 2019). Seaweeds are naturally enriched with numerous dietary minerals notably calcium, magnesium, iron, etc. and are a recognized dietary source of iodine (Rohani-Ghadikolaei *et al.*, 2012; Sánchez-Machado *et al.*, 2004; Tanna and Mishra, 2019; Wong and Cheung, 2000).

Not all seaweeds are edible (notably due to taste or texture, for example, calcified macroalgae), however, among edible algae certain reserves in their consumption have to be considered. As of all food, a certain moderation in the consumption of seaweed is advised to avoid any potential detrimental consequences of mass consumption of it. However, the daily addition of seaweed in the diet does not represent any health issues. Concerns over heavy metal and potential toxicity can be raised when seaweeds are grown in a highly polluted area, however, this remains an exception and not the general case (Almela *et al.*, 2002; Tanna and Mishra, 2019). Current and new regulations are applied to seaweeds to ensure their safety for consumption (Mohamed *et al.*, 2012; Tanna and Mishra, 2019). With the increasing consumption of seaweeds, concerns over allergy have been raised but there is no evidence of acute allergy to seaweed in any population as of today (Polikovsky *et al.*, 2018; Tanna and Mishra, 2019). People developing seaweed allergy should avoid seaweed products, however, seaweed is safe for people allergic to seafood according to the current knowledge.

Seaweed is also processed in various products, notably pasta, « bacon-alternative », flakes, mixed with salt or as chips. The list could be endless as new companies are adding seaweed-based products every year. We, therefore, chose to focus here on seaweed-extracted substances that are used for food applications. Seaweed polysaccharides are currently the main components extracted and used at a commercial scale. Indeed, this represents a worldwide market worth several billions of dollars from the selling of agar and carrageenan from red seaweed and alginate from brown seaweed (Alba, 2018). Applications of those three polysaccharides are from biotechnology to food including biomedical purposes. In food, those polysaccharides are usually used as texturizers as they are excellent at binding water and affecting the viscosity of solution even at a low level, sometimes even producing hydrogel (Alba, 2018; Tanna and Mishra, 2019). As a consequence, most of the population in the Western world has eaten seaweed ingredients as a food additive. Their addition to food products are currently growing as they are usually considered as healthy, vegan, natural (bio-based), and functional food ingredients, while a wide range of seaweed polysaccharides remain unexplored in the food industry (Alba, 2018; Barrow and Shahidi, 2008; Dawczynski *et al.*, 2007; Holdt and Kraan, 2011; Tanna and Mishra, 2019).

Current trends in food products are the use of natural, non-animal, high-quality, and ecological products. Seaweed components are among the potential candidates for new ingredients. Recently, the boom in research and development of a new source of protein and meat alternative has increased the interest in seaweed polymers (see above) and seaweed proteins, but their commercial applications are still at their infancy (Alba, 2018; Bleakley and Hayes, 2017; Cofrades *et al.*, 2011; Fleurence, 1999). Only a few other seaweed components are considered for food applications, such as mineral extracts (Magnusson *et al.*, 2016), the carotenoid fucoxanthin (Miyashita *et al.*, 2011), or starch (Prabhu *et al.*, 2019), as most of the current research and development both in academia and industry are focusing on seaweed polysaccharides and protein.

Finally, it should be noticed that the health benefits of seaweed depend on the bioavailability of seaweed dietary components, the effect of seaweed matrix on the bioavailability of other food components, their susceptibility to digestion, and their relationships with gut microbiomes. And all those aspects are far from being thoroughly explored from a nutritional, nutraceutical, or nutrigenomic point of view (Wells *et al.*, 2017). Notably additional parameters such as food processing, seasonal, and geographic variations can significantly affect how seaweed can benefit the human diet (Wells *et al.*, 2017). However, current literature and historical uses are depicting seaweed and seaweed ingredients as an excellent addition to the human diet. As the world is struggling to produce more food of better quality and sustainability, there is not a single doubt that seaweed will play an emerging but steadily increasing role in the future of food.

8.3 Pharma/Medicine

Macroalgae are a rich and varied source of pharmacologically bioactive natural products. The use of macroalgae for therapeutic purposes has a long history; however, until the 1950s, their medicinal properties were mainly restricted to traditional and folk medicine (Milledge *et al.*, 2016; Smit, 2004). The scientific research on their therapeutic property began during the 1970s when scientists successfully isolated chemical compounds from brown macroalgae that showed anticancer and antitumor activity (Pal *et al.*, 2014). Between 1977–1987, algae have been the source of ~35% of the newly discovered chemicals (Ireland *et al.*, 1993; Smit, 2004). In the last three decades, the discovery of natural metabolite and bioactive compounds with potential pharmaceutical and nutraceutical applications from macroalgae has increased significantly.

Macroalgae are considered as an excellent source for bioactive chemicals. In addition to their abundance, great diversity (number of different species), and primary metabolites, which are required for healthy growth, macroalgae produce many secondary metabolites (Murphy *et al.*, 2014; Pal *et al.*, 2014). These metabolites which are produced in response to a wide range of fluctuating environmental pressures (as competition over space, high salinity, dangerous levels of irradiance, water movement resulting in thallus breakage and wound formation, etc.), can be potentially useful for humans and may result in new technologies, natural chemicals, or drugs (Smit, 2004).

There are numerous reports of macroalgae derived compounds that have a broad range of biological activities, such an antibiotic, antimicrobial, and antibacterial (Ballantine *et al.*, 1987; Rizzo *et al.*, 2017; Wahidi *et al.*, 2014), antiviral (Damonte *et al.*, 2004), anti-cancer (Murphy *et al.*, 2014), anti-oxidants (Messyasz *et al.*, 2018; Wu, 2017), anti-diabetic, anti-neurodegenerative diseases (Barbosa *et al.*, 2014; Grosso *et al.*, 2014), anti-blood coagulation, and anti-inflammatory (Smit, 2004). Among these compounds are: lectins, polyphenols, sulfated polysaccharides, terpenes, fatty acids, proteins, carotenoids, alginates, vitamins, and sterols. The following will provide a few examples from the remarkable array of bioactive materials in current use or development. The list of compounds provided is by no means exhaustive, but it will allow a glimpse of the breadth of biological activities exhibited by macroalgae natural products.

Antioxidants: Oxidation is a chemical reaction caused by reactive oxygen species, unstable, and highly active molecules. These molecules can promote a chain reaction and thereby damage macromolecules and cells, such as lipid membrane, protein, or formation of DNA adducts that can cause cancer-promoting mutations or cell death. Antioxidants have multiple functions in biological systems, including defending against oxidative damage and participating in the major signaling pathways of cells (Ahmed *et al.*, 2014; Devi *et al.*, 2011).

Several studies have shown that marine macroalgae extracts, notable polyphenols, have antioxidant activity (Kuda *et al.*, 2005; Ng, 2002). The major active compounds in different macroalgae have been reported to be phlorotannins and fucoxanthin (Devi *et al.*, 2011). Brown algae generally contain higher amounts of polyphenols than red and green algae. Most brown algae contain fucoxanthin (carotenoid pigment) that besides its antioxidant qualities also has significant anti-inflammatory, anti-cancer, and UV-preventative activities (Milledge *et al.*, 2016). Other examples are fucoidan and lambda carrageenan, which was reported (De Souza *et al.*, 2007) to have the highest antioxidant activity among sulfated polysaccharides extracted from brown and red algae (Ahmed *et al.*, 2014).

Antiviral: The use of marine compounds for antiviral activity has enormous potential, and it is widely accepted as it usually has low cost with low cytotoxicity (Ahmed *et al.*, 2014). Several macroalgae sulfated polysaccharides have shown significant antiviral activities towards human infectious diseases, primarily against enveloped viruses (as human immunodeficiency virus (HIV), herpes simplex virus (HSV), human cytomegalic virus, dengue virus, and respiratory syncytial virus (RSV)) (Damonte *et al.*, 2004; Smit, 2004). Their antiviral activity depends on the molecular weight, sulfate content, and constituent sugar (Adhikari *et al.*, 2006) and are based on the formation of formally similar complexes that block the interaction of the viruses with the cells (Damonte *et al.*, 2004).

Fucoidan, for example, has shown antiviral properties against HIV-1 strain (Sugawara *et al.*, 1989; Trinchero *et al.*, 2009) and high inhibitory activity against HSV 1 and 2 (Feldman *et al.*, 1999; Ponce *et al.*, 2003; Trinchero *et al.*, 2009). Carraguard, a carrageenan-based vaginal microbicide, has been shown to block HIV and other sexually transmitted diseases in vitro. In 2003, it entered phase Ш clinical trials (https://journals.plos.org/plosone/article?id=10.1371/journal.pone.0003162, 2003; Smit, 2004). Other examples are sulfated galactan from *Schizymenia binderi* (Matsuhiro *et al.*, 2005) and Xylomannan from *Nothogenia fastigiata* (Mandal *et al.*, 2008), who showed antiviral activities against HSV-1 and 2. Besides algal polysaccharides, other compounds exhibit similar characteristics, as Kahalalide F, a small natural peptide from the species *Bryopsis*, being studied as a possible agent against HIV and cancer (Hamann *et al.*, 1996; Sewell *et al.*, 2005; Suárez *et al.*, 2003).

Anticancer: Extensive reviews have been published on the potential use of macroalgae concentrated extracts for cancer chemotherapy

(Murphy *et al.*, 2014; Stonik and Fedorov, 2014). This potential is due to considerable evidence that several molecules in macroalgae have cytotoxic properties. The presence of cytotoxins in macroalgae is not surprising since they need to protect against herbivory and encroachment of other seaweeds into their habitat (Murphy *et al.*, 2014). Among the molecules, groups that have shown antitumor and anti-cancer properties are Quinones, Carotenoids, Fucoidans, Ulvan, Porphyran, Lactones, and Lipids (Atashrazm *et al.*, 2015; Murphy *et al.*, 2014; Stonik and Fedorov, 2014).

Even though there is extensive evidence of the potential of marine macroalgae as a source of anti-cancer, only one derived from macroalgae, Kahalalide F, has entered clinical trials. It is useful in controlling tumors that cause lung, colon, and prostate cancer (Hamann *et al.*, 1996; Smit, 2004; Suárez *et al.*, 2003), and has been patented for use as a possible active substance for the treatment of human lung carcinoma (Scheuer *et al.*, 2000).

Antibiotic: Chemicals responsible for an antibiotic, antimicrobial, and antibacterial are widespread in macroalgae. Marine algal compounds able to inhibit the growth of bacterial pathogens may represent future alternatives to common antibiotics. The increasing resistance of pathogens to antibiotics represents a priority for exploring and developing effective natural antimicrobial agents (Rizzo *et al.*, 2017). Interesting compounds present in macroalgae are sterols, heterocyclic, phenolic, and halogenated compounds such as halogenated alkanes, haloforms, alkenes, alcohol, aldehyde, hydroquinone, and ketone (Lincoln *et al.*, 1991).

Despite macroalgae's considerable potential as a source of therapeutic compounds, translation to clinically useful preparations is almost non-existent (Murphy *et al.*, 2014). Few critical points in the process of drug development from marine organisms, and particularly from macroalgae, limit the possibility of successful marketing. Preclinical and clinical trials in the process of drug development require permanent availability and supply of a sufficient amount of organisms and compounds, which can markedly affect the marine ecosystem (Barbosa *et al.*, 2014; Lindequist, 2016). For marine drugs to have an opportunity on the market, the supply problem should be addressed either by sustainable, ecological collection from natural environments or by the novel, greener processes of marine biotechnology (Barbosa *et al.*, 2014; Lindequist, 2016).

Furthermore, there is a great diversity in macroalgae composition, with many compounds being unique to a specific group. There is also natural variability in the composition of bioactive compounds in the same species and even within different parts of the same thallus (Murphy *et al.*, 2014). Bioactivity can be affected by several factors, such as algae size, age, tissue type, salinity, season, nutrient levels, the intensity of herbivory, light intensity, and water temperature. Therefore, a better understanding of living conditions in the natural environment is necessary to develop alternative cultivation methods that will not only prevent the overexploitation of the natural population but also maintain the desired metabolite production over a long time (Barbosa *et al.*, 2014; Lindequist, 2016; Murphy *et al.*, 2014). Developing these novel cultivation techniques and efficient methods for controlling the process will contribute significantly to fulfilling the enormous potential of macroalgae in the pharmaceutical field.

8.4 Chemical Industry

The unique chemical composition and high growth rate of macroalgae offer a wide range of opportunities to use them for chemical production. One of the first examples for this use is California's kelp industry which was active during World War I. The war created both a significant demand for chemicals and a severe shortage of supplies to meet the demand, especially for explosives (Neushul, 1989). Responding to this wartime opportunity, a new industry was built to extract both potash and acetone from California's giant kelp, an ideal source of materials for explosives (Kelly and Dworjanyn, 2008; Neushul, 1989; Van Hal *et al.*, 2014). The potash served as an ingredient of gun powder and acetone as a critical component of cordite, a smokeless powder used for gun and artillery shells (Kelly and Dworjanyn, 2008; Neushul, 1989; Van Hal *et al.*, 2014). The potash was also used as fertilizer to improve sandy soils and in the manufacture of glass, soap, matches, and various dyes (Neushul, 1989). The largest plant during that time, built by Hercules Powder Co., was capable of producing 54 chemicals (as alkali, sodium carbonate) from macroalgae (Kelly and Dworjanyn, 2008; Milledge *et al.*, 2016). The plant was closed shortly after the war when demand fell and alternative cheaper supplies became available (Kelly and Dworjanyn, 2008).

In recent decades, awareness of the economic and environmental importance of biomass-based chemicals has increased significantly, and as a result, also the manufacture of chemicals from algae. The biofuel industry is an excellent example of that. Biofuel and bio-oils (as methane, butanol, ethanol, and acetone) produced by anaerobic fermentation from macroalgae are also commercially used in the bulk chemical market (van der Wal *et al.*, 2013). Moreover, the fermentation process also produces many byproducts, such as glycerol and organic acids (e.g. acetate, succinate), which are profitable and enhance the economic value of the seaweed fuel production process. Organic acids, for example, are a high-demand chemical feedstock for producing deicing salts and food additives. Glycerol has various applications in the manufacturing of food, pharmaceutical, and personal care products (Na Wei *et al.*, 2013).

Macroalgae can also be used to produce anti-microfouling chemicals. Biofouling is an undesirable accumulation of microorganisms, plants, and animals on unprotected surfaces exposed to water or air. This phenomenon appears in a wide variety of applications, e.g. colonization of pipe surfaces in the food and water industries, metal corrosion due to sulfate-reducing bacteria in the shipping and oil industries, in medical practice when tissues and prosthetic implants may become infected, and in aquaculture facilities (e.g. fish-cage, boats fouling) (Plouguerne *et al.*, 2008; Sargassum *et al.*, 2010).

Macroalgae produce a wide variety of chemically active metabolites, potentially as an aid to protect themselves against other settling organisms and biofouling. These metabolites can be used as natural, environmentally friendly antifouling compounds (Plouguerne *et al.*, 2008). Chloroform and ethanol extracted from *Sargassum muticum* and dichloromethane from *Grateloupia turuturu* have shown antimicrobial activity against several bacterial, fungal, and microalgal strains involved in the marine biofilm formation and could be the basis of an antifouling product (Milledge *et al.*, 2016; Plouguerne *et al.*, 2008; Sargassum *et al.*, 2010).

Also, algal biomass is an excellent renewable source for the production of novel polymers. For example, mannitol can be used to produce rigid polyurethane foams. It can be converted into intermediates to produce detergents, as well as isomannide (an isomer of isosorbide). These isohexides can be used as building blocks for polymers, fuel additives, and plasticizers (Rose and Palkovits, 2012). Alginate (derivatives of alginic acid or alginic acid itself from the cell wall of brown algae) can be converted to furan dicarboxylic acid (FDCA) or similar diacids that can be used for polyesters (Van Hal *et al.*, 2014). Biomass-based polyester has great economic and environmental importance. Biodegradable polyester blends are used in tissue engineering and controlled drug delivery systems (Noreen *et al.*, 2016).

Alginates are also used in the textile printing and paper industries. In textile printing, alginates are used as thickeners for the paste containing the dye. Textile printing accounts for about 50% of the global alginate market. In the paper industry, their primary use is in surface sizing. Alginate added to the standard starch sizing gives a smooth continuous film and a surface with less fluffing (McHugh, 2003).

8.5 Construction

The construction industry is responsible for the consumption of a significant part of all produced materials. However, only recently has this industry started to worry about its environmental impact (Pacheco-Torgsl *et al.*, 2014). Greenhouse emission and energy-intensive processes involved in the manufacture of materials served as a growing incentive for the development of new building technologies (Dove, 2014). The use of green, natural materials has been identified as a potential means for reducing the environmental impact and positively contributing to sustainable design strategies (Dove, 2014).

Bio admixtures are functional molecules used in building products to optimize material mechanical properties. They include natural or modified biopolymers and biotechnological and biodegradable products (Pacheco-Torgal *et al.*, 2016; Plank, 2004). One of the principles uses of these renewable bio-based additives is as alternative stabilizers to cement and lime in the earth and clay-based construction (Dove, 2014). Examples for biopolymers used in concrete, cement, or mortar are lignosulfonate, derivatives of starch and cellulose, chitosan, pine root extract, protein hydrolysates, vegetable oils, and water-soluble polysaccharides (Pacheco-Torgal *et al.*, 2016; Plank, 2004). Although there is a clear potential for natural biopolymers to be used as stabilizers for construction materials, further research is carried out to improve materials' mechanical properties. The study focuses on the appropriate amount of additives, the interaction between clay–biopolymer, and new types of biopolymers (Dove, 2014).

Few studies have discussed the use of macroalgae constitutes, mostly polysaccharides, for building products. For example, Galán-Marín *et al.* (2010) who referred to the use of alginate (polysaccharide found in the cell wall of brown macroalgae) in soil blocks. Dove (2014), who also used four different alginates from *Laminaria Hyperborea* mixed with soil to produce unfired clay bricks. Other examples are Achenza *et al.* (2006), who described the use of 10 mm "seaweed fibers" for stabilization of adobe bricks for earth construction while Susilorini *et al.* (2014) have investigated the addition of polysaccharide (agarans and carrageenans) from *Eucheuma cottonii* (as a gel) and *Gracilaria* sp. (as powder) to mortar. *Eucheuma cottonii* contains kappa carrageenan, which has an advantage as an emulsifier, suspensor, condenser, and stabilizer. While *Gracilaria* sp. contains agarose and agar pectin which make strong gels. Both products were used to increase the bonding mechanism of mortar.

All mentioned studies have demonstrated that the use of macroalgae products improves mechanical properties. Achenza *et al.* (2006), for an instant, showed an increase of 75% in compression strength, reduction in earth porosity, and improvement in earth resistance to water. Susilorini *et al.* (2014) (Susilorini *et al.*, 2014) showed that mortar modified with *Gracilaria* sp. powder performed well, showing great values for compressive strength and splitting tensile. Furthermore, the general advantages of construction materials made of algae are self-extinguishing and resistance to mold.

Additional use of dried macroalgae mass or fibers is insulating construction material for roofs, exterior and interior walls, and basement ceilings, which provide thermal and noise insulation. The beneficial thermal insulation properties are caused by small cavities within the algae mass. These cavities are small enough not to allow any significant convective heat transfer (Heinz, 2011). A recent experimental study, by Boulaoued *et al.* (2016) have shown that insulating materials made of at least 20% seaweed fibers in a cement matrix, present thermal properties (thermal conductivity and diffusivity) that are generally comparable to synthetic commercial insulating materials.

However, improvement in both thermal and mechanical properties of construction materials by macroalgae additives depends on multiple parameters, including type and amount (Dove, 2014; Susilorini *et al.*, 2014) of additive, a form of addition (gel, powder, fiber, etc.), the species of algae from which it was produced, and the environmental factors in which the algae were grown (temperature, irradiance, water hardness, etc.). Therefore, further research is required.

The construction industry has become a major potential field for the use of biopolymers. In 2000, an estimated $2 billion in sales was made at the manufacturer's level, and it is expected to continue to grow in the next years (Pacheco-Torgal *et al.*, 2016). Nonetheless, the research and use of macroalgae products in this industry are still in their infancy. More investigation should be done on the use of additive materials from different types of macroalgae and their effect on the properties of the products, as thermal performance, mechanical properties, and long-term durability.

8.6 Agriculture

The green revolution combined traditional agricultural practices with controlled irrigation and mechanized farming systems, and above all, heavy use of both fertilizers and pesticides (Dutta, 2012). In recent decades, this has included crop improvement by using novel genetic technologies (Pingali, 2012). Yet, it has been predicted that an additional increase of 35% in yield will be necessary by 2025 to maintain current world caloric requirements and to keep up with the expected rise in the global human population. However, many green-revolution crops have experienced a plateau in production gains, which suggests that we have reached a biological barrier to yields with current technologies.

The Green Revolution, and especially the increasing use of strong synthetic agrochemicals such as fertilizers and pesticides, have exacerbated urgent environmental problems related to water, soil, and air pollution. This has also caused detrimental effects on non-target species and endangers human health through exposure to toxic materials (Carvalho, 2006).

Alternative technologies to increase production are needed to support the world's growing population and increasing affluence. Biostimulants are a promising technology for protecting plants against environmental stress and increasing yields (du Jardin, 2015).

Since ancient times, seaweeds and seaweed crude extracts have been used as biostimulants in the field and proven to be effective for multiple crops. Many species of seaweeds and numerous extraction methods have been used over the years (Van Oosten *et al.*, 2017). For example, applying brown seaweed *Ascophyllum nodosum* extract to winter rapeseed (*Brassica napus*) crops were shown to increase nitrogen uptake (Jannin *et al.*, 2013). Enhanced germination and growth of tomatoes were recorded following biostimulation with different seaweed extracts such as *Ulva lactuca, Caulerpa sertularioides, Padina gymnospora*, and *Sargassum liebmannii* (Hernández-Herrera *et al.*, 2014). Although multiple studies have confirmed the plant-growth stimulating properties of

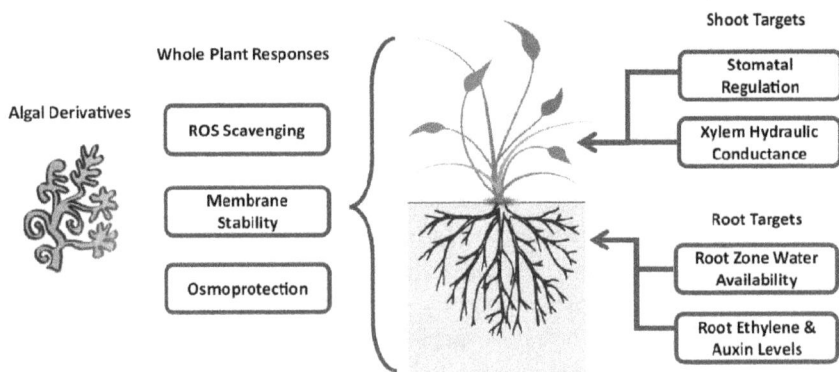

KEY MECHANISMS TARGETED BY ALGAL BASED BIOSTIMULANTS

Fig. 8.6   Illustrates a few possible pathways for the biostimulant mechanisms in sea-weeds. Current studies deal with the identification of specific active fractions that can be then used in formulations for precise agricultural uses.

Sources: Algal extracts target several pathways to increase tolerance under stress. Adapted with permission from Van Oosten *et al.*, 2017.

seaweed extracts, no specific mechanisms or functional molecules have been identified. Moreover, comparative studies have been hampered by the lack of standardization of extracts within and between species of sea-weeds (see Fig. 8.6).

8.7   Energy

While biomass rich in lipids will be suitable for the production of bio-diesel, bioethanol can be obtained more easily from sugar-rich sources. In this part, Tables 8.1, 8.2, and 8.3 list the different yields obtained for etha-nol, butanol, and biodiesel conversion, respectively, the type of macroal-gae, and the process. We can note that the low fraction of lignin in macroalgae ease the degradation and for that reason, pre-treatments do not need to be as harsh and energy-consuming as for more resistant bio-mass (Briand and Morand, 1997; Wargacki *et al.*, 2012). For ethanol and butanol production, before hydrolysis, algae are washed, dried, and put into smaller pieces (for instance via milling, grinding, comminution,

Table 8.1 Macroalgae conversion into ethanol: pre-treatment, hydrolysis, and fermentation processes.

Algal species	Mode	Hydrolysis/pre-treatment	Fermentation	Yield (ethanol)	Reference
Eucheuma cottonii	SHF	5% H_2SO_4, 100°C, 2 h[a]	*Saccharomyces cereviceae*, 28-30°C, 5-6 days	4.6%	Candra et al. (2011)
Kappaphycus alvarezii	SHF	2.5% H_2SO_4, 100°C, 1 h	*Saccharomyces cerevisiae* (NCIM 3523), 96 h	2.46%	Khambhaty et al. (2012)
Kappaphycus alvarezii	SHF	0.2 M H_2SO_4, 130°C, 15 min	*Saccharomyces cerevisiae*, 24 h	1.1-1.9 g/L	Meinita et al. (2012)
Gracilaria Salicornia[1]	SHF	2% H_2SO_4, 120°C, 30 min. Cellulase, 40°C[c]	*Escherichia coli* (KO11), 30°C, 30 h	79.1 g/1 kg	X. Wang et al. (2011)
Gelidium elegans	SHF	2% H_2SO_4, 121°C, 30 min. Meicelase, 50°C, 120 h[c]	*Saccharomyces cerevisiae* (IAM 4178), 30°C, 48 h	55 g/L	Yanagisawa et al. (2011)
Ulva pertusa	SHF	Meicelase, 50°C, 120 h (twice)[d]	*Saccharomyces cerevisiae* (IAM 4178), 30°C, 24 h	30 g/L	Yanagisawa et al. (2011)
Alaria crassifolia	SHF	Meicelase, 50°C, 120 h (twice)[d]	*Saccharomyces cerevisiae* (IAM 4178), 30°C, 24 h	34.4 g/L	Yanagisawa et al. (2011)
Laminaria hyperborea	SHF	Water pH 2, 65°C, 1 h. pH 4.5, 121°C, 20min	*Pichia angophorae*, 30°C, 75 h, surface aeration[g]	0.43 g/g substrate	Horn et al. (2000)
Saccharina latissima	SSF	Laminarinase, 32°C, 36 h	*Saccharomyces cerevisiae* (Ethanol Red), 32°C, 36 h	0.45% (v/v)	Adams et al. (2009)
Laminaria digitata[2]	SSF	2M HCl + 0.2 M HCl and laminarinase, 24°C, 69 h	*Pichia angophorae* (CBS 5830), 24°C, 69 h	167 mL/kg algae	Adams et al. (2011)
Enteromorpha intestinalis	SHF	75 mM H_2SO_4, 121°C, 60 min. Celluclast 1.5 L and Viscozyme L, 55°C, 54 h	*Saccharomyces cerevisiae* (KCTC 1126), 30°C, 48 h	8.6 g/L	Cho et al. (2013)
Enteromorpha intestinalis	SSF	75 mM H_2SO_4, 121°C, 60 min.[f] Celluclast 1.5 L and Viscozyme L, 30°C, 36 h	*Saccharomyces cerevisiae* (KCTC 1126), 30°C, 36 h	7.6 g/L	Cho et al. (2013)
Sargassum spp.	SHF	4% H_2SO_4, 115°C, 1.5 h. Cellulase and cellobiase, 50°C, 96 h	*Saccharomyces cerevisiae*, 40°C, 48 h	2.79 g/L	Borines et al. 2013)

(Continued)

Table 8.1 (*Continued*).

Algal species	Mode	Hydrolysis/pre-treatment	Fermentation	Yield (ethanol)	Reference
Gracilaria verrucosa	SHF	Cellulase and cellobiase, 50°C, 36 h	*Saccharomyces cerevisiae* (HAU), 30°C, 16 h	0.43 g/g sugars	Kumar *et al.* (2013)
Gelidium amansii[3]	SHF	2% H_2SO_4, 150°C, 3-3.5bar, 40L/h[b]	*Brettanomyces custersii* (KTCT18154P), 30°C, 39 h	27.6 g/L (0.38 g/g sugar)	J. H. Park *et al.* (2012)
Gelidiella acerosa	SHF	5% H_2SO_4 (1:4 ratio), boiling and then room temperature overnight. 2% cellulase 22086, 45°C, 36 h	*Saccharomyces cerevisiae* (MTCC no. 180), 28°C, 12 h	418 mg/g sugars	R. S. R. S. Baghel *et al.* (2015)
Gelidium pusillum	SHF	5% H_2SO_4 (1:4 ratio), boiling and then room temperature overnight. 2% cellulase 22086, 45°C, 36 h	*Saccharomyces cerevisiae* (MTCC no. 180), 28°C, 12 h	416 mg/g sugars	R. S. R. S. Baghel *et al.* (2015)
Gracilaria dura	SHF	5% H_2SO_4 (1:4 ratio), boiling and then room temperature overnight. 2% cellulase 22086, 45°C, 36 h	*Saccharomyces cerevisiae* (MTCC no. 180), 28°C, 12 h	411 mg/g sugars	R. S. R. S. Baghel *et al.* (2015)
Saccharina japonica[4]	CBP	Engineered *Escherichia coli* (BAL1611), 25-30°C, 150 h[i]		4.7% v/v (0.281 g/g dry algae)	Wargacki *et al.* (2012)

Notes: SHF: Separate Hydrolysis and Fermentation; SSF: Simultaneous Saccharification and Fermentation; CBP: Consolidated Bioprocessing.

M: molar (mol/L); N: equivalent concentration (=M/f_eq with f_eq the equivalence factor).

[a]100 g seaweed gel coming from 25 g of algae; [b]Combination of acid hydrolysis and enzymatic hydrolysis; [c]Successive saccharification with enzyme: two steps; [d]Air rate of 100 mL/min corresponding to an oxygen transfer coefficient (kLa) of 0.053 1/min; [e]Acid hydrolysis first and then simultaneous enzymatic hydrolysis and fermentation; [f]Oxygen rate is 5.9 mmol O_2/L/h; [g]Continuous hydrolysis reactor; [h]The engineered platform simultaneously allows hydrolysis and fermentation.

[1]Collected on a reef ; [2]Samples were harvested from wild stock; [3]No origin cited; [4]No origin cited.

Source: Adapted with permission from Fernand *et al.* (2016).

Table 8.2   Macroalgal conversion into butanol: pre-treatment, hydrolysis and fermentation processes.

Type of algae	Hydrolysis/ pre-treatment	Fermentation	Yield (butanol)	Reference
Ulva lactuca	1% H_2SO_4, 125°C, 30 min	*Clostridium beijerinckii.* *Clostridium saccharoperbutylacetonicum*[a]	0.29 g/g sugar	Potts *et al.* (2012)
Saccharina spp.		*Clostridium acetobutylicum* (ATCC 824), 209 h	0.12 g/g	Huesemann *et al.* (2012)
Ulva lactuca	Water, 150°C, 10 min Cellulase (GC220), 50°C, 24 h	*Clostridium beijerinckii* (NCIMB 8052), 140 h	3 g/L (0.35 g ABE/g sugar)	van der Wal *et al.* (2013)

Notes: ABE: Acetone Butanol Ethanol.
[a] Two-step fermentation: first with *Clostridium beijerinckii* and second with *Clostridium saccharoperbutylacetonicum*.
Source: Adapted with permission from Fernand *et al.* (2016).

chopping, beating, or extrusion). Pre-treatment and hydrolysis steps aim to improve and fasten the conversion process. For ethanol and butanol production, in most of the cases, the biomass undergoes thermal acidic or enzymatic hydrolysis, which produces sugars from carbohydrates. These sugars are then transformed into ethanol or butanol through fermentation. The fermentation organisms are mostly yeasts, but bacteria can also be used. An interesting, completely engineered platform allowing both hydrolysis and fermentation has been developed using synthetic biology tools by Wargacki *et al.* (2012). As a part of EU H2020 project, a rel car was fueled with seaweed derived biofuels (10% mix of ethanol and ABE) and drove 80 km (Fig. 8.7).

In the case of biodiesel production, the process is very different. There is an interesting relation between the harsh extraction procedures and the metabolite profile. The softer the extraction procedure, the less affected will be the chemical structures (Chaturvedi and Verma, 2013). In general, the lipid fraction in macroalgae first undergoes homogenization and extraction. A couple of extraction protocols have been developed based on solvents or supercritical fluids as well as on mechanical (i.e. by ultrasonication) or biological (i.e. by cell wall degrading enzymes)

Table 8.3   Conversion into biodiesel: pre-treatment, extraction and transesterification.

Type of algae	Pre-treatment	Extraction/transesterification	Oil content % DW	Yield of methyl esters	References
Ulva lactuca	Dried, ground	Chloroform/methanol (2:1, v/v), 6 h. NaOH, methanol, 16 h	4.2	3.8% from oil content	El-Moneim *et al.* (2010)
Asparagopsis taxiformis	Dried, ground	Chloroform/methanol (2:1, v/v), 6 h. NaOH, methanol, 16 h	4.1	3.64% from oil content	El-Moneim *et al.* (2010)
Mix of Pelvetia canaliculata, Fucus spiralis	Dried, crushed	Methanol-oil molar ratio 300:1, 1% mass g NaOH/g algae, 2.5 L n-hexane/kg algae, 60°C, 11 h[a]	2	11.42% from oil content	Maceiras *et al.* (2011)
Enteromorpha compressa	Dried, pulverized, sieved. Ultrasonication, 24 kHz, 50°C, 5 min	1.5% H_2SO_4, methanol-oil ratio 12:1, 400 rpm, 60°C, 90min. 1% NaOH, methanol-oil ratio 9:1, 600 rpm, 60°C, 70min[b]	11.14	90.6% from oil content	Suganya *et al.* (2013)
Himanthalia elongata	Dried, crushed	Algae-methanol ratio (1:15, w/v), NaOH (2 wt%), microwave (800 W), 3 min[a]	2.9	0.034% directly from dry biomass	Cancela *et al.* (2012)

Notes: FAME: Fatty Acid Methyl Esters.
[a]Simultaneous extraction and transesterification.
[b]Two steps: acid esterification followed by base transesterification.
Source: Adapted with permission from Fernand *et al.* (2016).

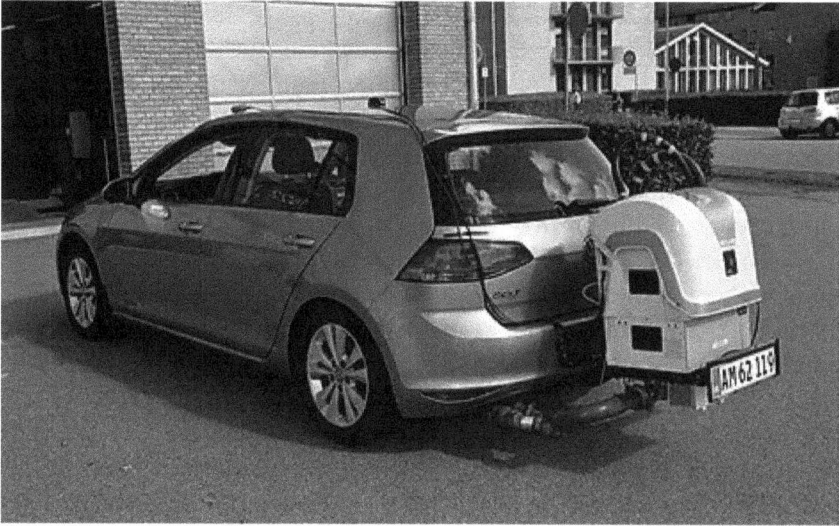

Fig. 8.7   A test car fueled with E10 produced from a mixture of fossil petrol with 10 weight % MacroFuels, that is ethanol and ABE. The car drove 80 km for emission tests by Danish Technological Institute as a part of H2020 MacroFuels project.

Source: MacroFuels H2020 project, https://www.macrofuels.eu/deliverables-1.

approaches. Base-catalyzed transesterification transforms fatty acids derived from triglycerides, phospholipids, and galactolipids but also free fatty acids into the respective methyl esters for further biodiesel production. As significant amounts of fatty acids might be transformed into breakdown products upon cell disruption as shown in several macroalgae, extraction and transesterification can also be achieved simultaneously in a one-pot reaction (Maceiras *et al.*, 2011). The distribution of fatty acids in lipids between brown, red, and green algae differs strongly, and variations within the genus level of, for example, *Ulva* can be observed and have to be considered as well (Alsufyani *et al.*, 2014; Pereira *et al.*, 2012). The biodiesel yield from macroalgae is thus strongly species-dependent.

8.8   Bioremediation

Macroalgae can be used for bioremediation of various contaminants in different techniques, combining environmental improvement with the production of potentially useful biomass. Here, we divided bioremediation into two main categories. The first, which can be coined nutrient bioremediation or eutrophication bioremediation, is the bioremediation via the consumption of nutrients (mostly nitrate, ammonium, and phosphate) which are needed for the macroalgae but may disturb the ecological balance and in high concentrations may pose an environmental hazard. Nutrient bioremediation may be applied as a pollution prevention step, such as in the IMTA concept, where macroalgae are integrated into farming of higher trophic levels, i.e. fish, thus consuming the effluent nutrients and minimizing environmental effects (Kang *et al.*, 2008). Alternatively, nutrient bioremediation may be applied to water bodies that are already contaminated and eutrophicated (Fei, 2004; Xu *et al.*, 2011). A second category is the bioremediation of industrial contaminants, mostly heavy metals, which may be hazardous even in low concentrations but can be accumulated by macroalgae nonetheless, or adsorbed on dead macroalgae biomass (He and Chen, 2014; Zeraatkar *et al.*, 2016).

8.8.1 *Eutrophication*

Eutrophication, a phenomenon associated with algal blooms, was defined by Nixon (1995) as "an increase in the rate of supply of organic matter to an ecosystem". Eutrophication may be caused, intuitively, by an increase in the input of inorganic nutrients, but also by a decrease in the turbidity of the water, a change in the hydraulic residence time of the water, a decline in grazing pressure, and other factors. Eutrophication may lead to species change in some cases (i.e. increases in non-visual predators and low-oxygen tolerant species such as jellyfish (Dong *et al.*, 2010; Lo *et al.*, 2008), and severe cases also to hypoxia and even fish kills (Nixon, 1995; Oswald, 2003)). Although increased productivity of micro- or macroalgae is initially associated with increased oxygen production, during night times photosynthetic processes cease and the accumulated primary producers become major oxygen consumers. Therefore, during night times, hypoxic or even anoxic conditions may develop, threatening all species which are not adapted to lack of oxygen (Diaz and Breitburg, 2009). Also, some microalgae species emit toxins, which in high concentrations may be hazardous to the ecosystem (Blackburn, 2004).

Eutrophication is an environmental concern that was first recognized only during the 1950s, with a significant increase in scientific attention only during the 1970s (Nixon, 1995). Although different factors can cause eutrophication, the most common single factor promoting eutrophication in freshwater and marine environments is an increase in the amounts of nitrogen and phosphorus that they receive (Oswald, 2003). Thus, it is not surprising that the increase in eutrophication processes and the awareness of the phenomenon have followed the dramatic global increase in phosphate rock mining and synthesis of nitrogen fertilizers since the 1960s (Nixon, 1995). Large-scale production of nitrogen synthetic fertilizers, for example, became possible after the development of the Haber-Bosch process for NH_3 synthesis in 1913, but due to political and economic reasons expanded globally only during the 1950s (Nixon, 1995). Eventually, when so many fertilizers are produced, large amounts of nutrients inevitably reach the coastal environment, in many ways — accidental release during production and transportation, natural wash away from farm fields into surface and groundwater, incorporation into plant tissues which are directly or indirectly consumed by people and eventually released to the environment by treated or untreated wastewater (Blackburn, 2004; Oswald, 2003), and others (Nixon, 1995).

8.8.2 *Eutrophication mitigation*

Different techniques were developed to prevent or mitigate eutrophication processes. These techniques include, for example, precision agriculture technologies (Zhang *et al.*, 2002b) and farming regulations regarding excess fertilization in the agricultural sector (Van Grinsven *et al.*, 2012), and advanced wastewater treatment technologies (Morse *et al.*, 1998; Paredes *et al.*, 2007), which have been followed by effluent water quality regulations (Epa, 2018), in the municipal sector. Wastewater treatment technologies for nutrient reduction include, among others, biotechnological treatments such as nitrification-denitrification processes that transform inorganic soluble nitrogen ions back to atmospheric inert gaseous nitrogen (Paredes *et al.*, 2007), and phosphorous direct precipitation or precipitation after assimilation in bacteria, which are later removed and disposed of (Morse *et al.*, 1998). However, in many cases, implementing these techniques is not enough. In developing countries, for example, economic considerations and lack of funding may prevent appropriate wastewater treatment (Kivaisi, 2001), and in developed countries, intensively fertilized agriculture may apply large amounts of nutrients which are very challenging to manage and control (Parris, 2011). Nutrient removal using current technologies is a resource demanding process, imposing an economic burden also on rich countries, which consequently compromise on partial wastewater treatment (for example, remove ammonium but not nitrate) (Neethling and Kennedy, 2018). Therefore, sometimes eutrophication remediation of the already polluted water source or coastal environment is the only solution.

8.8.3 *Macroalgae eutrophication bioremediation*

Macroalgae are a natural candidate for bioremediation of eutrophicated regions, as they are naturally stimulated by excess nutrients, yet large enough for easy harvesting. Harvesting is a key point of the bioremediation process, as most adverse effects of eutrophication occur due to the accumulation of excessively dense biomass, rather than due to the high production rate or high nutrient concentrations (Valiela *et al.*, 1997). Thus, macroalgae can be cultivated in eutrophicated regions, utilizing the favorable conditions for rapid biomass production, and easily harvested and removed, preventing excessive organic load and the formation of hypoxia conditions (Valiela *et al.*, 1997). Algal blooms in nutrient-rich conditions are a natural phenomenon and an important part of the marine food web. For example, spring upwelling of deep, nutrient-rich water, is the base for phytoplanktonic blooms, which are followed by zooplankton blooms and onwards up the marine food web (Boero, 2013; Legendre, 1990). However, when these algal blooms take over an ecosystem instead of supporting its food web, they become a problem. Harmful blooms of macroalgae, as opposed to microalgae blooms, are naturally concentrated in nearshore areas or even washed out to the beach, thus being cheaper and simpler to harvest (Ansell *et al.*, 1998). Therefore, bioremediation by macroalgae can be performed simply by harvesting of natural stocks where the density exceeds sustainable and environmentally healthy values (Monagail *et al.*, 2017), or by active controlled cultivation and harvest in nutrient-enriched regions (Fei, 2004; Xu *et al.*, 2011), which can also prevent harmful microalgae blooms. One interesting example is in north China, where large-scale cultivation of *Laminaria japonica* has been encouraged for several decades, intending to balance the negative environmental effects of scallop cultivation (Fei, 2004). Besides, *Laminaria japonica* may also contribute to the reduction of other pollutants such as heavy metals that are disposed of in the marine environments (Xu *et al.*, 2011).

Large brown seaweeds (kelps) are of immense economic value and present several environmental benefits as well. In coastal marine waters, red and brown seaweed species have been cultivated as nutrient scrubbers for nutrient remediation capacity. The morphology and lifecycle of kelps allow for their cultivation of structures in the ocean. In principle, kelp spores, gametophytes, or juvenile sporophytes are seeded onto ropes or textiles and left to grow for 6–15 months in the top meters of the water column. Land-based nurseries are also frequently used. The most commonly cultivated kelp species are *Saccharina japonica, Undaria pinnatifida, Macrocystis pyrifera, Saccharina latissima, Alaria esculenta, Laminaria digitata,* and *Saccorhiza polyschides.* Kelp species assimilate inorganic nutrients such as N and P and hence function as nutrient scrubbers, regardless of the source of nutrients. The nutrient remediation capacity of seaweeds is determined by the number of nutrients removed from the sea by harvesting the seaweed biomass.

Consequently, the quantitative nutrient assimilation capacity is defined by the nutrient content of the harvested biomass (percentage of fresh or dry biomass) and the areal biomass production capacity (tonnes of fresh or dry biomass harvested per area). Biomass yields and quality of any species depend strongly on the growth environment, causing variation in productivity and biochemical composition on a temporal and spatial scale. Yields reported from Europe are somewhat lower, and most often yields are in linear meters (volume of biomass per meter of the seeded line). Often, yields are reported from small-scale research experiments, with no description of the construction of the overall cultivation structure (distance between seeding lines, the vertical extent of cultivation lines, etc.), complicating the actual areal efficiency of the bioremediation effect. The kelp growth strategy implicates an inverse relation between kelp biomass yield and tissue nutrient content since, at the time of maximal biomass yield, the kelp will typically have exploited the internal nutrient reserves for sustaining growth. Viewing this trade-off and in a bioeconomy perspective, the timing of biomass harvest is crucial.

8.8.4 *Integrated systems*

Similarly, macroalgae can be cultivated downstream fish cages and other aquaculture facilities, implementing the IMTA concept (see Chapter 6), thus sequestering excess nutrients at the source and preventing potential eutrophication (Kang *et al.*, 2008). Furthermore, macroalgae can be used for wastewater treatment. Various wastewater streams, including municipal, industrial, and agricultural, have been studied as potential nutrient sources for growing algae, reporting the recovery of 90–99% of ammonium and orthophosphate from municipal sewage and dairy manure (Adey *et al.*, 1993; Dahiya *et al.*, 2012; Wilkie and Mulbry, 2002; Woertz *et al.*, 2009). The ability of macroalgae to grow even in high concentrations of heavy metals enables the integration of nutrient reduction with heavy metal removal from industrial wastewater (Zeraatkar *et al.*, 2016). Finally, the preferable design of the macroalgal bioremediation project, which should deliver a real economic added value, is the integration of pollution removal with the production of a valuable product (Nielsen *et al.*, 2012). Two suggested integrated systems are the integration of wastewater treatment and biofuel production (Dahiya *et al.*, 2012; Zeraatkar *et al.*, 2016) and the integration of dairy manure treatment by macroalgae with the production of methane and proteinaceous animal feed or fertilizer, thus providing a holistic solution to nutrient management problems in dairy farms (Wilkie and Mulbry, 2002).

8.8.5 *Industrial wastewater*

Industrial wastewater (IW) differ from municipal or agricultural wastewater in their composition, as they tend to contain unconventional materials which may be hazardous even in low concentrations, such as organic compounds, synthetic chemicals, nutrients, organic matter, and heavy metals (Mwangi and Ngila, 2012; Zeraatkar *et al.*, 2016). Specifically, heavy metal contaminations have become a global issue of concern as they are highly toxic, non-biodegradable, bioaccumulate in the food chain and the human body, and are carcinogenic (He and Chen, 2014). Therefore, Maximum Contaminant Levels (MCLs) were set by the World Health Organization (WHO), U.S. Environmental Protection Agency (USEPA), and many government environmental protection agencies for the concentrations of lead (Pb), mercury (Hg), chromium (Cr), arsenic (As), cadmium (Cd), zinc (Zn), copper (Cu), and nickel (Ni), which are the most common surface and groundwater contaminants (He and Chen, 2014).

The conventional methods of IW treatment involve precipitation, ion exchange, filtration, reverse osmosis, evaporation recovery, and other membrane or electrochemical methods (Ahluwalia and Goyal, 2007; He and Chen, 2014; Mata *et al.*, 2009; Yu *et al.*, 2008). These conventional treatment methods may be ineffective or expansive, especially in low metal concentrations (below 100 mg l⁻¹), due to incomplete removal of heavy metal ions and the high costs of chemicals at industrial scales (Nourbakhsh *et al.*, 1994; Yu *et al.*, 2008; Zeraatkar *et al.*, 2016). Moreover, increasing regulatory restrictions necessitate a change to alternative methods (Zeraatkar *et al.*, 2016).

One non-conventional heavy metal remediation method is based on biosorption, which is the ability of biological materials to accumulate heavy metals from wastewater streams by either metabolically mediated or purely physico-chemical pathways of uptake (Yu *et al.*, 2008). The phenomenon of biosorption was observed in the early 1970s when algae were found to accumulate radioactive elements and heavy metals in wastewater released from a nuclear power station (He and Chen, 2014). Studies conducted since then demonstrated the potential of macro- and microalgae biomass to offer cheaper and more sustainable remediation solutions. One application of the ability to accumulate hazardous elements is the use of macroalgae as "bioindicators" or for "biomonitoring" of environmental concentrations of heavy metals and radioactive elements (Chakraborty *et al.*, 2014; Jitar *et al.*, 2015; Saâ Nchez-Rodrõ Â Guez *et al.*, 2001; Torres *et al.*, 2005). A second application is environmental remediation by controlled biosorption. The algae biosorption technology possesses a few advantages, including high metal sorption and concentration capacities, high detoxifying efficiency also for low heavy metals concentrations, low operating costs and high cost-effectiveness, reusability and renewability, abundance in seawater and freshwater, and more (He and Chen, 2014; Mata *et al.*, 2009; Torres *et al.*, 2005; Zeraatkar *et al.*, 2016; Zeroual *et al.*, 2003). However, the heavy metal bioremediation concept is still considered theoretical, as no pilot or demonstration scale studies were conducted yet, and no detailed techno-economic analysis of the feasibility of such a process has been performed (Zeraatkar *et al.*, 2016).

8.8.6 *Biosorption mechanisms*

A prerequisite for effective biosorption is the diffusion of positive charge ions to the algae cell, which is negatively charged due to the ionization of functional groups (Zeraatkar *et al.*, 2016). Heavy metal biosorption is mostly based on an ion-exchange mechanism, enabling cations with higher affinity to replace low-affinity cations (mainly Ca^{2+} and Mg^{2+}) adsorbed to negative charged functional groups on the surface and inside algal cells (He and Chen, 2014). Functional groups, such as $OH-$, $SH-$, $COO-$, PO_4^{-3}, SO_3-, NO_3-, RNH_2-, $RS-$ and $RO-$ are present in the cell wall, in the cytoplasm and the vacuoles (He and Chen, 2014; Zeraatkar *et al.*, 2016). In the cell wall, for example, the adsorption capabilities are associated mostly with the alginic acid (alginate) for brown seaweed and to sulfated polysaccharides such as fucoidan, galatians, and ulvan, for brown, red, and green seaweed, respectively (He and Chen, 2014; Kanno *et al.*, 2014). The uptake mechanism can be described as initial adsorption to the cell wall, which is passive and occurs in a time scale of minutes (Zeraatkar *et al.*, 2016), followed by a slower active cytosolic protein transfer into the cell and finally to storage in the vacuoles that serve as an accumulation organelle (Zeraatkar *et al.*, 2016). Two other biosorption mechanisms are electrostatic interactions and complex formations (Zeraatkar *et al.*, 2016).

Macroalgae bioremediation processes can be divided into those that use living algae and those that use non-living algae biomass. Some authors refer only to the passive adsorption to non-living biomass as biosorption and use the term bioaccumulation for the active process, which requires the metabolic activity of living organisms (Davis *et al.*, 2003).

8.8.7 *Live macroalgae bioremediation*

For living algae, avoiding toxicity is a major issue. Monteiro *et al.* (2012) described a common microalgae intracellular metal detoxification mechanism: metal-binding peptides or proteins, namely class III metallothioneins or phytochelatins, bind to metal ions and transfer them into vacuoles, thus facilitating control over cytoplasmic concentrations and preventing toxic effects. However, this mechanism alone is not always enough, as excessive heavy metal concentrations may lead to protein denaturation, depletion of essential elements, or damage to the oxidative balance of the live algae, which depend also on the content of oxidized proteins and lipids in the cell (Zeraatkar *et al.*, 2016). Therefore, heavy metal ion concentrations should be optimized for efficient algal growth (Zeraatkar *et al.*, 2016). The growth rate is important as it continuously increases the biosorption capacity while decreasing the internal metal concentrations, slowing down toxic effects (Zeraatkar *et al.*, 2016). These toxic effects, which are caused by specific metals, can be moderated by dilution or mixing of the IW with other waste streams (Abinandan and Shanthakumar, 2015). Therefore, IW should be characterized to determine the pollution type and nutrient availability, which can both affect the algae's growth rate. The biosorption mechanism of living algae includes also an active, slower, stage (Zeraatkar *et al.*, 2016), emphasizing the importance of the contact time factor for the process. Lamai *et al.* (2005) investigated the toxicity of Pb and Cd ions to common filamentous green macroalgae, *Cladophora fracta,* by increasing exposure times and concentrations. This study found a chlorophyll loss followed by cell wall disintegration and reduced growth, which were more prominent for Cd than for Pb. This means that the accumulation potential, and therefore the bioremediation suitability of *Cladophora fractal,* is higher for Pb than for Cd.

Adsorption by living algae can be optimized by modification of the process parameters, including the concentration of metal ions and algae biomass, temperature, and the presence of competing ions (Zeraatkar *et al.*, 2016). However, parameter modification is limited by the need to maintain appropriate physiological conditions. Biosorption capacity correlates to the concentration of the metal ions directly, while biosorption efficiency correlates to the concentration of the metal ions inversely. For example, Monteiro *et al.* (2009) reported a 10-fold increase in Zn^{2+} biosorption by the microalgae *Scenedesmus obliquus* (19 to 209.6 mg/g DW), when increasing initial Zn^{2+} concentrations from 10 to 50 mg l^{-1}, followed by a decrease in the sorption efficiency. Increasing algal biomass density increases the total removal but decreases the specific sorption capacity per gram of biomass (Zeraatkar *et al.*, 2016). Literature regarding temperature effects is not consistent. One explanation is that optimum temperatures for biological reactions are usually a narrow range, and temperature variations cause different biosorption behaviors in various algal strains with different metal ions (Zeraatkar *et al.*, 2016). Nutrients and light are required for the bioremediation process, and an increase in those parameters increases the algae's growth rates and metal ion removal capacity (Taziki *et al.*, 2016). Therefore, some IW streams require fertilization or mixing with nutrient streams for a successful, continuous, bioremediation process. Finally, a better understanding and optimization of these and other parameters is necessary for the successful implementation of living-algae biosorption treatments (Zeraatkar *et al.*, 2016).

8.8.8 *Non-living macroalgae bioremediation*

At the same time, non-living algae have some significant advantages. These advantages include, first of all, biosorption capacities several times greater compared to living algae (He and Chen, 2014; Mata *et al.*, 2009; Zeraatkar *et al.*, 2016). Furthermore, dead biomass can be successfully reused in successive adsorption–desorption cycles (Areco *et al.*, 2012; Zeraatkar *et al.*, 2016) and can achieve improved performances by different chemical or physical treatments (before, during, or after the adsorption activity). In addition, non-living biomass does not require constant cultivation activities (harvesting, cleaning, fertilizing, etc.) and costs, and may be available as waste or byproduct (He and Chen, 2014; Mata *et al.*, 2009). The main parameter that affects the adsorption capacity is the solution pH. pH affects the charge of the biomass functional groups, on the one hand, and the solution chemistry of the heavy metal ions, on the other hand (Pavasant *et al.*, 2006). Increasing pH has a positive effect on biosorption as it is followed by more negatively charged functional groups and fewer protons which can compete with the heavy metal ions for the binding sites (He and Chen, 2014; Zeraatkar *et al.*, 2016). However, when pH increases too much, it affects the chemistry of the metal ions and causes precipitation. Studying the exact effects of pH on different functional groups and algae species, and on different metal ions, is most important for the optimization of metal removal of specific ions by specific algal biomass (Zeraatkar *et al.*, 2016). For example, Pavasant *et al.* (2006) studied metal removal by dried marine green macroalga *Caulerpa lentillifera* and found optimal removal capacities of 90% Pb at pH 5.5, 82% Cu at pH 6, 87% Zn at pH 8, and 87% Cd at pH 8.5. Other important parameters are contact time, biomass concentration, initial metal concentration, and biomass pre-treatment. The biosorption capacities of 14 different heavy metal ions using different micro- and macroalgal strains under optimal conditions are presented in Table 1 in the review by Zeraatkar *et al.* (2016).

8.8.9 *Biomass pre-treatment*

Pre-treatment of the dead biomass can be performed for two main reasons: (1) prevention of secondary pollution, and (2) enhancement of adsorption capacity. Different physical pre-treatments, such as heating, boiling, freezing, crashing, and drying, usually destruct cell membranes and provide increased surface area and enhanced adsorption capacity. Chemical pre-treatments with $CaCl_2$, formaldehyde, glutaraldehyde, NaOH, and HCl improve the ion-exchange performance and the availability of bio-mass binding sites in various mechanisms (Zeraatkar *et al.*, 2016). Besides, some chemical pre-treatment can prevent potential secondary pollutants, for example, by encapsulation or cell wall modification (He and Chen, 2014). Secondary pollution is the leaching of metal ions or organic molecules such as the green pigment chlorophyll during the bio-remediation treatment (Mwangi and Ngila, 2012). The secondary pollu-tion can affect the taste and the color of the water in mild cases and significantly increase the total dissolved solids and hardness of the water in other cases, and therefore is undesired. For example, Mwangi and Ngila (2012) have modified the seaweed *Caulerpa serrulate* with ethylenedi-amine (EDA) and reported the great potential of the modified seaweed for the removal of Cu, Pb, Cd, and dissolved organic carbon in polluted water. The modified seaweed showed an adsorption capacity of 5.27, 2.12, and 2.16 mg/g for Cu, Cd, and Pb, respectively, compared to 3.29, 4.57, and 1.05 mg/g^{-1} with the unmodified seaweed. These results offer a significant increase for Cu and Pb sorption and a decrease in Cd adsorption. In addi-tion, modification minimized the leaching of organic matter and improved the thermal stability of the material. Zhao *et al.* (1994) studied the effects of different treatments (HNO_3, HCl, NaOH, acetone, and 60°C water) on the biosorption of different ion metals (Pb, Cu, Zn, Cd, Cr, Mn, Ni, Co, Hg, Au, and Ag), using six species of marine algae, and found that all treatments improved the biosorption capacity.

8.8.10 *Macroalgae biomass derivatives*

Another alternative is the use of macroalgae biomass derivatives for bio-sorption needs. Alginate, a biopolymer that can be extracted from brown seaweed, can be modified chemically to efficiently remove anionic con-taminants from water solutions or even complexed with magnetite, thus adapting magnetic properties (LIM *et al.*, 2008). Dealginate, which is the left-over of seaweed alginate extraction, can be used as a biosorbent as it has an array of weakly acidic and basic functional groups, which enable them to bind metal ions (Romero-González *et al.*, 2000). This cheap mate-rial is especially useful as it is physically stable for long periods (Romero-González *et al.*, 2000). Dealginate has demonstrated the ability to adsorb about 90% of the metals Cd^{2+}, Cr^{3+}, Cu^{2+}, and Pb^{2+} in less than five minutes, and can be used as a column packing material for preconcentration and quantification (Romero-González et al., 2000).

After heavy metal biosorption in non-living biomass, a desorption stage is applied to elute the adsorbed ions and reuse the sorbent (Yu et al., 2008). Yu and Kaewsarn (2001) examined 11 eluting agents on Cu^{2+} adsorbed to the biomass of marine alga *Durvillaea potatorum* and found HCl, H_2SO_4, and EDTA to be the most efficient. The optimal elute composition was found to be 0.35 M HCl and 0.5 M $CaCl_2$, which kept desorption capacity close to the original after the fifth cycle. Zeroual et al. (2003) found *Ulva Lactuca* to be an effective Hg biosorbent in pH 7, maintaining similar performances after continuous cycles of adsorption–desorption with 0.3 N H_2SO_4 and regeneration with distilled water. A clean alternative for metal elution from the sorbent is using the Supercritical Fluid Extraction (SFE) technique which converts the charged metal species into neutral metal chelates soluble in supercritical fluids. A suitable choice for SFE is CO_2, because of its moderate critical constants (Tc of 32°C and Pc of 73 atm), and it can be modified with methanol and a suitable ligand for the elution task (WANG and WAI, 1996).

8.8.11 *Metals recovery*

High volume recovery of precious metals such as gold and silver or radionuclides such as uranium from waste streams could provide significant economic benefits to the bioremediation process in mines, electronic waste leachates, and relevant industries (Dahiya et al., 2012; Mata et al., 2009; Wang and Chen, 2009). Torres et al. (2005) examined the effectiveness of calcium alginate gels as a biomass-derived immobilization matrix for the uptake of Au and Ag and found that the optimum pH was 2 for Au and 4 for Ag. Furthermore, alginate presented a good biosorption capacity for dilute gold solutions (290 mg/g) and an acceptable capacity for silver (38 mg/g). Also, alginate (Mata et al., 2009; Torres et al., 2005) and dealginate (Cui and Zhang, 2008; Romero-González et al., 2003) have the ability to reduce metals such as gold and silver, thus offering an economical alternative for mining and industrial effluents treatment, based on a synthesis of nanoparticles of precious metals (gold, silver, platinum, and palladium) for use in the electronic and medical industries.

8.9 CO$_2$ Capture

Seaweeds are regarded as highly productive organisms with photosynthetic and growth rates frequently overpassing most terrestrial plants. Indeed, energy conversion efficiency from sunlight to stable intracellular compounds in most terrestrial plants is in the range of 0.5–0.8%, while seaweeds can attain 4.3–6.5%. Seaweeds are then viewed as palliative solutions for the sequestration of increasing amounts of seawater inorganic carbon (Ci) that restrain climate change, while simultaneously delivering valuable biomass to be used in a biorefinery.

The marine environments are relatively stable and seaweeds have adapted to the much higher diversified and challenging habitats, encountered in the intertidal zone. Current atmospheric CO$_2$ concentrations have increased by ca. 40% in only 50-y and will likely double in a few decades. Dissolved CO$_2$ in shallow coastal marine areas are expected to be in equilibrium with atmospheric CO$_2$ and, thus, will co-vary directly with atmospheric values. In seawater, the dominant forms of inorganic carbon (Ci) are CO$_2$, HCO$_3$-, and CO$_3$-, and vary with pH and temperature, altogether around 2.2 mm. Seaweeds can utilize inorganic carbon (Ci) through various routes that share the use of HCO$_3$-, which is naturally found at concentrations 200 times higher than dissolved CO$_2$. In terms of photosynthesis, increasing CO$_2$ should favor seaweeds from Ci potential limitations with an expected consequent increase of plant productivity, provided that light, other nutrients, and water are kept at plentiful levels. Ci limitations may occur because the enzyme that fixes CO$_2$ in the photosynthetic carbon reduction (PCR) cycle, ribulose-1,5-bisphosphate carboxylase/oxygenase (Rubisco), has a low affinity for CO$_2$ and also fixes O$_2$. Hence, photosynthesis would be severely limited if it were depended only on the diffusional entry of CO$_2$ from the medium to the site of fixation. The reasons for CO$_2$ limitation in seawater include (1) the rather low concentrations of CO$_2$ (ca. 30% lower (v/v) than in air, depending on temperature), (2) the low diffusion coefficient in liquid media (four orders of magnitude slower diffusion than in air), and (3) the slow uncatalyzed rates of formation of CO$_2$ from HCO$_3$-. Under such conditions, and given the high levels of HCO$_3$-, most marine macroalgae utilize the ionic Ci form. Further, seaweeds have developed operating carbon concentrating mechanism (CCM). In terrestrial plants, the mechanism underlying the CCM is C$_4$ photosynthesis. With very few distinct exceptions, photosynthesis of macroalgae is of the C$_3$ type (Israel and

Hophy, 2002), indicating that HCO_3^- utilization is the basis for their CCM.

In practice, the utilization of HCO_3^- can take place by (1) extracellular dehydration to form CO_2 before Ci uptake and (2) direct uptake through the plasma membrane. Since the establishment of equilibrium between CO_2 and HCO_3^- is a slow process, plants that employ the first method of HCO_3^- utilization do so by the aid of the enzyme carbonic anhydrase (CA). This seems to be a fairly common way of HCO_3^- utilization that was first described for the red algae *Chondrus crispus* (Smith and Bidwell, 1989) and subsequently found in some species of *Ulva* (Björk *et al.*, 1992) and several other genera (Beer and Björk, 1994). The second method of HCO_3^- utilization is through direct uptake that would require the presence of a transport protein to enhance its penetration through the membranes (Drechsler *et al.*, 1994). Whether or not a marine algae will respond to external elevated CO_2 levels will depend both on the method and degree of HCO_3^- utilization and the environmental conditions under which Ci enrichment is imposed. As with the CCM, adaptations of Rubisco to increasing CO_2 in marine macroalgae are not well known yet. Also, modulation in the amount of Rubisco is affected by nitrogen nutrition. Extrapolation of short-term macroalgae responses to globally high CO_2 is complex, and long-term growth trials under CO_2 enrichment are particularly important as are the combined effects of additional anthropogenic factors.

The term Carbon Offsetting, namely receiving credit for reducing, avoiding, or sequestering carbon, has become part of the bulk of solutions to mitigate carbon emissions, and thus climate change, through policy and voluntary markets, primarily by land-based but shortly presumably also through offshore seaweed aquaculture. As land and freshwater are limiting, seaweed farming is rapidly growing. When summarizing the currently available scientific literature, we assess the extent and cost of scaling seaweed aquaculture to provide sufficient CO_2 sequestration for several climate change mitigation scenarios, with a focus on the food sector, a major source of greenhouse gases. Given known ecological constraints (nutrients and temperature), a suitable area would consist of ca. 48 million km^2 for seaweed farming, which is largely unfarmed. Seaweed could create a carbon-neutral aquaculture sector with just 14% (mean = 25%) of current seaweed production (0.001% of the suitable area). At a much larger scale, we find seaweed culturing extremely unlikely to offset global agriculture, in part due to production growth and cost constraints.

Yet, offsetting agriculture appears more feasible at a regional level, especially in areas with strong climate policy. Importantly, seaweed farming can provide other benefits to coastlines affected by eutrophic, hypoxic, and/or acidic conditions, creating opportunities for seaweed farming to act as "charismatic carbon" that serves multiple purposes. Seaweed offsetting is not the sole solution to climate change, but it provides an invaluable new tool for a more sustainable future.

Together with seaweeds and viewing the ongoing processes of global change on Earth, vegetated coastal ecosystems, such as seagrass beds, mangroves, and tidal salt marshes, make globally significant contributions to carbon storage in biomass and long-term sequestration in sediment deposition (Duarte *et al.*, 2013). The carbon sequestered in both living and non-living biomass in the ocean and coastal habitats has been termed "Blue carbon" (Howard *et al.*, 2017). Some carbon is released back into the atmosphere through respiration and oxidation, but a proportion of assimilated carbon remains in the form of living biomass, contributing to organic carbon stored in soils (Murray *et al.*, 2011). In this context, additional benefits of seaweed photosynthesis refer to marine ecosystem services such as climate regulation, which is considered a valuable asset to human welfare. In the context of the oceanic environment, climate regulation can be addressed as an ecosystem service rendered by marine ecosystems through the absorption and deposition of atmospheric carbon dioxide (CO_2) within deep oceanic layers by marine organisms, a process often referred to as the "biological pump" (Peled *et al.*, 2018). Following its formation by seaweeds (and other primary producers) during photosynthesis, organic carbon is exported below the euphotic layer (depths corresponding to 0.1–1% of sunlight reaching the surface layer), where it is subjected to remineralization and solubilization at various depths (Raven and Falkowski, 1999). The fraction of organic carbon that remains within the ocean is dependent on various biogeochemical processes and its residence time within deep oceanic layers dictates the duration of temporary reduction of atmospheric CO_2 concentrations. The benefit derived from the prolonged sequestration of atmospheric CO_2 is the moderation of adverse climate change phenomena, such as extreme weather.

Seaweed aquaculture beds (SABs) that support the production of seaweed and their diverse products cover extensive coastal areas, especially in the Asian-Pacific region, and provide many ecosystem services such as nutrient removal and CO_2 assimilation. It is worth mentioning that land-based seaweed aquaculture practices can also significantly add to the

SABs in coastal areas. The use of SABs in potential CO_2 mitigation efforts has been proposed with commercial seaweed production in China, India, Indonesia, Japan, Malaysia, Philippines, Republic of Korea, Thailand, and Vietnam, and more recently in Australia and New Zealand (Sondak *et al.*, 2017). An estimate of the total annual potential of SABs to fix anthropogenic CO_2 and reduce current atmospheric levels has increased substantially in the Asian-Pacific region, surpassing $2.61 \cdot 106$ t DW. This biomass production translates in a total of carbon accumulated annually equaling more than $0.78 \cdot 106$ t y^{-1}, equivalent to over $2.87 \cdot 106$ t CO_2 y^{-1} (Sondak *et al.*, 2017). Consequently, CO_2 drawdown can be enhanced by increasing the area available for SABs, biomass production, and carbon accumulation. The conversion of biomass to biofuel can reduce the use of fossil fuels and provide additional mitigation of CO_2 emissions. Yet this approach might only be feasible within the biorefinery vision, combined with additional target products. Contributions of seaweeds as carbon donors to other ecosystems could be significant in global carbon sequestration. The ongoing development of SABs would not only ensure that Asian-Pacific countries will remain leaders in the global seaweed industry but may also provide an added dimension of helping to mitigate the problem of excessive CO_2 emissions (Sondak *et al.*, 2017). Nonetheless, the overall contribution of seaweeds as CO_2 sequesters on a global scale and in the long-run might be questionable. It is believed that a rather small scale, local seaweed aquaculture may have sustainable favorable ecological impacts when it comes to CO_2 sequestration.

8.10 Geo and Climate Engineering

Seaweeds play an important role in the coastal biome, in which vegetation has a remarkable capacity to engineer both biotic and abiotic ecosystem components (Duarte *et al.*, 2013). A survey of ecosystem services of different biomes ranked the vegetated coastal habitats as one of the most valuable ecosystems in Earth's biosphere, pointing out its crucial contribution to the global nutrient cycling (Costanza *et al.*, 1997). Other ecosystem services of the vegetated coastal habitat include carbon sequestration, coastal protection, erosion control, water catchment and purification, and maintenance of beneficial species (Barbier *et al.*, 2011). Among the different vegetative coastal ecosystems (Salt marshes, Mangroves, Seagrasses, and Seaweed), seaweeds are the most abundant and globally spread, in both natural stocks and cultivation farms, due to their large number of species with different habitat requirements and different biochemical characteristics. Thanks to their capacity to engineer ecosystem components, when controlled and manipulated by human activities, seaweed can be used for small-scale geo- or climate-engineering applications. In other words, seaweed can be utilized as an important player in the marine ecological engineering, which aims to "design sustainable ecosystems that integrate human society with its natural environment for the benefit of both" (Mitsch and Jørgensen, 2003; Odum, 1962; Odum *et al.*, 1963).

8.10.1 *Ecological engineering*

The most known example of seaweed ecological engineering is the IMTA, in which seaweeds are added as nutrient-recyclers and used to design aquaculture systems to be more sustainable. However, seaweed cultivation can achieve different objectives beyond the previously discussed biomass production (for various applications), CO_2 sequestration, and nutrient or metal bioremediation. Each naturally occurring ecosystem service (mentioned above) can be achieved artificially from seaweed cultivation. Thus, seaweed cultivation can be used for mitigating climate change by carbon sequestration (see Section 8.12), protecting the environment from pollution by water purification (see Section 8.8), promoting ecosystem restoration by offering "safe zones" from increasing pH, and for proliferation and recruitment of many species and protecting the coast from erosion.

8.10.2 *Ecosystem restoration*

Seaweed cultivation can be used for ecosystem restoration, as dense seaweed canopies can offer protection from predation to some species while allowing refuge from desiccation at low tide for others (Monagail *et al.*, 2017; Phillippi *et al.*, 2014). Furthermore, seaweed cultivation may be used to mitigate ocean acidification locally, as large-scale marine photosynthesis can increase the pH significantly (Buapet *et al.*, 2013), thus offering "safe zones" to calcifying organisms, such as shrimps, oysters, and the particularly vulnerable corals (Gattuso *et al.*, 2018; Mcleod *et al.*, 2013). A model by Unsworth *et al.* (2012) suggests that in shallow water reef environments, *Scleractinian coral* calcification downstream of sea-grass can potentially increase by about 18% compared to an environment without seagrass. Based on these results, Unsworth *et al.* (2012) offered to consider the future use of seagrass as a useful tool in marine park management at a local scale. Similarly, Mcleod *et al.* (2013) suggested to use local resources such as seaweed cultivation to influence the pH and carbonate system to be reef favorable. Following these ideas, the role of reef-associated habitats of seagrass and mangroves was examined by Camp *et al.* (2016), with results supporting the hypothesis that seagrass can potentially buffer near-by corals reefs from ocean acidification.

8.10.3 *Coastal protection*

Another possible application of seaweed cultivation, which relates to their effects on the abiotic environment, is coastline erosion prevention. Naturally, seaweed canopies dissipate wave energy, increase bottom shear stress, and reduce flow and turbulence, thus reducing tidal surge and waves which cause most coastline erosion (Duarte *et al.*, 2013; Mendez and Losada, 2004; Monagail *et al.*, 2017). At the same time, increased sedimentation rates, which are the opposite of erosion, together with high burial rates of seaweed biomass raise the seafloor, buffering the impacts of rising sea level and wave action that are associated with climate change (Duarte *et al.*, 2013). Theoretically, this property of energy dissipation can be utilized for large-scale eco-engineering coastal protection projects.

Laboratory experimental studies by Løvås and Tørum (2001), aimed to better understand the consequences of kelp harvesting along the Norwegian coast, found that kelp reduces the coastal erosion by reducing the time till waves fade, equilibrium is obtained, and erosion ends. Earlier laboratory and field experiments by Price *et al.* (1969) reported that arti-ficial seaweed can lead to net onshore sediment transport and beach build-up. Studies found that wavelength-to-water-depth ratio has a significant effect on the efficiency of the wave energy damping (Duarte *et al.*, 2013; Løvås and Tørum, 2001), together with the specific properties of species. *Laminaria Hyperborea*, for example, was found especially suitable for this task (Løvås and Tørum, 2001).

Currently, applied coastal protection eco-engineering solutions are still quite rare and usually consist of other species rather than seaweed. One mega-scale example is the Dutch coastline in which erosion of sandy beaches and dunes is prevented by using various civil- and eco-engineering solutions. These eco-engineering solutions include sand fixation with different vegetation types, large wetlands that serve as buffers against flooding, willow trees that reduce wave impacts on the dikes, and more (Borsje *et al.*, 2011). Finally, the utilization of natural and artificial seaweed beds for various eco-engineering solutions, including CO_2 sequestration, nutrient recycling, ecosystem conservation or restoration, and coastal protection, offer an interesting, cost-effective, and mostly untapped, strategy for environmental improvement and climate change mitigation and adaptation.

Chapter 9

Economics of Macroalgae Biorefineries

The desire for economic growth, coupled with growing concern about the environmental impact of development, has led to the concept of "sustainable development". The "bioeconomy" has been referred to as the set of economic activities that relate to the invention, development, production, and use of biological products and processes. Novel biorefinery technologies allow utilizing renewable natural resources and introduce new industries of bioeconomy. Seaweed, also called marine macroalgae, has the potential to be a valuable feedstock for biorefineries. Depending on seaweed type and species, it is possible to extract different fatty acids, oils, natural pigments, antioxidants, high-value biological components, and other substances, which can be potentially used in an industrial production system. The seaweed biorefinery framework presents a conceptual model for the production of high-value products along with low-value biofuels, either fluid or gaseous. Yet, the implementation of industrial level macroalgae biorefinery for the commercial production of renewable chemicals and fuels is currently very limited. This is because of high capital and operating expenditures, irregularities in biomass supply chain, technical process immaturity, and scale-up challenges. There is a need for methodologies for economic and policy analysis of novel biorefinery technologies, taking into account environmental side effects and physical and economic uncertainties. The question is not only how the microalgae-based biorefinery can be developed, but also how it can be developed sustainably in terms of economic and environmental concerns. Studies point at both the scientific and economic challenges that require multidisciplinary effort to develop viable technologies, cost-effective harvesting techniques, and equipment, processing facilities, supporting infrastructure, and sustainable supply chains, as well as creating markets for novel macroalgae-based goods.

The bioeconomy describes the global industrial transition of sustainably utilizing renewable aquatic and terrestrial resources in energy, intermediates, and final products for economic, environmental, social, and national security benefits (Golden *et al.*, 2015). The bioeconomy consists of complex supply chains that include biomass production, transportation, conversion into products at biorefineries, and distribution. Biomass is regarded as a renewable carbon source, offering multi-faceted benefits, from carbon sequestration to the production of bioenergy and bio-products (Ubando *et al.*, 2020). Macroalgae have the potential to be valuable feedstock for marine biorefineries that convert the macroalgae into a range of bio-based chemicals, e.g. materials, chemicals, and energy. The microalgae-based biorefinery was introduced for the production of high-valued bio-products from seaweeds (Ingle *et al.*, 2018). Macroalgae-based fuels and products have the potential of decreasing the negative environmental impact and creating products with higher added value. The chapter will discuss the economic potential and the challenges of macroalgae-based biorefineries.

Between 75 and 85% of worldwide seaweed production is used for direct human consumption in Asia (van den Burg *et al.*, 2019). The second important application of seaweed is for the production of thickeners (such as alginate and carrageenan) used in multiple food and non-food products. The seaweed feedstock to biorefineries is processed into bio-based chemicals. Bio-based chemicals can be defined as those classes of chemicals that are produced by using natural feedstock and have minimal impact on the environment. Examples for bio-based chemicals include (but are not limited to) carboxylic acids, Polylactic Acid (PLA), Fatty Acids, isoprene, Biosolvents (e.g. bio-ethanol), amino acids, vitamins, bio-pesticides, bio-fertilizers, antioxidants, sterols, and even industrial enzymes (Golden *et al.*, 2015). Globally, many different new markets are considered interesting. The bioactive compounds in seaweed may be applied in a processed or isolated form in food additives and pharmaceuticals (see Chapter 8). Seaweed species may also contain other chemicals or chemical precursors and microchemicals (starch, various polysaccharides, and proteins) to be used in chemical production (Bikker *et al.*, 2016a) or animal feed (Seghetta *et al.*, 2016a). There are indications that specific species may even reduce enteric fermentation in ruminants thus reducing climate change impact of the production of beef or mutton (van den Burg *et al.*, 2019). Seaweed-based products for these markets are subject to study or already commercially available at a small scale.

Although there is already a market for several bio-based plastics and other bio-based chemicals, the demand is considerably lower in comparison to traditional fossil fuel-based plastics (Stichnothe *et al.*, 2016). However, the market for bio-based chemicals is anticipated to increase significantly shortly. Global algae (macro- and microalgae) products' market stood at $9.9 billion in 2018 and is projected to grow at a compound annual growth rate (CAGR) of over 7% during 2019–2024 to reach $14.99 billion by 2024 (ReportLinker, 2019). The major driving factors that are expected to boost the demand include, among others:

- the availability of raw materials at a reduced cost;
- increasing the use of algae products in health-food and natural nutrient supplements, for example, as a substitute for animal-based nutrition;
- increasing consumer awareness towards bio-based products and subsequent demand;
- government initiatives to promote "green" products.

Various chemical, nutritional components, and elements are also highly abundant in algal biomass and can be extracted and processed as additives, fertilizers, cosmetics, medicines, and biofuels (see Chapters 7 and 8). However, current extraction and purification technologies are often limited to produce single primary bioproducts alone, and oftentimes make the biomass underutilized for other higher valued applications (see Chapter 7). Nevertheless, research findings suggest that better and more suitable technologies would emerge, which together with the discovery of more appropriate algal species would drive industry investment for upcoming biorefineries.

The major drivers for the deployment of biorefineries are:

(i) Sustainable and renewable energy supply — as biorefineries utilize renewable feedstock.
(ii) Saving foreign exchange reserves — required alternatively for importing fossil fuels and other chemicals.
(iii) Reduced dependency on imported crude petroleum and other chemicals — due to locally grown feedstock for biorefineries.

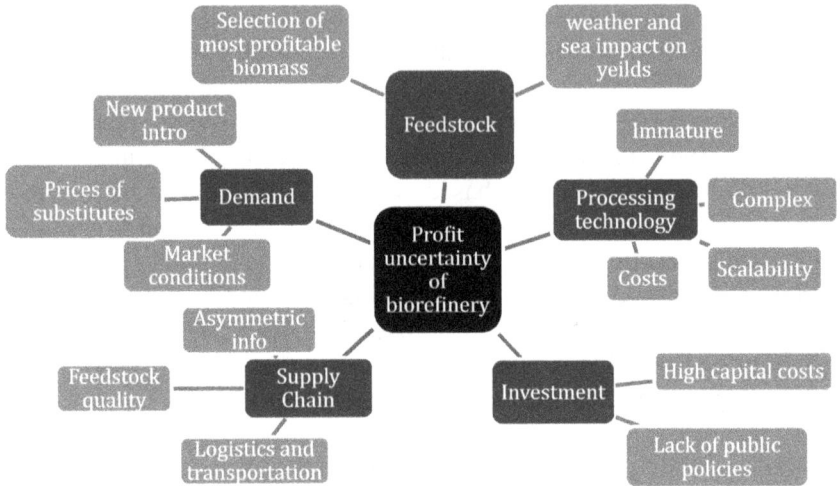

Fig. 9.1   Major sources of uncertainty for marine biorefinery deployment.

(iv) Establishment of the carbon-neutral and circular economy — allowed by low carbon footprint and net positive environmental impacts of biorefineries.

There are numerous challenges associated with the successful deployment of marine biorefinery operations, as summarized in Fig. 9.1. The profitability of marine biorefineries is subject to various sources of uncertainty, such as that of feedstock supply, processing technology, investment, contracting, and demand (Palatnik and Zilberman, 2017).

9.1 Feedstock Uncertainty

Selection of right biomass: There are numerous species of macro- and microalgae (Golden *et al.*, 2015; Jong *et al.*, 2012). Each species differs in terms of chemical structure and therefore differs in the biobased chemicals that can be produced. The entrepreneur should determine which algae-based activities are profitable under a multi-dimensional uncertainty outlined below.

Yields affected by weather and sea conditions: The rate of feedstock growth shows a wide range of values. Studies report the range of

6–108 tonnes/ha per year (Valderrama *et al.*, 2013). This uncertainty in feedstock yield has a major impact on the cost-effectiveness of the technology. Feedstock growth depends on saturation kinetics by light intensity, ambient dissolved inorganic nutrient concentrations, and temperature (Buschmann *et al.*, 2004). Cultivation uncertainty is exacerbated by stochastic weather, seasonal variability between regions, within years and between years. Studies point at biomass productivity as the main constraint against being competitive with other energy and protein-producing technologies (Seghetta *et al.*, 2016a).

9.2 Processing Technology Uncertainties

Anaerobic digestion, fermentation, transesterification, liquefaction, and pyrolysis can convert algal biomass into proteins and sugars that can be further processed into food, chemicals, and biofuels. At each stage of the production process, the entrepreneur should decide between various options that ultimately affect the irreversible (sunk) and variable costs of the production, the productivity, and the output, therefore affecting the total profitability. Yet, the biorefinery yields are highly uncertain (Table 9.1) signaling the immaturity of the technology. The upper value can be ten times larger than the lower one, significantly affecting the potential profitability of the process.

Scalability: Numerous studies have focused on the effort to evaluate future costs of the process that is currently available mostly in small (lab) scale (e.g. Korzen *et al.*, 2015b; Seghetta *et al.*, 2016a). This effort is

Table 9.1 Biorefinery conversion factors (Yields from *Ulva* Seaweed).

Biomass dry weight derived product	Conversion factor	Reference
Ethanol [g m^{-2}]	0.03–0.23	Nikolaisen *et al.* (2011a)
Buthanol	0.03–0.06	van der Wal *et al.* (2013)
Ethanol	0.03–0.23	van der Wal *et al.* (2013)
Acetone	0.01–0.02	Potts *et al.* (2012); van der Wal *et al.* (2013)
Methane [m^3/ton DW]	10–96	Bruhn *et al.* (2011b)

Source: Adapted with permission from Lehahn *et al.* (2016a).

remarkable and should not be underestimated. However, the studies mostly lack (or do not report) a structured production function that leads to a cost function. The common assumption is a linear approximation. This assumption should be treated cautiously and verified against actual data when production is scaled up.

9.3 Investments

The development of a new biorefinery, its design, and construction, requires huge investments (Stichnothe *et al.*, 2016). The strategy about the capacity of the biorefinery may change over time; the innovator may experiment by starting at a small scale. Once the production system is established, the innovator may either expand operations or reach out to cooperatives to provide it with inputs.

Moreover, introducing and perfecting innovations is a random process, and the economic conditions that face technology vary over time. Learning takes time, and the dynamics of knowledge accumulation affect the timing of the introduction of innovations, their refinement, and their commercialization. Timing can also affect the decision regarding both the capacity of innovation and the extent of reliance on external sources.

Lack of public policies supporting the biorefinery sector limits the long-term investment decision required. There are various strategies, but there are no distinct policy drivers for the utilization of bio-based chemicals, in direct contrast to the biofuels industry where various national regulations are driving rapid growth (Hermann *et al.*, 2011).

9.4 Price Uncertainties

The impact of price variation should be analyzed in several aspects: the price uncertainties that the aquafarmer faces, the price uncertainties of feedstock for biorefineries, and the price uncertainty of competitive outputs (backstop technology). A seaweed industry that contains many small-scale price-takers is especially prone to boom–bust cycles. For example, the strong demand from China drove the price of dry cottonii in the Philippines from \$900/tonne in 2007 to almost \$3,000/tonne in 2008, causing the Philippines' production to double from 1.5 million tonnes (wet weight) in 2007 to 3.3 million tonnes in 2008. The "seaweed rush" lasted only one year — the price dropped to $1,300/tonne in 2009 (Hurtado, 2013).

Table 9.2 exemplifies the range of prices as well as annual growth rates for one of the macroalgae species — Kappaphycus, and two possible biorefinery outputs: carrageenan and proteins, in the years 1991–2016. Generally, when strong demand for dry seaweeds drives up the price, seaweed farmers tend to increase their planting efforts and/or harvest immature crops. A likely result would then be that feedstock supply exceeds demand, which might cause a consequent collapse in price. However, if the price is low, seaweed farmers tend to reduce production, which creates sourcing difficulties for the biorefineries. On the other hand, biorefineries would tend to reduce demand as prices rise by substituting with cheaper feedstock alternatives (McHugh, 2006).

Table 9.2   Examples for costs of macroalgae production and commodities' prices.

Description	Average value	Range	Source
First unit cost of *carrageenan*	4500 $/ton 2014	4000–6500 $/ton 2010	Brown (2015)
First unit cost of feedstock seaweed Kappaphycus	1600 $/ton 2016	600–7000 $/ton 2010	Calculated based on Valderrama (eds.) (2013)
Price of protein	5000 $/ton	1000–15000 $/ton 2016	Price calculated from value and quantity world 2016[a]
Annual growth of the price of protein	7%	−43–92% in 1991–2016	Price calculated from value and quantity Philippines and USA export 1991–2016[b]
Price of *carrageenan*	5500 $/ton	3000–6000 $/ton 2016	https://www.alibaba.com/trade/search?fsb=y&IndexArea=product_en&CatId=&SearchText=carrageenan
Annual growth of the price of *carrageenan*	4%	−11–53% in 1991–2016 S. D. 16%	Price calculated from value and quantity Philippines export[c]

Notes: [a]UN COMTRADE; commodity 210610 protein; concentrates and textured protein substances).
[b]UN COMTRADE; commodity 210610 protein; concentrates and textured protein substances.
[c]UN COMTRADE; commodity HS130239 (mucilages and thickeners).

9.4 Demand

The economics of biorefinery-based products depends heavily on drop-in versus non-drop-in (existing demand and infrastructure). Therefore, demand may be very strong or very weak, leading to general uncertainty. It is difficult to know, for example, if an investment in the bio-based supply chain will make economic sense. It might not be possible to sell the produced bio-based chemicals at a price necessary to make the investment profitable. Of course, these are the kinds of decisions that all businesses face, but the reliance of biomass markets on policy measures and the lack of long-term signals regarding biomass, for example, in the EU policy, means that uncertainties are unusually high. In addition to the production cost, the value of biorefinery products when reaching end users may also reflect the expenses on research and development (R&D), formulation, marketing, etc. (Valderrama *et al.*, 2013) Specific information on these aspects is generally lacking.

9.5 Supply Chains

Within the bioeconomy, the biorefinery is one stage of the supply chain. To "implement" the bioeconomy, the optimal supply chain should be designed in terms of procurement of feedstock (intermediate inputs), production and processing, and marketing. The producers must decide how much to produce, what segments of the supply chain to undertake in-house versus sourcing externally, and what institutions, such as contracts and standards, they will use to coordinate the suppliers assuring its external sourcing (Du *et al.*, 2016). The general structure for deciding upon a sustainable supply chain is outlined in Fig. 9.2.

Established supply chains for seaweeds used for food production do exist (Valderrama *et al.*, 2013). However, supply chains for biorefinery outputs are still to be developed. Decisions about the scale of operation and the division of supply of inputs between in-house and external operations are key in the design of a basic supply chain (Du *et al.*, 2016). These decisions are affected by the investor's financial situation, the political and social system, the technology available, etc.

As supply chains increasingly encompass far-flung markets and supply sources, manufacturers and retailers are susceptible to various types of supply chain risks (Tang, 2006). There are diverse supply chain risks associated with disruptions or delays (Chopra and Sodhi, 2004) that could

natural resource

- 1st stage: feedstock farming

biobased chemicals

- 2nd stage: biorefinery

product

- for each molecule identify profitable economic activities

market

- for each product, identify market potential
- Logistics and transportation

sustainable supply chain

- uncertainty and environmental externalities

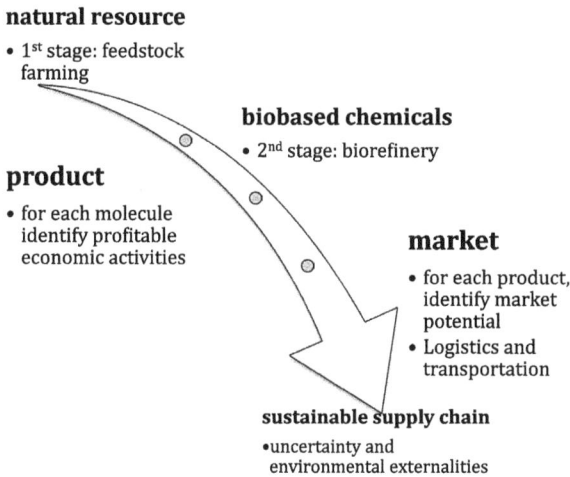

Fig. 9.2   General structure of bio-based chemical supply chain.

be categorized into supply risks, process risks, demand risks, intellectual property risks, behavioral risks, and political/social risks (Tang and Tomlin, 2008).

Supply chain contract uncertainties may occur due to asymmetric information (Du *et al.*, 2016). That is, the innovator may not observe the ability and effort being devoted by the contracted supplier, or the quality of his product.

Entrepreneurs may invest in protective measures to increase the resilience of their supply chains to extreme weather risks. For example, they may geographically diversify their external sources of feedstock to reduce exposure to weather shocks (Reardon and Zilberman, 2018). Therefore, incorporating risk considerations may increase the cost of investment in implementing innovation, especially if the enterprise is constrained by credit.

The supply chain design of biorefinery-based industries may require determining strategies for the production of feedstocks, and for processing the feedstocks to produce multiple products (Du *et al.*, 2016). Corn is used to produce ethanol as well as the residue product, Distillers Dried Grains (DDGs), which are being sold as animal feedstock (Popp *et al.*, 2016). Many of the agrifood innovations increased the value-added of agricultural resources either by identifying non-food uses of agricultural products and residues as part of the bioeconomy or by producing differentiated products by increasing their convenience and quality (Du *et al.*, 2016).

In some regions, new biomass supply chains develop in a hostile economic/political environment as the climatic benefit is challenged, and potential negative environmental impacts are highlighted (Bentsen, 2019). Also, the lack of transparency or public awareness of the environmental impacts of fossil fuels and biomass-based supply chains contributes to social barriers and constraints to mobilization. Successful deployment of biomass-based supply chains requires a social license to operate, and communication, knowledge transfer, and a broad societal stakeholder consensus seem to help overcome some of the barriers. A learning-by-doing approach to deployment can help overcome some of the chicken-and-egg problems in initiating and scaling up a new supply chain (Smith *et al.*, 2017).

A comprehensive review of optimization-oriented biomass supply-chain designs shows numerous prior works that addressed various important conditions for a profitable supply chain (Ghaderi *et al.*, 2016). This review of 146 studies concluded that researchers have been mostly orientated towards single-feedstock, single-product, single-period, single-objective, and deterministic models, without considering all the dimensions of sustainability.

Another recent study developed a methodology to assess the performance of the integrated two-stage supply chain — feedstock farming and processing into multiple outputs (Palatnik *et al.*, 2018). The results from the non-linear dynamic model clarify that learning in a multi-stage supply chain creates a positive externality of co-outputs. Moreover, if the learning rate is faster than the cost increase, then output grows faster than prices. Next, they demonstrated the application of the modeling framework on macroalgae (seaweed) farming and processing in the biorefinery into crude proteins and polysaccharides (carrageenan). The results indicated that for average prices of proteins and carrageenan, and average costs of investment in cultivation farm and the biorefinery, macroalgae utilization is cost-efficient. The study indicates that using near-future aquaculture technologies, offshore cultivation of macroalgae has the potential to provide some of the basic products required for human society in the coming decades. However, the profitability of this supply chain is fragile due to the high volatility of outputs' prices, as well as a wide range of feedstock growth rate and chemical composition. Notably, the research identifies the first stage of the supply chain, namely macroalgae marine cultivation, as the main constraint for commercialization.

In this chapter, we discuss the economic opportunities and challenges of marine biorefineries. Studies outline the state of the art of technological and economic abilities of feedstock conversion. The recent developments in biorefineries show the potential to produce not only food and coloring, but also sugars for biofuels, proteins, and high-value biochemicals. Biorefineries have evolved into successful commercial endeavors in several tropical countries endowed with clear, unpolluted intertidal environments and protected beach locations. For example, carrageenan seaweed farming in Asia has minimum capital and technological requirements and, as such, produces feedstock at competitive prices. In developed countries, though, the deployment of macroalgae-based supply chains is still challenging. Production costs are high, compared to costs in developing economies and compared to prices paid for seaweed in the major global markets. Therefore, future production and harvesting systems will need to be highly automated to reduce high labor costs.

Consumer demand for seaweed products is currently low, and dietary changes are not easily realized. The environmental benefits of seaweeds compared with other products are not easily quantified, not only due to lack of data but also because comparison of land-based and sea-based farming systems is inherently difficult. Moreover, uncertainties and random factors that need to be taken into account are the demand for final products, the cost of the refining technology, the cost of feedstock production in-house, or the reliability and performance of various external suppliers (Du *et al.*, 2016).

In response to these challenges, it is often argued that upscaling in the developed countries is needed (Golberg *et al.*, 2014) assuming that more production, more research, and more demand will suffice.

Another solution might be combining macroalgae cultivation with other sea-related economic activities (Buck *et al.*, 2008; Korzen *et al.*, 2015b; Prabhu *et al.*, 2020; Zollmann *et al.*, 2019b). Co-management with other offshore systems like wind farms and fisheries to increase economic and environmental benefits, and to diversify the revenue sources, should be considered. Integrated feedstock production of mussels and/or fish caves and biorefineries producing bio-based products using bioenergy produced on-site and exporting just the bioenergy surplus is a promising way forward (see Chapter 7). Yet, integrated biorefineries still have to prove that they can become cost-competitive.

The integration of different pre-treatment and conversion technologies in biorefineries can maximize the use of all biomass components and improve both the economic and ecological efficiency of the whole value chain. However, eco-efficient bio-based value chains call for different ways of thinking about agriculture, energy infrastructure, processing industry, and rural economic development. A key challenge for developing and growing a commercially secure biorefinery sector is a reliable and consistent feedstock supply.

Another way to diversify revenue is the co-production in the stage of feedstock conversion (to e.g. biofuels and food, Chapter 8). The variation in shares of co-products between, for example, butanol-acetone, ethanol, and methane, as well as protein, may affect substantially the net benefits of the production process. More research on the key aspects of co-production that leads to increased profitability of biorefineries is crucial.

Next, investments in macroalgae utilization are risky not only due to the uncertainty in feedstock cultivation, but also in processing technology, contracting, and demand. Considering uncertainty is most pertinent when a new processing technology, such as new bio-refining technology, is invented, the design of a sustainable biorefinery, which will generate sustainable food, fuels, and biochemicals, is a complex task and is largely influenced by local raw material supplies, advances in multiple technologies, and socio-economic conditions (Fatih Demirbas, 2009). Also, comprehensive scientific studies on the question of whether the novel biorefinery can increase the yield by the order of 10 in a rigid manner to assure a profitable process should be undertaken.

Therefore, more research on economies of scale in macroalgae cultivation and the refinery is crucial for establishing the industry.

Investments in production capacity or consumption infrastructure are also susceptible to market uncertainties from, for example, fluctuations in energy prices (Rajagopal *et al.*, 2009) and demand uncertainty that is often associated with new product introduction. Similarly to traditional crops, the price paid to seaweed farmers is determined in part by the complexity of the supply chain and partly by the quality of the macroalgae. But, crops destined for conversion into bio-fuels have prices determined in large part by the ethanol market, which is linked to the volatile gasoline market (Tyner, 2008).

Energy crop price volatility is likely to be aggravated as ethanol shifts in and out of status as a cost-effective fuel substitute for gasoline, based on the relative prices of petroleum and corn grain, the leading current ethanol feedstock in the United States. However, more investigation on the impact of output price variability on technology adoption decisions is essential.

The example of the impact of uncertainty of daily growth rates (DGR) of a green seaweed *Ulva* cultivated in the shallow waters near Tel Aviv (Fig. 9.3) and the uncertainty of its biomass conversion to bioethanol (Fig. 9.4) on the area required to provide ethanol as fuel to Israel economy is shown in Fig. 9.4 (Chemodanov *et al.*, 2017a). The impact of additional uncertainty of ethanol prices on the carbon sequestration potential and revenues of potential offshore seaweed farms is shown in Fig 9.5 (Chemodanov *et al.*, 2017a).

Price volatility is also compounded by the absence of relevant, reliable, and timely production statistics and market intelligence. Unlike for some agricultural commodities such as coffee or tea, there are no organized markets to provide benchmarking international prices for seaweeds (Tinne *et al.*, 2006). Unavailability of reliable information is especially detrimental to uninformed seaweed farmers who are at the lowest end of

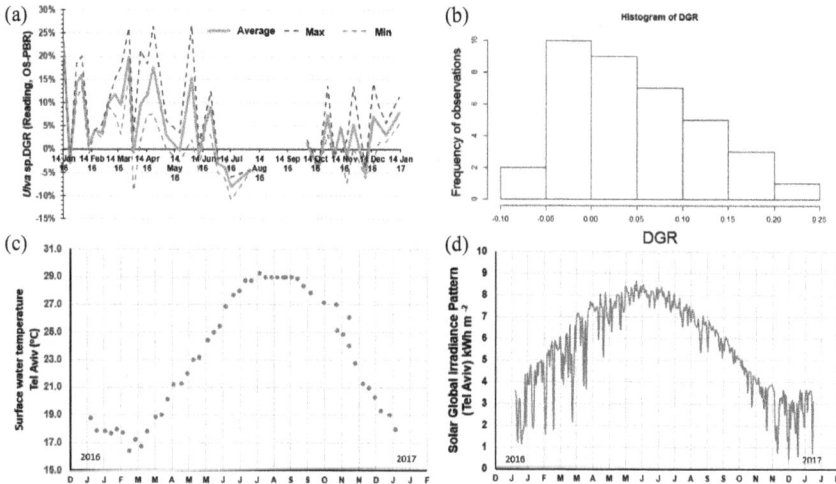

Fig. 9.3 Annual growth rates and environmental conditions at the cultivation site. (a) Measured annual daily growth rate (%DGR) of *Ulva* biomass at Reading (N = 3 for each point). (b) Histogram of DGR observed during the year; (c) Annual profile of surface water temperature (°C) in Tel Aviv; (d) Annual global irradiance (kwh m^{-2}) at Tel Aviv.

Source: Adapted with permission from Chemodanov *et al.* (2017a).

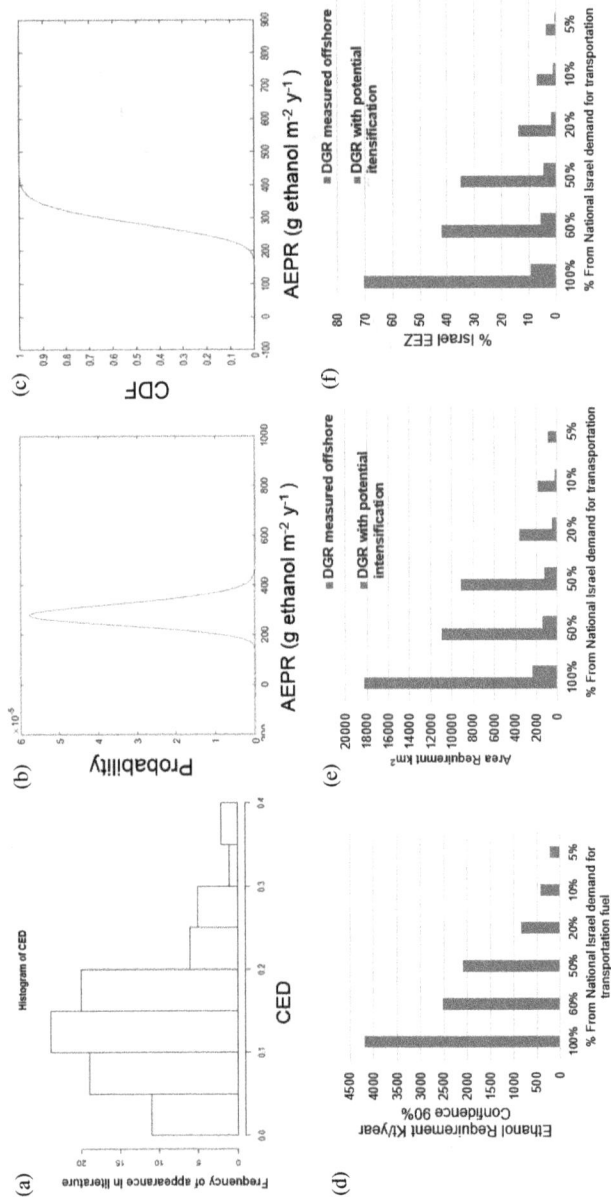

Fig. 9.4 Bioethanol production estimation from the offshore-cultivated biomass. (a) Conversion efficiency distribution (CED) based on meta-data analysis of conversion literature. (b) Annual ethanol production rate (AEPR) probability density function. (c) Annual ethanol production rate (AEPR) cumulative density function; (d) Ethanol requirements (Kton/year) to supply a fraction of Israel transportation fuels. Comparison between areas required for biomass grown in a single layer photobioreactor with no intensification as observed in the offshore experiment with biomass grown under nutrients saturation as observed in laboratory experiments (e–f). (e) Offshore area allocation (km²) requirement to displace oil in transportation in Israel. (f) The fraction of Israel Exclusive Economic Zone (EEZ) required to displace a fraction of oil in transportation in Israel.

Source: Adapted with permission from Chemodanov *et al.* (2017a).

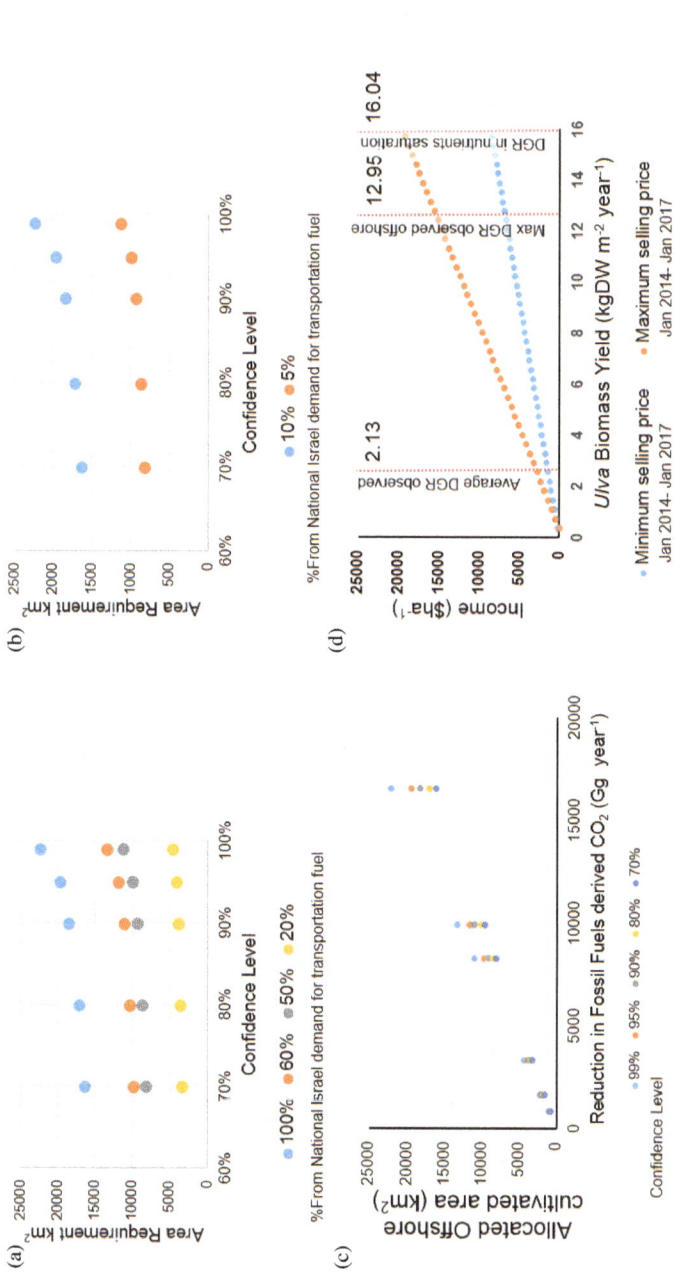

Fig. 9.5 Allocated offshore areas sensitivity analysis for *Ulva* sp. biomass cultivated in a single layer photobioreactor with no intensification. (a) To supply biomass for bioethanol production to displace 100%, 60%, 50%, and 20% of the oil used in the transportation sector in Israel. (b) To supply biomass for bioethanol production to displace 5% and 10% of the oil used in the transportation sector in Israel. (c) Allocated for the offshore cultivation area required to produce biomass for bioethanol to reduce new CO_2 emissions from fossil fuels on national levels in Israel. (d) Estimated annual income from the production of bioethanol derived from offshore cultivated macroalgae.

Source: Adapted with permission from (Chemodanov *et al.*, 2017a).

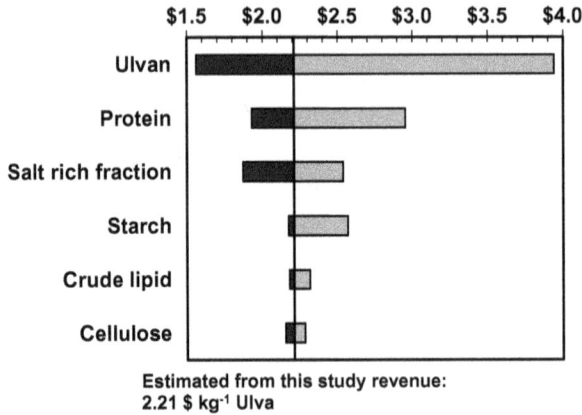

Fig. 9.6   One-way sensitivity analysis of total revenue from *Ulva* biomass when sold as fractionated ingredients (Prabhu *et al.*, 2020).

the seaweed value chain and often forced to accept whatever price is offered.

Moreover, it is essential to identify the biochemical compounds that may provide higher value than the ethanol. As of now, macroalgae-based bioethanol cannot compete with corn-ethanol or sugarcane-based ethanol. To generalize, rather than competing with existing goods, the scientific challenge can be the investigation of the potential to utilize macroalgae for unique foods, high-value chemicals, and fuels. An example of such an income analysis, based on the fractionation of green seaweed *Ulva* sp and market prices of 2019, as shown in Figs. 4.3 and 4.4, is shown in Fig. 9.6 (Prabhu *et al.*, 2020). The modeled revenue from *Ulva* fractions alone is higher than ethanol that could be produced from the same biomass.

Besides, transparency and multidisciplinary interaction will increase the learning curve and will make the biorefinery sector more structured and efficient. Alternative specifications for biorefinery and cultivation processes in a transparent way would allow replication and induce the improvement of methodologies. A multidisciplinary effort is required for improving the knowledge of production and cost functions to lead to the establishment of economic models.

Most of the studies assessing biorefineries used the conveniently available calculation module on large-scale ethanol production from corn stover by NREL (Humbird *et al.*, 2011). Indeed, the popularity of this research signals that more up-to-date studies with a transparent open-source tool would increase our understanding of the economic viability of the novel biorefinery technologies.

Nor less important is the uninvestigated effect that mass cultivation of offshore algae might have on the environment. On the one hand, the transition from small to big-scale algae cultivation involves direct and external effects that may completely reshape the process. On the other hand, if algae-based biofuel crowds-out the use of fossil fuels and crop-based bioethanol, it mediates the environmental externalities, as well as negative effects on agricultural supply and land use (Zilberman *et al.*, 2017). Further analysis of macroalgae external costs and benefits is required for accurate policy intervention. The analysis of the technological prospects of macroalgae biorefineries should evaluate the social net benefit too. Consequently, the recommendation upon optimal fuel mix is to be based on social (versus private) costs.

At present, the bioeconomy is a political concept, and efforts from all actors are required for its realization (Stichnothe *et al.*, 2016). The pace of large-scale implementation of biorefineries depends not just on the environmental and economic performance of pre-treatment and conversion technologies, but also governmental policy frameworks and societal preferences. It is unlikely that a single solution or policy will determine the extent to which biorefineries are incorporated into a global bio-based economy. Rather, it is likely that a robust portfolio of solutions, appropriate for site-specific conditions (e.g. feedstock availability, supply chain infrastructure, rural development strategies, etc.), will be needed to sustainably address the future material and energy demand. The rapid growth in bio-feedstock utilization will require changes in the supply chain infrastructure and may also need more efficient socio-economic and policy frameworks (Richard, 2010). For example, an expanded policy that includes bio-based products provides added flexibility without compromising GHG targets (Posen *et al.*, 2015).

Bibliography

Abdelhamid, A., Jouini, M., Bel Haj Amor, H., Mzoughi, Z., Dridi, M., Ben Said, R., and Bouraoui, A. (2018). Phytochemical analysis and evaluation of the antioxidant, anti-inflammatory, and antinociceptive potential of phlorotannin-rich fractions from three mediterranean brown seaweeds. *Mar. Biotechnol.*, https://doi.org/10.1007/s10126-017-9787-z.

Abideen, Z., Ansari, R., and Khan, M.A. (2011). Halophytes: Potential source of ligno-cellulosic biomass for ethanol production. *Biomass. Bioenerg.*, **35**, 1818–1822. https://doi.org/http://dx.doi.org/10.1016/j.biombioe.2011.01.023.

Abinandan, S. and Shanthakumar, S. (2015). Challenges and opportunities in application of microalgae (Chlorophyta) for wastewater treatment: A review. *Renew. Sustain. Energ. Rev.*, **52**, 123–132. https://doi.org/10.1016/J.RSER.2015.07.086.

Achenza, M. and Fenu, L. (2006). On earth stabilization with natural polymers for earth masonry construction. *Mater. Struct. Constr.*, **39**, 21–27. https://doi.org/10.1617/s11527-005-9000-0.

Adams, G.D.J., Cook, I., and Ward, K.R. (2015). The principles of freeze-drying, in: *Cryopreservation and Freeze-Drying Protocols, Methods in Molecular Biology.* Springer Science+Business Media New York, pp. 121–143.

Adams, J.M., Gallagher, J.A., and Donnison, I.S. (2009). Fermentation study on saccharina latissima for bioethanol production considering variable pre-treatments. *J. Appl. Phycol.*, **21**, 569–574. https://doi.org/10.1007/s10811-008-9384-7.

Adams, J.M.M., Bleathman, G., Thomas, D., and Gallagher, J.A. (2017). The effect of mechanical pre-processing and different drying methodologies on bioethanol production using the brown macroalga Laminaria digitata (Hudson) JV Lamouroux. *J. Appl. Phycol.*, **29**, 2463–2469. https://doi.org/10.1007/s10811-016-1039-5.

Adams, J.M.M., Toop, T.A., Donnison, I.S., and Gallagher, J.A. (2011). Seasonal variation in Laminaria digitata and its impact on biochemical conversion routes to biofuels. *Bioresour. Technol.*, **102**, 9976–9984. https://doi.org/10.1016/j.biortech.2011.08.032.

Adey, W., Luckett, C., and Jensen, K. (1993). Phosphorus removal from natural waters using controlled algal production. *Restor. Ecol.*, **1**, 29–39. https://doi.org/10.1111/j.1526-100X.1993.tb00006.x.

Adhikari, U., Mateu, C.G., Chattopadhyay, K., Pujol, C.A., Damonte, E.B., and Ray, B. (2006). Structure and antiviral activity of sulfated fucans from Stoechospermum marginatum. *Phytochemistry*. **67**, 2474–2482. https://doi.org/10.1016/j. phytochem.2006.05.024.

Ahluwalia, S.S. and Goyal, D. (2007). Microbial and plant derived biomass for removal of heavy metals from wastewater. *Bioresour. Technol.*, **98**, 2243–2257. https://doi.org/10.1016/J.BIORTECH.2005.12.006.

Ahmed, A.B.A., Adel, M., Karimi, P., and Peidayesh, M. (2014). Pharmaceutical, cosmeceutical, and traditional applications of marine carbohydrates, 1st ed, *Advances in Food and Nutrition Research*. Elsevier Inc. https://doi.org/10.1016/B978-0-12-800268-1.00010-X.

Ahuja, D.B., Ahuja, U.R., Singh, S.K., and Singh, N. (2015). Comparison of Integrated Pest Management approaches and conventional (non-IPM) practices in late-winter-season cauliflower in Northern India. *Crop Prot.*, **78**, 232–238. https://doi.org/10.1016/j.cropro.2015.08.007.

Ainis, A., Vellanoweth, R., Lapeña, Q., and Thornber, C.S. (2014). Using non-dietary gastropods in coastal shell middens to infer kelp and seagrass harvesting and paleoenvironmental conditions. *J. Archaeol. Sci.*, **49**, 343–360.

Alavanja, M.C.R., Ross, M.K., and Bonner, M.R. (2013). Increased cancer burden among pesticide applicators and others due to pesticide exposure. CA. *Cancer J. Clin.* https://doi.org/10.3322/caac.21170.

Alejandro, B. and Patricio, G. (1993). Interaction mechanisms between *Gracilaria chilensis* (Rhodophyta) and epiphytes. *Hydrobiologia*, **260**, 345–351. https://doi.org/10.1007/BF00049039.

Alexandratos, N. and Bruinsma, J. (2012). World agriculture towards 2015/2030: The 2012 revision. ESA Working paper (Vol. 12, No. 3), ESA Working paper. https://doi.org/10.1016/S0264-8377(03)00047-4.

Alexandru, L., Cravotto, G., Giordana, L., Binello, A., and Chemat, F. (2013). Ultrasound-assisted extraction of clove buds using batch- and flow-reactors: A comparative study on a pilot scale. *Innov. Food Sci. Emerg. Technol.*, **20**, 167–172. https://doi.org/10.1016/J.IFSET.2013.07.011.

Alsufyani, T., Engelen, A.H., Diekmann, O.E., Kuegler, S., and Wichard, T. (2014). Prevalence and mechanism of polyunsaturated aldehydes production in the green tide forming macroalgal genus *Ulva* (*Ulvales, Chlorophyta*).

Chem. Phys. Lipids., **183**, 100–109. https://doi.org/10.1016/j.chemphyslip. 2014.05.008.

Alsufyani, T.A., Weiss, A., Engelen, A., and Kwantes, M. (2020). Macroalgal — bacterial interactions: Identification and role of thallusin in morphogenesis of the seaweed *Ulva* (*Chlorophyta*). *J. Exp. Bot.* https://doi.org/10.1093/jxb/eraa066.

Alveal, K., Romo, H., Werlinger, C., and Oliveira, E. (1997). Mass cultivation of the agar-producing alga *Gracilaria chilensis* (Rhodophyta) from spores. *Aquaculture*, **148**, 77–83.

Alvira, P., Ballesteros, M., and Negro, M.J. (2010). Pretreatment technologies for an efficient bioethanol production process based on enzymatic hydrolysis: A review. *Bioresour. Technol.*, **101**, 4851–4861. https://doi.org/10.1016/j. biortech.2009.11.093.

Amamou, S., Sambusiti, C., Monlau, F., and Dubreucq, E. (2018). Mechano-enzymatic deconstruction with a New Enzymatic cocktail to enhance enzymatic hydrolysis and bioethanol fermentation of two macroalgae species. *Molecules*, **23**,174. https://doi.org/10.3390/molecules23010174.

Amarasekara, A.S. (2019). Ionic liquids in biomass processing. *Isr. J. Chem.*, **59**, 789–802. https://doi.org/10.1002/ijch.201800140.

Andersson, E. and Hahn-Hägerdal, B. (1990). Bioconversions in aqueous two-phase systems. *Enzym. Microb. Technol.*, **12**, 242–254. https://doi.org/10. 1016/0141-0229(90)90095-8.

Angell, A.R., Mata, L., de Nys, R., and Paul, N.A. (2016). The protein content of seaweeds: a universal nitrogen-to-protein conversion factor of five. *J. Appl. Phycol.*, **28**, 511–524. https://doi.org/10.1007/s10811-015-0650-1.

Anicia, H. and Alan, C. (2006). Occurrence of Polysiphonia epiphytes in Kappaphycus farms at Calagas Is., Camrines Norte, Philippines. *J. Appl. Phycol.* https://doi.org/10.1007/978-1-4020-5670-3.

Ansell, A., Gibson, R., Barnes, M., and Press, U. (1998). Ecological impact of green macroalgal blooms. Oceanography and Marine Biology: An Annual Review, 36th ed.

Appels, L., Lauwers, J., Degrève, J., Helsen, L., Lievens, B., Willems, K., Van Impe, J., and Dewil, R. (2011). Anaerobic digestion in global bio-energy production: Potential and research challenges. *Renew. Sustain. Ener. Rev.*, **15**, 4295–4301. https://doi.org/http://dx.doi.org/10.1016/j.rser.2011. 07.121.

Areco, M.M., Hanela, S., Duran, J., and dos Santos Afonso, M. (2012). Biosorption of Cu(II), Zn(II), Cd(II) and Pb(II) by dead biomasses of green alga Ulva lactuca and the development of a sustainable matrix for adsorption implementation. *J. Hazard. Mater.*, **213–214**, 123–132. https://doi.org/10. 1016/J.JHAZMAT.2012.01.073.

Asadullah, M. (2014). Barriers of commercial power generation using biomass gasification gas: A review. *Renew. Sustain. Ener. Rev.*, **29**, 201–215. https://doi.org/10.1016/j.rser.2013.08.074.

Asghari, A., Bothast, R., Doran, J., and Ingram, L. (1996). Ethanol production from hemicellulose hydrolysates of agricultural residues using genetically engineered *Escherichia coil* strain KOll. *J. Ind. Microbiol.*, **16**, 42–47.

Ashkenazi, D.Y., Israel, A., and Abelson, A. (2018). A novel two-stage seaweed integrated multi-trophic aquaculture. *Rev. Aquac.*, **11**, 246–262. https://doi.org/10.1111/raq.12238.

Association of Official Analytical Chemists (AOAC). (1990). *Off. Methods Anal.* 15th Edn., 556.

Atashrazm, F., Lowenthal, R.M., Woods, G.M., Holloway, A.F., and Dickinson, J.L. (2015). Fucoidan and Cancer: A Multifunctional Molecule with Anti-Tumor Potential 2327–2346. https://doi.org/10.3390/md13042327.

Ausubel, J.H., Wernick, I.K., and Waggoner, P.E. (2013). Peak farmland and the prospect for land sparing. *Popul. Dev. Rev.* https://doi.org/10.1111/j.1728-4457.2013.00561.x.

Aziz, M., Oda, T., and Kashiwagi, T. (2014). Advanced energy harvesting from macroalgae — Innovative integration of drying, gasification and combined cycle. *Energies*, **7**, 8217–8235. https://doi.org/10.3390/en7128217.

Baghel, R.S., Trivedi, N., Gupta, V., Neori, A., Reddy, C.R.K., Lali, A., and Jha, B. (2015). Biorefining of marine macroalgal biomass for production of biofuel and commodity chemicals. *Green. Chem.*, **17**, 2436–3443.

Baghel, R.S., Trivedi, N., and Reddy, C.R.K. (2016). A simple process for recovery of a stream of products from marine macroalgal biomass. *Bioresour. Technol.*, **203**, 160–165. https://doi.org/10.1016/j.biortech.2015.12.051.

Baghel, R.S.R.S., Trivedi, N., Gupta, V., Neori, A., Reddy, C.R.K.R.K., Lali, A., and Jha, B. (2015). Biorefining of marine macroalgal biomass for production of biofuel and commodity chemicals. *Green. Chem.*, **17**, 2436–3443. https://doi.org/10.1039/ C4GC02532F.

Ballantine, D.L., Gerwick, W.H., Velez, S.M., Alexander, E., and Guevara, P. (1987). Antibiotic activity of lipid-soluble extracts from Caribbean marine algae. *Hydrobiologia*, **469**, 463–469.

Barbarino, E. and Federal, U. (2005). An evaluation of methods for extraction and qualification of protein from marine macro- and microalgae marine macro- and microalgae. *J. Appl. Phycol.*, **17**, 447–460. https://doi.org/10.1007/s10811-005-1641-4.

Barbarino, E. and Lourenço, S.O. (2005). An evaluation of methods for extraction and quantification of protein from marine macro- and microalgae. *J. Appl. Phycol.*, **17**, 447–460. https://doi.org/10.1007/s10811-005-1641-4.

Barbera, C., Bordehore, C., Borg, J.A., Glémarec, M., Grall, J., Hall-Spencer, J.M., de la Huz, C., Lanfranco, E., Lastra, M., Moore, P.G., Mora, J., Pita,

M.E., Ramos-Esplá, A.A., Rizzo, M., Sánchez-Mata, A., Seva, A., Schembri, P.J., and Valle, C. (2003). Conservation and management of northeast Atlantic and Mediterranean maerl beds. *Aquat. Conserv. Mar. Freshw. Ecosyst.*, **13**, S65–S76. https://doi.org/10.1002/aqc.569.

Barbier, E.B., Hacker, S.D., Kennedy, C., Koch, E.W., Stier, A.C., and Silliman, B.R. (2011). The value of estuarine and coastal ecosystem services. *Ecol. Monogr.*, **81**, 169–193. https://doi.org/10.1890/10-1510.1.

Barbosa, M., Valentão, P., and Andrade, P.B. (2014). Bioactive compounds from macroalgae in the New Millennium: Implications for neurodegenerative diseases. *Mar. Dru.*, **12**, 4934–4972.

Barilotti, D.C. and Zertuche-González, J.A. (1990). Ecological effects of seaweed harvesting in the Gulf of California and Pacific Ocean off Baja California and California. *Hydrobiologia*, **204**, 35–40.

Baum, Z., Palatnik, R.R., Kan, I. and Rapaport-Rom, M. (2016). Economic impacts of water scarcity under diverse water salinities. *Water Econ. Policy*, 2. https://doi.org/10.1142/S2382624X15500137.

Beer, S. and Björk, M. (1994). Photosynthetic properties of protoplasts, as compared with Thalli, of *Ulva Fasciata* (*Chlorophyta*). *J. Phycol.* https://doi.org/10.1111/j.0022-3646.1994.00633.x.

Beer, S. and Levy, I. (1983). Effects of photon fluence rate and light spectrum composition on growth, photosynthesis and pigment relations. In *Gracilaria* sp. *J. Phycol.* https://doi.org/10.1111/j.0022-3646.1983.00516.x.

Bennett, A.J., Bending, G.D., Chandler, D., Hilton, S., and Mills, P. (2012). Meeting the demand for crop production: The challenge of yield decline in crops grown in short rotations. *Biol. Rev.* https://doi.org/10.1111/j.1469-185X.2011.00184.x.

Bentsen, N.S. (2019). Biomass for biorefineries: Availability and costs, In: Bastidas-Oyanedel, J., Schmidt, J. (Eds.), *Biorefinery. Cham*: Springer, pp. 37–48.

Bermejo, R., Vergara, J.J., and Hernández, I. (2012). Application and reassessment of the reduced species list index for macroalgae to assess the ecological status under the water framework directive in the Atlantic coast of Southern Spain. *Ecol. Indic.* https://doi.org/10.1016/j.ecolind.2011.04.008.

Bernstein, B. and Jung, N. (1979). Selective pressures and coevolution in a Kelp Canopy Community in Southern California. *Ecol. Monogr.*, **49**, 335–355.

Bikker, P., van Krimpen, M.M., van Wikselaar, P., Houweling-Tan, B., Scaccia, N., van Hal, J.W., Huijgen, W.J.J., Cone, J.W., López-Contreras, A.M., Scaccia, N., van Hal, J.W. vanhal, wurnl W., Wouter J Huijgen, W.J.J., and Cone, J.W.W. (2016). Biorefinery of the green seaweed *Ulva lactuca* to produce animal feed, chemicals and biofuels. *J. Appl. Phycol.*, **28**, 3511–3525. https://doi.org/10.1007/s10811-016-0842-3.

Bikker, P., van Krimpen, M.M.M., van Wikselaar, P., Houweling-Tan, B., Scaccia, N., van Hal, J.W.W., Huijgen, W.J.J., Cone, J.W.W., López-Contreras, A.M., Lopez-Contreras, A.M., Krimpen, M.M., Wikselaar, P., Houweling-Tan, B., Scaccia, N., Hal, J.W., Huijgen, W.J.J., Cone, J.W.W., López-Contreras, A.M., van Krimpen, M.M.M., van Wikselaar, P., Houweling-Tan, B., Scaccia, N., van Hal, J.W.W., J Huijgen, W.J., Cone, J.W.W., López-Contreras, A.M., Scaccia NazarenoScaccia, N., Jaap van Hal vanhal, wurnl W., Wouter J Huijgen, ecnnl J., Cone johncone, J.W., Krimpen, M.M., Wikselaar, P., Houweling-Tan, B., Scaccia, N., Hal, J.W., Huijgen, W.J.J., Cone, J.W.W., López-Contreras, A.M., van Krimpen, M.M.M., van Wikselaar, P., Houweling-Tan, B., Scaccia, N., van Hal, J.W.W., J Huijgen, W.J., Cone, J.W.W., López-Contreras, A.M., Scaccia NazarenoScaccia, N., Jaap van Hal vanhal, wurnl W., Wouter J Huijgen, ecnnl J., Cone johncone, J.W., and Lopez-Contreras, A.M. (2016c). Biorefinery of the green seaweed *Ulva lactuca* to produce animal feed, chemicals and biofuels. *J. Appl. Phyco.* https://doi.org/10.1007/s10811-016-0842-3.

Bixler, H.J. and Porse, H. (2011). A decade of change in the seaweed hydrocolloids industry. *J. Appl. Phycol.*, **23**, 321–335. https://doi.org/10.1007/s10811-010-9529-3.

Björk, M., Haglund, K., Ramazanov, Z., Garcia-Reina, G., and Pedersén, M. (1992). Inorganic-carbon assimilation in the green seaweed *Ulva rigida* C.Ag. (Chlorophyta). *Planta.* https://doi.org/10.1007/BF00201637.

Blackburn, S. (2004). Water pollution and bioremediation by microalgae: Eutrophication and water poisoning. *Handb. Microalgal Cult.* 417–429. https://doi.org/10.1002/9780470995280.ch24.

Bligh, E.G. and W.J.D. (1959). A rapid method of total lipid extraction and purification. *Can. J. Biochem. Physiol.*, **37**, 911–917.

Blitzer, H. and Redman, C.L. (2001). Human impact on ancient environments. *Am. J. Archaeol.* https://doi.org/10.2307/507334.

Bocken, N.M.P., Short, S.W., Rana, P., and Evans, S. (2014). A literature and practice review to develop sustainable business model archetypes. *J. Clean. Prod.* https://doi.org/10.1016/j.jclepro.2013.11.039.

Boero, F. (2013). Review of jellyfish blooms in the mediterranean and black sea. *General Fisheries Commission for the Mediterranean.* FAO studies and reviews 2013. N 92.

Bolton, J.J. (1994). Global seaweed diversity: Patterns and anomalies. *Bot. Mar.* https://doi.org/10.1515/botm.1994.37.3.241.

Bolton, J.J. (2019). The problem of naming commercial seaweeds. *J. Appl. Phycol.* https://doi.org/10.1007/s10811-019-01928-0.

Bono, A., Farm, Y.Y., Yasir, S.M., Arifin, B., and Jasni, M.N. (2011). Production of fresh seaweed powder using spray drying technique. *J. Appl. Sci.*, **11**(13), 2340–2345.

Borines, M.G., de Leon, R.L., and Cuello, J.L. (2013). Bioethanol production from the macroalgae *Sargassum* spp. *Bioresour. Technol.*, **138**, 22–29. https://doi.org/10.1016/j.biortech.2013.03.108.

Borines, M.G., McHenry, M.P., and de Leon, R.L. (2011). Integrated macroalgae production for sustainable bioethanol, aquaculture and agriculture in Pacific island nations. *Biofuels, Bioprod. Biorefining.* **5**, 599–608.

Borsje, B.W., van Wesenbeeck, B.K., Dekker, F., Paalvast, P., Bouma, T.J., van Katwijk, M.M., and de Vries, M.B. (2011). How ecological engineering can serve in coastal protection. *Ecol. Eng.*, **37**, 113–122. https://doi.org/10.1016/J.ECOLENG.2010.11.027.

Bostock, J., McAndrew, B., Richards, R., Jauncey, K., Telfer, T., Lorenzen, K., Little, D., Ross, L., Handisyde, N., Gatward, I., and Corner, R. (2010). Aquaculture: Global status and trends. *Philos. Trans. R. Soc. B Biol. Sci.*, **365**, 2897–2912. https://doi.org/10.1098/rstb.2010.0170.

Boulaoued, I., Amara, I., and Mhimid, A. (2016). Experimental determination of thermal conductivity and diffusivity of new building insulating materials. *Int. J. Heat Technol.*, **34**, 325–331. https://doi.org/10.18280/ijht.340224.

Briand, X. and Morand, P. (1997). Anaerobic digestion of *Ulva* sp. 1. Relationship between Ulva composition and methanisation. *J. Appl. Phycol.*, **9**, 511–524. https://doi.org/10.1023/A:1007972026328.

Brittain, C. and Potts, S.G. (2011). The potential impacts of insecticides on the life-history traits of bees and the consequences for pollination. *Basic Appl. Ecol.* https://doi.org/10.1016/j.baae.2010.12.004.

Bruhn, A., Dahl, J., Nielsen, H.B., Nikolaisen, L., Rasmussen, M.B., Markager, S., Olesen, B., Arias, C., and Jensen, P.D. (2011). Bioenergy potential of *Ulva lactuca*: Biomass yield, methane production and combustion. *Bioresour. Technol.*, **102**, 2595–2604. https://doi.org/10.1016/j.biortech.2010.10.010.

Bruinsma, J. (2009). The resource outlook to 2050: By how much do land, water and crop yields need to increase by 2050? *FAO Expert Meeting on How to Feed the World in 2050.*

Buapet, P., Gullström, M., and Björk, M. (2013). Photosynthetic activity of seagrasses and macroalgae in temperate shallow waters can alter seawater pH and total inorganic carbon content at the scale of a coastal embayment. *Mar. Freshw. Res.*, **64**, 1040. https://doi.org/10.1071/MF12124.

Buck, B.H. and Buchholz, C.M. (2004). The offshore-ring: A new system design for the open ocean aquaculture of macroalgae. *J. Appl. Phycol.*, **16**, 355–368. https://doi.org/10.1023/B:JAPH.0000047947.96231.ea.

Buck, B.H. and Buchholz, C.M. (2005). Response of offshore cultivated Laminaria saccharina to hydrodynamic forcing in the North Sea. *Aquaculture*, **250**, 674–691. https://doi.org/10.1016/J.AQUACULTURE.2005.04.062.

Buck, B.H., Krause, G., Michler-Cieluch, T., Brenner, M., Buchholz, C.M., Busch, J.A., Fisch, R., Geisen, M., and Zielinski, O. (2008). Meeting the quest for spatial efficiency: Progress and prospects of extensive aquaculture within offshore wind farms. *Helgol. Mar. Res.*, **62**, 269–281. https://doi.org/10.1007/s10152-008-0115-x.

Buck, B.H., Krause, G., and Rosenthal, H. (2004). Extensive open ocean aquaculture development within wind farms in Germany: The prospect of offshore co-management and legal constraints. *Ocean Coast. Manag.*, **47**, 95–122. https://doi.org/10.1016/J.OCECOAMAN.2004.04.002.

Buck, B.H., Troell, M.F., Krause, G., Angel, D.L., Grote, B., and Chopin, T. (2018). State of the art and challenges for offshore integrated multi-trophic aquaculture (IMTA). *Front. Mar. Sci.*, **5**, 165. https://doi.org/10.3389/fmars.2018.00165.

Budarin, V.L., Zhao, Y., Gronnow, M.J., Shuttleworth, P.S., Breeden, S.W., Macquarrie, D.J., and Clark, J.H. (2011). Microwave-mediated pyrolysis of macro-algae. *Gree. Chem.*, **13**, 2330–2333. https://doi.org/10.1039/C1GC15560A.

Buschmann, A., Camus, C., Infante, J., Neori, A., Israel, Á., Hernández-González, M.C., Pereda, S.V., Gomez-Pinchetti, J.L., Golberg, A., Tadmor-Shalev, N., and Critchley, A. (2017). Seaweed production: overview of the global state of exploitation, farming and emerging research activity. *Eur. J. Phycol.*, **52**, 391–406. https://doi.org/10.1080/09670262.2017.1365175.

Buschmann, A.H., López, D.A., and Medina, A. (1996). A review of the environmental effects and alternative production strategies of marine aquaculture in Chile. *Aquac. Eng.*, **15**, 397–421. https://doi.org/10.1016/s0144-8609(96)01006-0.

Buschmann, A.H., Varela, D., Cifuentes, M., del Carmen Hernández-González, M., Henríquez, L., Westermeier, R., and Correa, J.A. (2004). Experimental indoor cultivation of the carrageenophytic red alga *Gigartina skottsbergii*. *Aquaculture*, **241**, 357–370. https://doi.org/10.1016/J.AQUACULTURE.2004.08.026.

Chan, J., Cheung, P.C.-K., and Ang, J.P.O. (1997). Comparative studies on the effect of three drying methods on the nutritional composition of seaweed *Sargassum hemiphyllum* (Turn) C. Ag. †. *J. Agric. Food Chem.*, **45**, 3056–3059. https://doi.org/10.1021/jf9701749.

Cabrera, R., Umanzor, S., Díaz-Larrea, J., and Araújo, P.G. (2019). *Kappaphycus alvarezii* (*Rhodophyta*): New record of an exotic species for the Caribbean Coast of Costa Rica. *Am. J. Plant Sci.* https://doi.org/10.4236/ajps.2019.1010133.

Cabrita, A.R.J., Maia, M.R.G., Sousa-Pinto, I., and Fonseca, A.J.M. (2017). Ensilage of seaweeds from an integrated multi-trophic aquaculture system. *Algal Res.*, **24**, 290–298. https://doi.org/10.1016/J.ALGAL.2017.04.024.

Camp, E.F., Suggett, D.J., Gendron, G., Jompa, J., Manfrino, C., and Smith, D.J. (2016). Mangrove and seagrass beds provide different biogeochemical services for corals threatened by climate change. *Front. Mar. Sci.* **3**, 52. https://doi.org/10.3389/fmars.2016.00052.

Campbell, I., Macleod, A., Sahlmann, C., Neves, L., Funderud, J., Øverland, M., Hughes, A.D., and Stanley, M. (2019). The environmental risks associated with the development of seaweed farming in Europe — prioritizing key knowledge gaps. *Front. Mar. Sci.*, https://doi.org/10.3389/fmars.2019. 00107.

Cancela, A., Maceiras, R., Urrejola, S., and Sanchez, A. (2012). Microwave-assisted transesterification of macroalgae. *Energies*, **5**, 862–871. https://doi.org/10.3390/en5040862.

Candra, K.P. and Sarwono, S. (2011). Study on bioethanol production using red seaweed Eucheuma cottonii from Bontang sea water. *J. Coast. Dev.*, **15**, 45–50.

Cao, L., Yu, I.K.M., Cho, D.-W., Wang, D., Tsang, D.C.W., Zhang, S., Ding, S., Wang, L., and Ok, Y.S. (2019). Microwave-assisted low-temperature hydro-thermal treatment of red seaweed (Gracilaria lemaneiformis) for production of levulinic acid and algae hydrochar. *Bioresour. Technol.*, **273**, 251–258. https://doi.org/10.1016/J.BIORTECH.2018.11.013.

Card, J.C. and Farrell, L.M. (1982). Separation of alcohol-water mixtures using salts. Oak Ridge National Lab., TN (USA).

Carlos, M.D., Iñigo, J.L., Iris E.H., Mazarrasa, I., and Marbà, N. (2013). The role of coastal plant communities for climate change mitigation and adaptation. *Nat. Clim. Chang.* https://doi.org/10.1038/NCLIMATE1970.

Carpenter, S.R., Caraco, N.F., Correll, D.L., Howarth, R.W., Sharpley, A.N., and Smith, V.H. (1998). Nonpoint pollution of surface waters with phosphorus and nitrogen. *Ecol. Appl.* https://doi.org/10.1890/1051-0761(1998)008 [0559:NPOSWW]2.0.CO;2.

Carrington, E. (1990). Drag and dislodgment of an intertidal macroalga: Consequences of morphological variation in Mastocarpus papillatus Kützing. *J. Exp. Mar. Bio. Ecol.* https://doi.org/10.1016/0022-0981(90)90146-4.

Carvalho, F.P. (2006). Agriculture, pesticides, food security and food safety. *Environ. Sci. Policy*, **9**, 685–692. https://doi.org/10.1016/j.envsci.2006.08.002.

Cesaro, A. and Belgiorno, V. (2015). Combined biogas and bioethanol produc-tion: Opportunities and challenges for industrial application. *Energies*, **8**, 8121–8144.

Chakraborty, S., Bhattacharya, T., Singh, G., and Maity, J.P. (2014). Benthic macroalgae as biological indicators of heavy metal pollution in the marine environments: A biomonitoring approach for pollution assessment. *Ecotoxicol. Environ. Saf.*, **100**, 61–68.

Chandrasekaran, S., Arun Nagendran, N., Pandiaraja, D., Krishnankutty, N., and Kamalakannan, B. (2008). Bioinvasion of Kappaphycus alvarezii on corals in the Gulf of Mannar, India. *Curr. Sci.*, **94**, 1167–1172.

Charrier, B., Abreu, M.H., Araujo, R., Bruhn, A., Coates, J.C., De Clerck, O., Katsaros, C., Robaina, R.R., and Wichard, T. (2017). Furthering knowledge of seaweed growth and development to facilitate sustainable aquaculture. *New Phytol.* https://doi.org/10.1111/nph.14728.

Charrier, B., Wichard, T., and Reddy, C.R.K. (Eds.) (2018). *Protocols for Macroalgae Research.* CRC Press. Boca Raton. FL.

Chatel, G., De Oliveira Vigier, K., and Jérôme, F. (2014). Sonochemistry: What potential for conversion of lignocellulosic biomass into platform chemicals? *ChemSusChem.*, 7, 2774–2787. https://doi.org/10.1002/cssc.201402289.

Chaturvedi, V. and Verma, P. (2013). An overview of key pretreatment processes employed for bioconversion of lignocellulosic biomass into biofuels and value added products. *3 Biotech.*, **3**, 415–431. https://doi.org/10.1007/s13205-013-0167-8.

Chemat, F., Vian, M.A., and Cravotto, G. (2012). Green extraction of natural products: Concept and principles. *Int. J. Mol. Sci.*, **13**, 8615–8627. https://doi.org/10.3390/ijms13078615.

Chemodanov, A., Jinjikhashvily, G., Habiby, O., Liberzon, A., Israel, A., Yakhini, Z., and Golberg, A. (2017a). Net primary productivity, biofuel production and CO_2 emissions reduction potential of *Ulva* sp. (Chlorophyta) biomass in a coastal area of the Eastern Mediterranean. *Ener. Convers. Manag.*, **148**, 1497–1507. https://doi.org/10.1016/j.enconman.2017.06.066.

Chemodanov, A., Robin, A., and Golberg, A. (2017b). Design of marine macroalgae photobioreactor integrated into building to support seagriculture for biorefinery and bioeconomy. *Bioresour. Technol.*, **241**, 1084–1093. https://doi.org/10.1016/j.biortech.2017.06.061.

Chemodanov, A., Robin, A., Jinjikhashvily, G., Yitzhak, D., Liberzon, A., Israel, A., and Golberg, A. (2019). Feasibility study of *Ulva* sp. (Chlorophyta) intensive cultivation in a coastal area of the Eastern Mediterranean Sea. *Biofu. Bioprod. Biorefining.* https://doi.org/10.1002/bbb.1995.

Chen, J., Li, J., Zhang, Z., and Ni, S. (2014). Long-term groundwater variations in Northwest India from satellite gravity measurements. *Glob. Planet. Change.* https://doi.org/10.1016/j.gloplacha.2014.02.007.

Chen, W.-T., Ma, J., Zhang, Y., Gai, C., and Qian, W. (2014). Physical pretreatments of wastewater algae to reduce ash content and improve thermal decomposition characteristics. *Bioresource Technology.* https://doi.org/10.1016/j.biortech.2014.07.076.

Chen, Y. and Mu, T. (2019). Application of deep eutectic solvents in biomass pretreatment and conversion. *Green Ener. Environ.*, **4**, 95–115. https://doi.org/10.1016/J.GEE.2019.01.012.

Cheng J. (2017). *Biomass to Renewable Energy Processes.* CRC Press. Boca Raton, FL 437p.

Cho, M., Kang, I.-J., Won, M.-H., Lee, H.-S., and You, S. (2010). The antioxidant properties of ethanol extracts and their solvent-partitioned fractions from various green seaweeds. *J. Med. Food*, **13**, 1232–1239. https://doi.org/10.1089/jmf.2010.1124.

Cho, Y., Kim, M., and Kim, S. (2013). Ethanol Production from Seaweed, Enteromorpha intestinalis, by Separate Hydrolysis and Fermentation (SHF) and Simultaneous Saccharification and Fermentation (SSF) with Saccharomyces cerevisiae. *KSBB Journal*, 28, 366–371.

Choi, W.-Y.Y., Kang, D.-H.H., and Lee, H.-Y.Y. (2013). Enhancement of the saccharification yields of *Ulva pertusa* kjellmann and rape stems by the high-pressure steam pretreatment process. *Biotechnol. Bioproc. Eng.*, **18**, 728–735. https://doi.org/10.1007/s12257-013-0033-x.

Chopin, T., Buschmann, A.H., Halling, C., Troell, M., Kautsky, N., Neori, A., Kraemer, G.P., Zertuche-González, J.A., Yarish, C., and Neefus, C. (2001). integrating seaweeds into marine aquaculture systems: A key toward sustainability. *J. Phycol.*, **37**, 975–986. https://doi.org/10.1046/j.1529-8817.2001.01137.x.

Chopin, T., Cooper, J.A., Reid, G., Cross, S., and Moore, C. (2012). Open-water integrated multi-trophic aquaculture: Environmental biomitigation and economic diversification of fed aquaculture by extractive aquaculture. *Rev. Aquac.*, **4**, 209–220. https://doi.org/10.1111/j.1753-5131.2012.01074.x.

Chopin, T. and Sawhney, M. (2019). Seaweeds and their mariculture, In: *Encyclopedia. of Ocean Sciences.* https://doi.org/10.1016/B978-0-12-813081-0.00757-6.

Chopin, T. and Sawhney, M. (2009). Seaweeds and their Mariculture, in: *Encyclopedia of Ocean Sciences*: Second Edition. Elsevier Inc., pp. 317–326. https://doi.org/10.1016/B978-012374473-9.00757-8.

Chopin, T., Troell, M., Reid, G.K., Knowler, D., Robinson, S.M.C., Neori, A., Buschmann, A.H., and Pang, S. (2010a). Integrated multi-trophic aquaculture. Part I. Responsible Practice Provides Diversified Products, Biomitigation. *Glob. Aquac. Advocate.*

Chopin, T., Troell, M., Reid, G.K., Knowler, D., Robinson, S.M.C., Neori, A., Buschmann, A.H., and Pang, S. (2010b). Integrated Multi Trophic Aquaculture Part II. Increasing IMTA Adoption. *Glob. Aquac. Advoc.*, 17–20.

Chopra, S., and Sodhi, M. (2004). Managing Risk to Avoid Supply-Chain Breakdown. *Sloan Manage. Rev.* October 15, 2014. https://sloanreview.mit.edu/article/managing-risk-to-avoid-supplychain-breakdown/.

Cikoš, A.-M., Jokić, S., Šubarić, D., and Jerković, I. (2018). Overview on the Application of Modern Methods for the Extraction of Bioactive Compounds from Marine Macroalgae. *Mar. Drugs*, **16**, 348. https://doi.org/10.3390/md16100348.

Cocero, M.J., Cabeza, Á., Abad, N., Adamovic, T., Vaquerizo, L., Martínez, C.M., and Pazo-Cepeda, M.V. (2018). Understanding biomass fractionation in subcritical & supercritical water. *J. Supercrit. Flui.*, **133**, 550–565. https://doi.org/10.1016/J.SUPFLU.2017.08.012.

Collén, J. and Davison, I.R. (1999). Stress tolerance and reactive oxygen metabolism in the intertidal red seaweeds Mastocarpus stellatus and Chondrus crispus. *Plant, Cell Environ.* https://doi.org/10.1046/j.1365-3040.1999.00477.x.

Cook, J., Oreskes, N., Doran, P.T., Anderegg, W.R.L., Verheggen, B., Maibach, E.W., Carlton, J.S., Lewandowsky, S., Skuce, A.G., Green, S.A., Nuccitelli, D., Jacobs, P., Richardson, M., Winkler, B., Painting, R., and Rice, K. (2016). Consensus on consensus: A synthesis of consensus estimates on human-caused global warming. *Environ. Res. Lett.* https://doi.org/10.1088/1748-9326/11/4/048002.

Correa, J. (1991). Pigmented algal endophytes of Chondrus cripus stackhouse: host specificity, fine structure and effects on host performance in infections by Acrochaete operculata Correa and A. heteroclada Correa & Nielsen. Thesis. https://dalspace.library.dal.ca/handle/10222/55201.

Costanza, R., d'Arge, R., de Groot, R., Farber, S., Grasso, M., Hannon, B., Limburg, K., Naeem, S., O'Neill, R. V., Paruelo, J., Raskin, R.G., Sutton, P., and van den Belt, M. (1997). The value of the world's ecosystem services and natural capital. *Nature*, **387**, 253–260. https://doi.org/10.1038/387253a0.

Coste, O., Malta, E., López, J.C., and Fernández-Díaz, C. (2015). Production of sulfated oligosaccharides from the seaweed *Ulva* sp. using a new ulvan-degrading enzymatic bacterial crude extract. *Algal Res.*, **10**, 224–231.

Cottier-Cook, E.J., Nagabhatla, N., Badis, Y., Campbell, M.L., Chopin, T., Dai, W., Fang, J., He, P., Hewitt, C.L., Kim, G.H., Huo, Y., Jiang, Z., Kema, G., Li, X., Liu, F., Liu, H., Liu, Y., Lu, Q., Luo, Q., Mao, Y., Msuya, F.E., Rebours, C., Shen, H., Stentiford, G.D., Yarish, C., Wu, H., Yang, X., Zhang, J., Zhou, Y., and Gachon, C.M.M. (2016). Safeguarding the future of the global seaweed aquaculture industry. 12. https://doi.org/ISBN 978-92-808-6080-1.

Crowl, T.A., Crist, T.O., Parmenter, R.R., Belovsky, G., and Lugo, A.E. (2008). The spread of invasive species and infectious disease as drivers of ecosystem change. *Front. Ecol. Environ.*, **6**, 238–246. https://doi.org/10.1890/070151.

Cui, J. and Zhang, L. (2008). Metallurgical recovery of metals from electronic waste: A review. *J. Hazard. Mater.*, **158**, 228–256. https://doi.org/10.1016/J.JHAZMAT.2008.02.001.

Cvjetko Bubalo, M., Radošević, K., Radojčić Redovniković, I., Halambek, J., and Gaurina Srček, V. (2014). A brief overview of the potential environmental hazards of ionic liquids. *Ecotoxicol. Environ. Saf.*, **99**, 1–12. https://doi.org/10.1016/J.ECOENV.2013.10.019.

Dahiya, A., Todd, J.H., and McInnis, A. (2012). Wastewater treatment integrated with algae production for biofuel, in: Gordon, R., Seckbach, J. (Eds.), *The Science of Algal Fuels Phycology, Geology, Biophotonics, Genomics, and Nanotechnology*. Springer, Dordrecht, pp. 447–466. https://doi.org/10.1007/978-94-007-5110-1_24.

Dai, J., Chen, B., and Sciubba, E. (2014). Ecological accounting based on extended exergy: a sustainability perspective. *Environ. Sci. Technol.*, **48**, 9826–9833. https://doi.org/10.1021/es404191v.

Dalin, C., Wada, Y., Kastner, T., and Puma, M.J. (2017). Groundwater depletion embedded in international food trade. *Nature*. https://doi.org/10.1038/nature21403.

Dalsgaard, J.P.T., Lightfoot, C., and Christensen, V. (1995). Towards quantification of ecological sustainability in farming systems analysis. *Ecol. Eng.* https://doi.org/10.1016/0925-8574(94)00057-C.

Damonte, E.B., Matulewicz, M.C., and Cerezo, A.S. (2004). Sulfated Seaweed Polysaccharides as Antiviral Agents Sulfated Seaweed Polysaccharides as Antiviral Agents. *Curr. Med. Chem.*, **11**, 2399–2419. https://doi.org/10.2174/0929867043364504.

Daneshvar, S., Salak, F., Ishii, T., and Otsuka, K. (2012). Application of subcritical water for conversion of macroalgae to value-added materials. *Ind. Eng. Chem. Res.*, **51**, 77–84. https://doi.org/10.1021/ie201743x.

Das, A.K., Sharma, M., Mondal, D., and Prasad, K. (2016). Deep eutectic solvents as efficient solvent system for the extraction of κ-carrageenan from Kappaphycus alvarezii. *Carbohydr. Polym.*, **136**, 930–935.

Das, P., Ganesh, A., and Wangikar, P. (2004). Influence of pretreatment for deashing of sugarcane bagasse on pyrolysis products. *Biom. and Bioener.*, **27**, 445–457. https://doi.org/10.1016/j.biombioe.2004.04.002.

Dave, A., Huang, Y., Rezvani, S., McIlveen-Wright, D., Novaes, M., and Hewitt, N. (2013). Techno-economic assessment of biofuel development by anaerobic digestion of European marine cold-water seaweeds. *Bioresour. Technol.*, **135**, 120–127.

Davis, T.A., Volesky, B., and Mucci, A. (2003). A review of the biochemistry of heavy metal biosorption by brown algae. *Water Res.* https://doi.org/10.1016/S0043-1354(03)00293-8.

Dawes, C.J. and Lawrence, J.M. (1980). Seasonal changes in the proximate constituents of the seagrasses Thalassia testudinum, Halodule wrightii. and Syringodium filiforme. *Aquat. Bot.* https://doi.org/10.1016/0304-3770(80)90066-2.

Dawes, C.J., Teasdale, B.W., and Friedlander, M. (2000). Cell wall structure of the agarophytes Gracilaria tikvahiae and *G. cornea* (Rhodophyta) and penetration by the epiphyte *Ulva* lactuca (Chlorophyta) 567–575.

De Clerck, O., Guiry, M.D., Leliaert, F., Samyn, Y., and Verbruggen, H. (2013). Algal Taxonomy: A Road to Nowhere? *J. Phycol.* https://doi.org/10.1111/jpy.12020.

De Souza, C.M.R., Marques, C.T., Maria, C., Dore, G., Roberto, F., Alexandre, H., Rocha, O., and Leite, E.L. (2007). Antioxidant activities of sulfated polysaccharides from brown and red seaweeds. *Appl. Phycol.*, 153–160. https://doi.org/10.1007/s10811-006-9121-z.

de Vries, F.T., Liiri, M.E., Bjørnlund, L., Setälä, H.M., Christensen, S., and Bardgett, R.D. (2012). Legacy effects of drought on plant growth and the soil food web. *Oecologia.* https://doi.org/10.1007/s00442-012-2331-y.

DeBusk, T.A. and Ryther, J.H. (1984). Effects of seawater exchange, pH and carbon supply on the growth of G. tikvahiae in large scale cultures. *Bot. Mar.*, **27**, 357–362.

Delaney, A., Frangoudes, K., and Ii, S.-A. (2016). Society and Seaweed: Understanding the Past and Present, Seaweed in Health and Disease Prevention. Academic Press. https://doi.org/10.1016/B978-0-12-802772-1.00002-6.

Denis, C., Ledorze, C., Jaouen, P., and Fleurence, J. (2009). Comparison of different procedures for the extraction and partial purification of R-phycoerythrin from the red macroalga Grateloupia turuturu. *Bot. Mar.*, **52**, 278–281.

Devi, G.K., Manivannan, K., Thirumaran, G., Rajathi, A.A., and Anantharaman, P. (2011). In vitro antioxidant activities of selected seaweeds from Southeast coast of India. *Asian Pac. J. Trop. Med.*, **4**, 205–211. https://doi.org/10.1016/S1995-7645(11)60070-9.

Di Donato, P., Buono, A., Poli, A., Finore, I., Abbamondi, R.G., Nicolaus, B., and Lama, L. (2018). Exploring marine environments for the identification of extremophiles and their enzymes for sustainable and green bioprocesses. sustain. https://doi.org/10.3390/su11010149.

Diamond, A.D. and Hsu, J.T. (1992). Aqueous two-phase systems for biomolecule separation, in: *Bioseparation.* Springer, pp. 89–135.

Diaz, R.J. and Breitburg, D.L. (2009). Chapter 1 The hypoxic environment. *Fish Physiol.*, **27**, 1–23. https://doi.org/10.1016/S1546-5098(08)00001-0.

Dillehay, T.T.D., Ramirez, C., Pino, M., Collins, M.B., Rossen, J., and Pino-Navarro, J.D. (2008). Monte Verde: Seaweed, food, medicine, and the peopling of South America. *Science* (80-.). **320**, 784–786. https://doi.org/10.1126/science.1156533.

Döll, P., Müller Schmied, H., Schuh, C., Portmann, F.T., and Eicker, A. (2014). Global-scale assessment of groundwater depletion and related groundwater abstractions: Combining hydrological modeling with information from well observations and GRACE satellites. *Water Resour. Res.* https://doi.org/10.1002/2014WR015595.

Dong, Z., Liu, D., and Keesing, J.K. (2010). Jellyfish blooms in China: Dominant species, causes and consequences. *Mar. Pollut. Bull.*, **60**, 954–963. https://doi.org/10.1016/J.MARPOLBUL.2010.04.022.

Doornbosch, R. and Steenblik, R. (2007). *Round Table on Sustainable Development Biofuels: Is the Cure Worse Than the Disease?*. OECD report SG/SD/RT(2007)3/REV1 https://www.oecd.org/sd-roundtable/39411732.pdf.

Dove, C. (2014). The development of unfired earth bricks using seaweed bio-polymers, in: Brebbia, C.A., Pulselli, R. (Eds.), *Eco-Architecture V Harmonisation between Architecture and Nature*, Wittpress, pp. 219–230.

Drechsler, Z., Sharkia, R., Cabantchik, Z.I., and Beer, S. (1994). The relationship of arginine groups to photosynthetic HCO_3 — uptake in *Ulva* sp. mediated by a putative anion exchanger. *Planta*. https://doi.org/10.1007/bf00196395.

Dreywood, R. (1946). Qualitative test for carbohydrate material. *Ind. Eng. Chem.*, **18**, 499. https://doi.org/10.1021/i560156a015.

Drimer, N. (2016). Offshore Structures First principle approach to the design of an open sea aquaculture system. https://doi.org/10.1080/17445302.2016.1213491.

Druehl, L.D., Baird, R., Lindwall, A., Lloyd, K.E., and Pakula, S. (1988). Longline cultivation of some Laminariaceae in British Columbia, Canada. John Wiley & Sons, Ltd. https://doi.org/10.1111/j.1365-2109.1988.tb00428.x.

du Jardin, P. (2015). Plant biostimulants: Definition, concept, main categories and regulation. *Sci. Hortic.* (Amsterdam). https://doi.org/10.1016/j.scienta.2015.09.021.

Du, X., Lu, L., Reardon, T., and Zilberman, D. (2016). Economics of agricultural supply chain design: A portfolio selection approach, in: *American J. of Agricul. Econo.* https://doi.org/10.1093/ajae/aaw074.

Duarte, P. and Ferreira, J.G. (1997). A model for the simulation of macroalgal population dynamics and productivity. *Ecol. Modell.*, **98**, 199–214. https://doi.org/10.1016/S0304-3800(96)01915-1.

Dubois, M., Gilles, K.A., Hamilton, J.K., Rebers, P.A., and Smith, F. (1956). Colorimetric method for determination of sugars and related substances. *Anal. Chem.*, **28**, 350–356. https://doi.org/10.1021/ac60111a017.

Duffy, J.E. (1990). *Amphipods on Seaweeds: Partners or Pests? Oecologia* **83** 267–276.

Dutta, S. (2012). Green revolution revisited: The contemporary agrarian situation in Punjab, India. *Soc. Chan.*, **42**, 229–247. https://doi.org/10.1177/004908571204200205.

Eggert, A. (2012). *Seaweed Responses to Temperature*. https://doi.org/10.1007/978-3-642-28451-9_3.

Ehler, L. (2006). Integrated pest management (IPM): definition, historical development and implementation, and the other IPM. *Pest Manag. Sci.*, **62**, 787–789. https://doi.org/10.1002/ps.

El-Dalatony, M.M., Kurade, M.B., Abou-Shanab, R.A.I., Kim, H., Salama, E.S., and Jeon, B.H. (2016). Long-term production of bioethanol in repeated-batch fermentation of microalgal biomass using immobilized Saccharomyces cerevisiae. *Bioresour. Technol.*, **219**, 98–105. https://doi.org/10.1016/j.biortech.2016.07.113.

El-Moneim M. R. Afify, A., Shalaby, E. A., and Shanab, S.M.M. (2010). Enhancement of biodiesel production from different species of algae. *Grasas y Aceites*, **61**, 416–422. https://doi.org/10.3989/gya.021610.

El-Sayed, W.M.M., Ibrahim, H.A.H., Abdul-Raouf, U.M., and El-Nagar, M.M. (2016). Evaluation of bioethanol production from *Ulva lactuca* by Saccharomyces cerevisiae. *J Biotech. Biomat.*, **6**. https://doi.org/10.4172/2155-952X.1000226.

Elena Romero-González, M., Williams, C.J., and Gardiner, P.H.E. (2000). The application of dealginated seaweed as a cation exchanger foron-line preconcentration and chemical speciation of trace metals. *J. Anal. At. Spectrom.*, **15**, 1009–1013. https://doi.org/10.1039/B002737P.

Ellis, E.C. and Ramankutty, N. (2008). Putting people in the map: Anthropogenic biomes of the world. *Front. Ecol. Environ.* https://doi.org/10.1890/070062.

Emmett Duffy, J., Paul Richardson, J., and Canuel, E. (2003). Grazer diversity effects on ecosystem functioning in seagrass beds. *Ecol. Lett.*, **6**, 637–645. https://doi.org/doi:10.1046/j.1461-0248.2003.00474.x.

EPA, U.S. (2018). Final 2016 Effluent Guidelines Program Plan (April 2018).

Erlandson, J., Braje, T., Gill, K., and Graham, M.H. (2015). Ecology of the kelp highway: Did marine resources facilitate human dispersal from Northeast Asia to the Americas? *J. Isl. Coast. Archaeol.*, **10**, 392–411.

Euroview, E.U. (2010). Euroview Biorefinery: Current situation and potential of biorefinery concept in the EU-Strategic framework and guidelines for its development, 2007. Final Report Summary.

Fabrowska, J., Ibañez, E., Lęska, B., and Herrero, M. (2016). Supercritical fluid extraction as a tool to valorize underexploited freshwater green algae. *Algal. Res.*, **19**, 237–245. https://doi.org/10.1016/J.ALGAL.2016.09.008.

Fabrowska, J., Messyasz, B., Szyling, J., Walkowiak, J., and Leska, B. (2017). Isolation of chlorophylls and carotenoids from freshwater algae using different extraction methods. *Phycol. Res.* https://doi.org/10.1111/pre.12191.

FAO, IFAD, UNICEF, W., and W. (2019). Food Security and Nutrition in the World 2019, IEEE *J. Selected Topics in Applied Earth Observations and Remote Sensing.* https://doi.org/10.1109/JSTARS.2014.2300145.

FAO (2004). Risk analysis for quarantine pests, including analysis of environmental risks and living modified organisms.

FAO (2016). The State of World Fisheries and Aquaculture 2016.

FAO (2016). The state of world fisheries and aquaculture 2016. Contributing to food security and nutrition for all., Food and Agriculture Organization of the United Nations, Rome, 2016.

FAO, Kapetsky, J.M., Aguilar-Manjarrez, J., and Jenness, J. (2013). A global assessment of offshore mariculture potential from a spatial perspective.

Fatih, D.M. (2009). Biorefineries for biofuel upgrading: A critical review. *Appl. Ener.*, **86**, S151–S161. https://doi.org/10.1016/j.apenergy.2009.04.043.

Fayad, S., Nehmé, R., Tannoury, M., Lesellier, E., Pichon, C., and Morin, P. (2017). Macroalga Padina pavonica water extracts obtained by pressurized liquid extraction and microwave-assisted extraction inhibit hyaluronidase activity as shown by capillary electrophoresis. *J. Chromatogr. A.*, **1497**, 19–27. https://doi.org/10.1016/J.CHROMA.2017.03.033.

Fei, X. (2004). Solving the coastal eutrophication problem by large scale seaweed cultivation. *Hydrobiologia*, **512**, 145–151. https://doi.org/10.1023/B:HYDR. 0000020320.68331.ce.

Feinberg, D. and Hock, S. (1985). Technical and Economic Evaluation of Macroalgae Cultivation for Fuel Production (Draft).

Feldman, S., Reynaldi, S., and Stortz, C. (1999). Antiviral properties of fucoidan fractions from Leathesia difformis. *Phytomed.*, **6**, 335–340.

Fernand, F., Israel, A., Skjermo, J., Wichard, T., Timmermans, K.R., and Golberg, A. (2016). Offshore macroalgae biomass for bioenergy production: Environmental aspects, technological achievements and challenges. *Renew. Sustain. Ener. Rev.*, **75**, 35–45. https://doi.org/10.1016/j.rser.2016.10.046.

Fingerman, K.R., Torn, M.S., O'Hare, M.H., and Kammen, D.M. (2010). Accounting for the water impacts of ethanol production. *Environ. Res. Lett.*, **5**, 14020.

Fiset, C., Liefer, J., Irwin, A.J., and Finkel, Z. V. (2017). Methodological biases in estimates of macroalgal macromolecular composition. *Limnol. Oceanography: Methods*, 618–630. https://doi.org/10.1002/lom3.10186.

Fletcher, R.L. (1995). Epiphytism and fouling in Gracilaria cultivation: An overview. *J. Appl. Phycol.*, 7, 325–333. https://doi.org/10.1007/BF00004006.

Fleurence, J., Le Coeur, C., Mabeau, S., Maurice, M., and Landrein, A. (1995). Comparison of different extractive procedures for proteins from the edible seaweeds *Ulva rigida* and *Ulva rotundata*. *J. Appl. Phycol.*, 7, 577–582. https://doi.org/10.1007/BF00003945.

Folch, J., Lees, M., and G.S.-S., (1957). A simple method for the isolation and purification of total lipids from animal tissues. *J. Biol.*, **226**, 497–509.

Foley, J.A., Ramankutty, N., Brauman, K.A., Cassidy, E.S., Gerber, J.S., Johnston, M., Mueller, N.D., O'Connell, C., Ray, D.K., West, P.C., Balzer, C., Bennett, E.M., Carpenter, S.R., Hill, J., Monfreda, C., Polasky, S., Rockström, J., Sheehan, J., Siebert, S., Tilman, D., and Zaks, D.P.M. (2011). Solutions for a cultivated planet. *Nature*. https://doi.org/10.1038/nature10452.

Foreman, R.E. (1976). Physiological aspects of carbon monoxide production by the brown alga *Nereocystis luetkeana*. *Can. J. Bot.* https://doi.org/10.1139/b76-032.

Forster, J. and Radulovich, R. (2015). Seaweed and food security, in: *Seaweed Sustainability*. Elsevier, pp. 289–313.

Freitas, J.R.C., Salinas Morrondo, J.M., and Cremades, U. J. (2016). Saccharina latissima (Laminariales, Ochrophyta) farming in an industrial IMTA system in Galicia (Spain). *J. Appl. Phycol.*, **28**, 377–385. https://doi.org/10.1007/s10811-015-0526-4.

French, B. (1960). Some considerations in estimating assembly cost functions for agricultural processing operations. *J. Farm. Econ.*, **42**,767–778.

Friedlander, M. (1992). Gracilaria conferta and its epiphytes. The effect of culture conditions on growth. *Bot. Mar.*, **35**, 423–428.

Friedlander, M. (1991). Growth rate, epiphyte biomass and agar yield of *G. conferta* in an annual outdoor experiment. 1. Irradiance and nitrogen. *Bioresour. Technol.*, **38**, 203–208.

Friedlander, M. and Ben-amotz, A. (1991). The effect of outdoor culture conditions on growth and epiphytes of Gracilaria conferta. *Aquat. Bot.*, **39**, 315–333.

Friedlander, M. and Levy, I. (1995). Cultivation of gracilaria in outdoor tanks and ponds. *J. Appl. Phycol.*, **7**, 315–324.

Fu, H., Sun, Y., Teng, H., Zhang, D., and Xiu, Z. (2015). Salting-out extraction of carboxylic acids. *Sep. Purif. Technol.*, **139**, 36–42. https://doi.org/10.1016/J.SEPPUR.2014.11.001.

Fu, H., Wang, X., Sun, Y., Yan, L., Shen, J., Wang, J., Yang, S.-T., and Xiu, Z. (2017). Effects of salting-out and salting-out extraction on the separation of butyric acid. *Sep. Purif. Technol.*, **180**, 44–50. https://doi.org/10.1016/J.SEPPUR.2017.02.042.

Fujita, D., Ishikawa, T., Kodama, S., Kato, Y., and Notoya, M. (2006). Distribution and recent reduction of gelidium beds in Toyama Bay, Japan. *J. Appl. Phycol.*, **18**, 591–598. https://doi.org/10.1007/s10811-006-9060-8.

Gagnon, K., Sjöroos, J., Yli-Rosti, J., Stark, M., Rothäusler, E., and Jormalainen, V. (2016). Nutrient enrichment overwhelms top-down control in algal communities around cormorant colonies. *J. Exp. Mar. Bio. Ecol.* https://doi.org/10.1016/j.jembe.2015.12.007.

Gajaria, T.K., Suthar, P., Baghel, R.S., Balar, N.B., Sharnagat, P., Mantri, V.A., and Reddy, C.R.K. (2017). Integration of protein extraction with a stream of byproducts from marine macroalgae: A model forms the basis for marine bioeconomy. *Bioresour. Technol.*, **243**, 867–873. https://doi.org/10.1016/j.biortech.2017.06.149.

Galán-Marín, C., Rivera-Gómez, C., and Petric, J. (2010). Clay-based composite stabilized with natural polymer and fibre. *Constr. Build. Mater.*, **24**, 1462–1468. https://doi.org/10.1016/j.conbuildmat.2010.01.008.

Gallagher, Joe A., Turner, L.B., Adams, J.M.M., Barrento, S., Dyer, P.W., Theodorou, M.K., and Gallagher, J.A. (2018). Species variation in the effects of dewatering treatment on macroalgae. *J. Appl. Phycol.*, **30**, 2305–2316. https://doi.org/10.1007/s10811-018-1420-7.

Gallagher, J.A., Turner, L.B., Adams, J.M.M., Dyer, P.W., and Theodorou, M.K. (2017). Dewatering treatments to increase dry matter content of the brown seaweed, kelp (Laminaria digitata ((Hudson) JV Lamouroux)). *Bioresour. Technol.*, **224**, 662–669. https://doi.org/10.1016/j.biortech.2016.11.091.

Ganesan, M., Reddy, C.R.K., and Jha, B. (2015). Impact of cultivation on growth rate and agar content of *Gelidiella acerosa* (Gelidiales, Rhodophyta). *Algal Res.*, **12**, 398–404. https://doi.org/10.1016/j.algal.2015.10.001.

Gao, G., Clare, A.S., Rose, C., and Caldwell, G.S. (2017). Non-cryogenic preservation of thalli, germlings, and gametes of the green seaweed *Ulva rigida*. *Aquaculture*, **473**, 246–250. https://doi.org/10.1016/j.aquaculture.2017.02.012.

Gao, J., Chen, C., Wang, L., Lei, Y., Ji, H., and Liu, S. (2019). Utilization of inorganic salts as adjuvants for ionic liquid–water pretreatment of lignocellulosic biomass: enzymatic hydrolysis and ionic liquid recycle. *3 Biotech*, **9**, 264. https://doi.org/10.1007/s13205-019-1788-3.

Gao, K. and McKinley, K.R. (1994). Use of macroalgae for marine biomass production and CO_2 remediation: A review. *J. Appl. Phycol.*, **6**, 45–60. https://doi.org/10.1007/BF02185904.

Gaspar, R., Pereira, L., and Neto, J.M. (2012). Ecological reference conditions and quality states of marine macroalgae sensu water framework directive: An example from the intertidal rocky shores of the Portuguese coastal waters. *Ecol. Indic.* https://doi.org/10.1016/j.ecolind.2011.08.022.

Gattuso, J.P., Magnan, A.K., Bopp, L., Cheung, W.W.L., Duarte, C.M., Hinkel, J., Mcleod, E., Micheli, F., Oschlies, A., Williamson, P., Billé, R., Chalastani, V.I., Gates, R.D., Irisson, J.O., Middelburg, J.J., Pörtner, H.O., and Rau, G.H. (2018). Ocean solutions to address climate change and its effects on marine ecosystems. *Front. Mar. Sci.* https://doi.org/10.3389/fmars.2018.00337.

Geertz-Hansen, O., Sand-Jensen, K., Hansen, D.F., and Christiansen, A. (1993). Growth and grazing control of abundance of the marine macroalga, *Ulva lactuca L.* in a eutrophic Danish estuary. *Aquat. Bot.*, **46**, 101–109. https://doi.org/10.1016/0304-3770(93)90039-Y.

George, R., McFarlane, D., and Nulsen, B. (1997). Salinity threatens the viability of agriculture and ecosystems in Western Australia. *Hydrogeol. J.* https://doi.org/10.1007/s100400050103.

Gereniu, C.R.N.C.R.N., Saravana, P.S.P.S., and Chun, B.-S.B.-S. (2018). Recovery of carrageenan from Solomon Islands red seaweed using ionic liquid-assisted subcritical water extraction. *Sep. Purif. Technol.*, **196**, 309–317.

Ghaderi, H., Pishvaee, M.S., and Moini, A. (2016). Biomass supply chain network design: An optimization-oriented review and analysis. *Ind. Crops Prod.* https://doi.org/10.1016/j.indcrop.2016.09.027.

Ghadiryanfar, M., Rosentrater, K.A., Keyhani, A., and Omid, M. (2016). A review of macroalgae production, with potential applications in biofuels and bioenergy. *Renew. Sustain. Ener. Rev.*, 54, 473–481. https://doi.org/10.1016/j.rser.2015.10.022.

Ghatak, H.R. (2011). Biorefineries from the perspective of sustainability: Feedstocks, products, and processes. Renew. *Sustain. Ener. Rev.*, **15**, 4042–4052. https://doi.org/10.1016/J.RSER.2011.07.034.

Ghosh, S., Gnaim, R., Greiserman, S., Fadeev, L., Gozin, M., and Golberg, A. (2019). Macroalgal biomass subcritical hydrolysates for the production of polyhydroxyalkanoate (PHA) by *Haloferax mediterranei. Bioresour. Technol.* https://doi.org/10.1016/j.biortech.2018.09.108.

Gibbs, H.K., Ruesch, A.S., Achard, F., Clayton, M.K., Holmgren, P., Ramankutty, N., and Foley, J.A. (2010). Tropical forests were the primary sources of new agricultural land in the 1980s and 1990s. *Proc. Natl. Acad. Sci. U.S.A.* https://doi.org/10.1073/pnas.0910275107.

Gil, I.D., Uyazán, A.M., Aguilar, J.L., Rodríguez, G., and Caicedo, L.A. (2008a). Separation of ethanol and water by extractive distillation with salt and solvent as entrainer: Process simulation. *Brazilian J. Chem. Eng.* https://doi.org/10.1590/S0104-66322008000100021.

Gil, R., Zahavi, A., and Einav, R. (2008b). Seaweed communities on abrasion platforms along the Newe Yam Island, in the north of Israel. *Isr. J. Plant Sci.* https://doi.org/10.1560/IJPS.56.1-2.103.

Glasson, C.R.K., Sims, I.M., Carnachan, S.M., de Nys, R., and Magnusson, M. (2017). A cascading biorefinery process targeting sulfated polysaccharides (ulvan) from *Ulva ohnoi. Algal Res.*, **27**, 383–391. https://doi.org/10.1016/j.algal.2017.07.001.

Goh, C.S. and Lee, K.T. (2010). A visionary and conceptual macroalgae-based third-generation bioetanol (TGB) biorefinery in Sabah, Malaysia as an underlay for renewable and sustainable development. *Renew. Sust. Energ. Rev.*, **12**. https://doi.org/10.1016/j.rser.2009.10.001.

Golberg, A. (2015). Environmental exergonomics for sustainable design and analysis of energy systems. *Energy.* https://doi.org/10.1016/j.energy.2015.05.053.

Golberg, A. and Liberzon, A. (2015). Modeling of smart mixing regimes to improve marine biorefinery productivity and energy efficiency. *Algal Res.*, **11**, 28–32. https://doi.org/10.1016/j.algal.2015.05.021.

Golberg, A., Sack, M., Teissie, J., Pataro, G., Pliquett, U., Saulis, G., Stefan, T., Miklavcic, D., Vorobiev, E., and Frey, W. (2016). Energy-efficient biomass processing with pulsed electric fields for bioeconomy and sustainable development. *Biotechnol. Biofue.*, **9**, 94.

Golberg, A., Vitkin, E., Linshiz, G., Khan, S.A., Hillson, N.J., Yakhini, Z., and Yarmush, M.L. (2014). Proposed design of distributed macroalgal biorefineries: Thermodynamics, bioconversion technology, and sustainability implications for developing economies. *Biofuels, Bioprod. Biorefin.*, **8**, 67–82. https://doi.org/10.1002/bbb.1438.

Golberg, A., Zollmann, M., Prabhu, M., and Palatnik, R.R. (2020). Enabling Bioeconomy with Offshore Macroalgae Biorefineries, in: *Bioeconomy for Sustainable Development*. Springer, Singapore, pp. 173–200. https://doi.org/10.1007/978-981-13-9431-7_10.

Golden, J.S., Handfield, R.B., Daystar, J., and T.E.M. (2015). An Economic Impact Analysis of the U.S. Biobased Products Industry: A Report to the Congress of the United States of America. North Carolina State University: A Joint Publication of the Duke Center for Sustainability & Commerce and the Supply Chain Resource.

Greiserman, S., Epstein, M., Chemodanov, A., Steinbruch, E., Prabhu, M., Guttman, L., Jinjikhashvily, G., Shamis, O., Gozin, M., Kribus, A., and Golberg, A. (2019). Co-production of monosaccharides and hydrochar from green macroalgae *Ulva* (Chlorophyta) sp. with subcritical hydrolysis and carbonization. *BioEner. Res.*, **12**, 1090–1103. https://doi.org/10.1007/s12155-019-10034-5.

Grosso, C., Valentão, P., Ferreres, F., and Andrade, P.B. (2014). Bioactive marine drugs and marine biomaterials for brain diseases. *Mar. Dru.*, **12**, 2539–2589. https://doi.org/10.3390/md12052539.

Gunaseelan, V.N. (1997). Anaerobic digestion of biomass for methane production: A review. *Biom. and Bioene.*, **13**, 83–114.

Gunny, A.A.N., Arbain, D., Edwin Gumba, R., Jong, B.C., and Jamal, P. (2014). Potential halophilic cellulases for in situ enzymatic saccharification of ionic liquids pretreated lignocelluloses. *Bioresour. Technol.* https://doi.org/10.1016/j.biortech.2013.12.101.

Gunny, A.A.N., Arbain, D., Javed, M., Baghaei-Yazdi, N., Gopinath, S.C.B., and Jamal, P. (2019). Deep eutectic solvents-halophilic cellulase system: An efficient route for in situ saccharification of lignocellulose. *Proc. Biochem.*, **81**, 99–103.

Gupta, V., Trivedi, N., Simoni, S., and Reddy, C.R.K. (2018). Marine macroalgal nursery: A model for sustainable production of seedlings for large scale farming. *Algal Res.*, **31**, 463–468. https://doi.org/10.1016/J.ALGAL.2018.02.032.

Gutiérrez-Arnillas, E., Rodríguez, A., Sanromán, M.A., and Deive, F.J. (2016). New sources of halophilic lipases: Isolation of bacteria from Spanish and Turkish saltworks. *Biochem. Eng. J.*, **109**, 170–177. https://doi.org/10.1016/J.BEJ.2016.01.015.

Haberl, H., Erb, K.-H., Krausmann, F., Bondeau, A., Lauk, C., Müller, C., Plutzar, C., and Steinberger, J.K. (2011). Global bioenergy potentials from agricultural land in 2050: Sensitivity to climate change, diets and yields. *Biom. Bioen.*, **35**, 4753–4769. https://doi.org/10.1016/j.biombioe.2011.04.035.

Habiby, O., Nahor, O., Israel, A., Liberzon, A., and Golberg, A. (2018). Exergy efficiency of light conversion into biomass in the macroalga *Ulva* sp. (*Chlorophyta*) cultivated under the pulsed light in a photobioreactor. *Biotechnol. Bioeng.* https://doi.org/10.1002/bit.26588.

Habig, C., DeBusk, T.A., and Ryther, J.H. (1984). The effect of nitrogen content on methane production by the marine algae *Gracilaria tikvahiae* and *Ulva* sp. *Biomass*, **4**, 239–251.

Hach, C., Bowden, B., Kopelove, A., and Brayton, S. (1987). More powerful peroxide Kjeldahl digestion method. *J. Assoc. off. Anal. Chem.*, **70**, 783–787.

Haddad, N.M., Brudvig, L.A., Clobert, J., Davies, K.F., Gonzalez, A., Holt, R.D., Lovejoy, T.E., Sexton, J.O., Austin, M.P., Collins, C.D., Cook, W.M., Damschen, E.I., Ewers, R.M., Foster, B.L., Jenkins, C.N., King, A.J., Laurance, W.F., Levey, D.J., Margules, C.R., Melbourne, B.A., Nicholls, A.O., Orrock, J.L., Song, D.X., and Townshend, J.R. (2015). Habitat fragmentation and its lasting impact on Earth's ecosystems. *Sci. Adv.* https://doi.org/10.1126/sciadv.1500052.

Hafting, J.T., Craigie, J.S., Stengel, D.B., Loureiro, R.R., Buschmann, A.H., Yarish, C., Edwards, M.D., and Critchley, A.T. (2015). Prospects and challenges for industrial production of seaweed bioactives. *J. Phycol.* https://doi.org/10.1111/jpy.12326.

Hamann, M., Otto, C., Scheuer, P., and Dunbar, D. (1996). Kahalalides: Bioactive peptides from a marine Mollusk Elysia rufescens and its Algal Diet *Bryopsis* sp. *Org. Chemi.*, **61**, 6594–6600.

Hanisak, M. (1987). Cultivation of Gracilaria and other macroalgae in Florida for energy production. *Dev. Aquac. Fish. Sci.*, 191–218.

Hanisak, M.D. (1987). Cultivation of Gracilaria and other macroalgae in Florida for energy production, in: Bird, K.T., Benson, P.H. (Eds.), *Seaweed Cultivation for Renewable Resources*. Elsevier, Amsterdam, pp. 191–218.

Hannah, L., Roehrdanz, P.R., K. C., K.B., Fraser, E.D.G., Donatti, C.I., Saenz, L., Wright, T.M., Hijmans, R.J., Mulligan, M., Berg, A., and van Soesbergen, A. (2020). The environmental consequences of climate-driven agricultural frontiers. *PLoS One*, **15**, e0228305. https://doi.org/10.1371/journal.pone.0228305.

Harley, C.D.G., Anderson, K.M., Demes, K.W., Jorve, J.P., Kordas, R.L., Coyle, T.A., and Graham, M.H. (2012). Effects of climate change on global seaweed communities. *J. Phycol.* https://doi.org/10.1111/j.1529-8817.2012.01224.x.

Harrison, P.J. and Hurd, C.L. (2001). Nutrient physiology of seaweeds: Application of concepts to aquaculture. *Cah. Biol. Mar.*, **42**(1–2), 71–82.

Harte, J. and Newman, E.A. (2014). Maximum information entropy: A foundation for ecological theory. *Trends Ecol. Evol.*, **29**, 384–389. https://doi.org/10.1016/j.tree.2014.04.009.

He, J. and Chen, J.P. (2014). A comprehensive review on biosorption of heavy metals by algal biomass: Materials, performances, chemistry, and modeling simulation tools. *Bioresour. Technol.*, **160**, 67–78. https://doi.org/10.1016/j.biortech.2014.01.068.

He, Y., Fang, Z., Zhang, J., Li, X., and Bao, J. (2014). De-ashing treatment of corn stover improves the efficiencies of enzymatic hydrolysis and consequent ethanol fermentation. *Bioresour. Technol.*, **169**, 552–558. https://doi.org/10.1016/j.biortech.2014.06.088.

Heesch, S. and Peters, A.F. (1999). Scanning electron microscopy observation of host entry by two brown algae endophytic in *Laminaria saccharina* (*Laminariales, Phaeophyceae*), **47**, 1–5.

Hermann, B., Carus, M., Patel, M., and Blok, K. (2011). Current policies affecting the market penetration of biomaterials. *Biofuels, Bioprod. Biorefining*, **5**, 708–719. https://doi.org/10.1002/bbb.327.

Hermann, W.A. (2006). Quantifying global exergy resources. *Energy*, **31**, 1349–1366. https://doi.org/10.1016/j.energy.2005.09.006.

Hernández-Herrera, R.M., Santacruz-Ruvalcaba, F., Ruiz-López, M.A., Norrie, J., and Hernández-Carmona, G. (2014). Effect of liquid seaweed extracts on growth of tomato seedlings (Solanum lycopersicum L.). *J. Appl. Phycol.*, **26**, 619–628. https://doi.org/10.1007/s10811-013-0078-4.

Hernandez, R.R., Easter, S.B., Murphy-Mariscal, M.L., Maestre, F.T., Tavassoli, M., Allen, E.B., Barrows, C.W., Belnap, J., Ochoa-Hueso, R., Ravi, S., and Allen, M.F. (2014). Environmental impacts of utility-scale solar energy. *Renew. Sustain. Energy Rev.* https://doi.org/10.1016/j.rser.2013.08.041.

Herrero, M., Cifuentes, A., and Ibañez, E. (2006). Sub- and supercritical fluid extraction of functional ingredients from different natural sources: Plants, food-by-products, algae and microalgae: A review. *Food Chem.*, **98**, 136–148. https://doi.org/10.1016/J.FOODCHEM.2005.05.058.

Herrero, M. and Ibáñez, E. (2015). Green processes and sustainability: An overview on the extraction of high added-value products from seaweeds and microalgae. *J. Supercrit. Fluids*, **96**, 211–216. https://doi.org/10.1016/J.SUPFLU.2014.09.006.

Hoekstra, A.Y. and Mekonnen, M.M. (2012). The water footprint of humanity. *Proc. Natl. Acad. Sci. U.S.A.* https://doi.org/10.1073/pnas.1109936109.

Holdt, S.L. and Edwards, M.D. (2014). Cost-effective IMTA: A comparison of the production efficiencies of mussels and seaweed. *J. Appl. Phycol.*, **26**, 933–945. https://doi.org/10.1007/s10811-014-0273-y.

Holt, T.J. (1984). The development of techniques for the cultivation of Laminariales in the Irish Sea. PhD. Thesis. Liverpool Univ. (United Kingdom) 283p.

Horn, S.J., Aasen, I.M., and Ostgaard, K. (2000). Ethanol production from seaweed extract. *J. Ind. Microbiol. Biotechnol.*, **25**, 249–254. https://doi.org/10.1038/sj.jim.7000065.

Hosseini, S.E. and Wahid, M.A. (2014). Development of biogas combustion in combined heat and power generation. *Renew. Sustain. Energy Rev.*, **40**, 868–875.

Howard, J., McLeod, E., Thomas, S., Eastwood, E., Fox, M., Wenzel, L., and Pidgeon, E. (2017). The potential to integrate blue carbon into MPA design and management. *Aquat. Conserv. Mar. Freshw. Ecosyst.* https://doi.org/10.1002/aqc.2809.

Hu, Y., Wang, S., Wang, Q., He, Z., Lin, X., Xu, S., Ji, H., and Li, Y. (2017). Effect of different pretreatments on the thermal degradation of seaweed biomass. *Proc. Combust. Inst.*, **36**, 2271–2281. https://doi.org/https://doi.org/10.1016/j.proci.2016.08.086.

Huang, C., Wu, X., Huang, Y., Lai, C., Li, X., and Yong, Q. (2016). Prewashing enhances the liquid hot water pretreatment efficiency of waste wheat straw with high free ash content. *Bioresour. Technol.*, **219**, 583–588. https://doi.org/10.1016/j.biortech.2016.08.018.

Huesemann, M.H., Kuo, L.J., Urquhart, L., Gill, G.A., and Roesijadi, G. (2012). Acetone-butanol fermentation of marine macroalgae. *Bioresour. Technol.* **108**, 305–309. https://doi.org/10.1016/j.biortech.2011.12.148.

Hughes, A.D., Kelly, M.S., Black, K.D., and Stanley, M.S. (2012). Biogas from Macroalgae: Is it time to revisit the idea? *Biotechnol. Biofuels*, **5**, 86. https://doi.org/10.1186/1754-6834-5-86.

Humbird, D., Davis, R., Tao, L., Kinchin, C., and Hsu, D. (2011). Process design and economics for biochemical conversion of lignocellulosic biomass to ethanol. NREL Technical Report NREL/TP-5100-47764.

Hunter, M.C., Smith, R.G., Schipanski, M.E., Atwood, L.W., and Mortensen, D.A. (2017). Agriculture in 2050: Recalibrating targets for sustainable intensification. *Bioscience.* https://doi.org/10.1093/biosci/bix010.

Hupa, M. (2012). Ash-related issues in fluidized-bed combustion of biomasses: Recent research highlights. *Energy & Fuels*, **26**, 4–14. https://doi.org/10.1021/ef201169k.

Hurd, C.L. (2000). Water motion, marine macroalgal physiology, and production. *J. Phycol.*, **36**, 453–472. https://doi.org/10.1046/j.1529-8817.2000.99139.x.

Hurd, C.L., Harrison, P.J., Bischof, K., and Lobban, C.S. (2014). *Seaweed Ecology and Physiology*, Second Edition. https://doi.org/10.1017/CBO9781139192637.

Hurtado, A.Q. (2013). Social and economic dimensions of carrageenan seaweed farming in the Philippines. Rome: FAO.

Hurtado, A.Q., Critchley, A.T., Trespoey, A., and Bleicher, L.G. (2006). Occurrence of Polysiphonia epiphytes in Kappaphycus farms at Calaguas Is., Camarines Norte, Phillippines. *J. Appl. Phycol.*, **18**, 301–306.

Iida, T. (1998). Competition and communal regulations in the Kombu Kelp (*Laminaria angustata*) harvest. *Hum. Ecol.*, **26**, 405–423. https://doi.org/10.1023/A:1018704231913.

Ingle, K., Vitkin, E., Robin, A., Yakhini, Z., Mishori, D., and Golberg, A. (2017). Macroalgae biorefinery from *Kappaphycus alvarezii*: Conversion modeling and performance prediction for India and Philippines as examples. *BioEner. Res.*, 1–11. https://doi.org/10.1007/s12155-017-9874-z.

Ingle, K.N., Polikovsky, M., Chemodanov, A., and Golberg, A. (2018). Marine integrated pest management (MIPM) approach for sustainable seagriculture. *Algal Res.*, **29**, 223–232. https://doi.org/10.1016/j.algal.2017.11.010.

Ingle, K.N.K.N., Polikovsky, M., Chemodanov, A., and Golberg, A. (2018). Marine integrated pest management (MIPM) approach for sustainable seagriculture. *Algal Res.*, **29**, 223–232. https://doi.org/10.1016/j.algal.2017.11.010.

Intergovernmental Panel on Climate Change, Intergovernmental Panel on Climate Change (2015). Agriculture, Forestry and Other Land Use (AFOLU), in: *Climate Change 2014 Mitigation of Climate Change*. https://doi.org/10.1017/cbo9781107415416.017.

International Institute for Applied Systems Analysis (2012). Global Agro-Ecological Zones. IIASA 40th Anniv. Conference.

Ireland, C., Copp, B., Foster, M., McDonald, L., Radisky, D., and Swersey, J. (1993). Biomedical Potential of Marine Natural Products, in: *Pharmaceutical and Bioactive Natural Products*. Plenum Press, NY, pp. 1–43.

Isik, M., Sardon, H., and Mecerreyes, D. (2014). Ionic liquids and cellulose: Dissolution, chemical modification and preparation of new cellulosic materials. *Int. J. Mol. Sci.*, **15**, 11922–11940. https://doi.org/10.3390/ijms150711922.

Israel, A. and Einav, R. (2017). Alien seaweeds from the Levant basin (Eastern Mediterranean Sea), with emphasis to the Israeli shores in: *Isr. J. Plant Sci.*, **64**, 99–110.

Israel, A. and Hophy, M. (2002). Growth, photosynthetic properties and Rubisco activities and amounts of marine macroalgae grown under current and elevated seawater CO_2 concentrations. *Glob. Chang. Biol.* https://doi.org/10.1046/j.1365-2486.2002.00518.x.

Israel, A., Levy, I., and Friedlander, M. (2006). Experimental tank cultivation of Porphyra in Israel. *J. Appl. Phycol.* https://doi.org/10.1007/s10811-006-9024-z.

Israel, A., Martinez-Goss, M., and Friedlander, M. (1999). Effect of salinity and pH on growth and agar yield of *Gracilaria tenuistipitata var. liui* in laboratory and outdoor cultivation. *J. Appl. Phycol.* https://doi.org/10.1023/A:1008141906299.

Jannin, L., Arkoun, M., Etienne, P., Laîné, P., Goux, D., Garnica, M., Fuentes, M., Francisco, S.S., Baigorri, R., Cruz, F., Houdusse, F., Garcia-Mina, J.M., Yvin, J.C., and Ourry, A. (2013). Brassica napus growth is promoted by *Ascophyllum nodosum* (*L.*) Le Jol. seaweed extract: Microarray analysis and physiological characterization of N, C, and S metabolisms. *J. Pla. Grow. Regul.*, **32**, 31–52. https://doi.org/10.1007/s00344-012-9273-9.

Jiang, R., Ingle, K.N.K.N.K.N., and Golberg, A. (2016). Macroalgae (seaweed) for liquid transportation biofuel production: what is next? *Algal Res.*, **14**, 48–57. https://doi.org/10.1016/j.algal.2016.01.001.

Jiang, R, Linzon, Y., Vitkin, E., Yakhini, Z., Chudnovsky, A., and Golberg, A. (2016). Thermochemical hydrolysis of macroalgae *Ulva* for biorefinery: Taguchi robust design method. *Sci. Rep.*, **6**. https://doi.org/doi:10.1038/srep27761.

Jitar, O., Teodosiu, C., Oros, A., Plavan, G., and Nicoara, M. (2015). Bioaccumulation of heavy metals in marine organisms from the Romanian sector of the Black Sea. *N. Biotechnol.*, **32**, 369–378. https://doi.org/10.1016/j.nbt.2014.11.004.

Jmel, M.A., Anders, N., Yahmed, N.B., Schmitz, C., Marzouki, M.N., Spiess, A., and Smaali, I. (2017). Variations in physicochemical properties and bioconversion efficiency of *Ulva lactuca* polysaccharides after different biomass pretreatment techniques. *Appl. Biochem. Biotechnol.*, 1–17. https://doi.org/10.1007/s12010-017-2588-z.

John, V.S. (2000). Implementing precision agriculture in the 21st century. *J. Agric. Eng. Res.*, **76**, 267–275.

Johnson, C.R. and Mann, K.H. (1986). The importance of plant defense abilities to the structure of subtidal seaweed communities: The Kelp Longzcrurzs De La Pylaie survives grazing by the snail *Lacuna vzncta* (montagu) at high population densities. *J. Exp. Mar. Biol. Ecol.*, **91**, 231–267.

Jong, E., Higson, A., Walsh, P., and Wellisch, M. (2012). Bio-based chemicals value added products from biorefineries. *IEA Bioenergy*. Final Report Task 42. https://www.ieabioenergy.com/publications/bio-based-chemicals-value-added-products-from-biorefineries/.

Jordan, P. and Vilter, H. (1991). Extraction of proteins from material rich in anionic mucilages: partition and fractionation of vanadate-dependent bromoperoxidases from the brown algae *Laminaria digitata* and *L. saccharina* in aqueous polymer two-phase systems. *Biochim. Biophys. Acta (BBA)-General Subj.*, **1073**, 98–106.

Jørgensen, S.E. (1990). Ecosystem theory, ecological buffer capacity, uncertainty and complexity. *Ecol. Modell.* https://doi.org/10.1016/0304-3800(90)90013-7.

Jørgensen, S.E. (2007). An integrated ecosystem theory. *Ann. Eur. Acad. Sci.*, 2006–2007, 19–33.

Jørgensen, S.E. (2015). New method to calculate the work energy of information and organisms. *Ecol. Modell.*, **295**, 18–20. https://doi.org/10.1016/j.ecolmodel.2014.09.001.

Jørgensen, S.E., Ladegaard, N., Debeljak, M., and Marques, J.C. (2005). Calculations of exergy for organisms. *Ecol. Modell.*, **185**, 165–175. https://doi.org/10.1016/j.ecolmodel.2004.11.020.

Jørgensen, S.E. and Mejer, H. (1977). Ecological buffer capacity. *Ecol. Modell.* https://doi.org/10.1016/0304-3800(77)90023-0.

Jung, K.A.A., Lim, S.-R.R., Kim, Y., and Park, J.M.M. (2013). Potentials of macroalgae as feedstocks for biorefinery. *Bioresour. Technol.*, **135**, 182–190. https://doi.org/10.1016/j.biortech.2012.10.025.

Kamalakannan, B., Joyson Joe Jeevamani, J., Arun Nagendran, N., Pandiaraja, D., and Chandrasekaran, S. (2014). Impact of removal of invasive species Kappaphycus alvarezii from coral reef ecosystem in Gulf of Mannar, India. *Curr. Sci.* https://doi.org/10.18520/cs/v106/i10/1401-1408.

Kamermans, P., Malta, E.-J., Verschuure, J.M., Schrijvers, L., Lentz, L.F., and Lien, A.T.A. (2002). Effect of grazing by isopods and amphipods on growth of *Ulva* spp. (Chlorophyta). *Aquat. Ecol.*, **36**, 425–433. https://doi.org/10.1023/A:1016551911754.

Kang, S.-R., Choi, J.-H., Kim, D.-W., Park, S.-E., Sapkota, K., Kim, S., and Kim, S.-J. (2016). A bifunctional protease from green alga *Ulva pertusa* with anti-coagulant properties: Partial purification and characterization. *J. Appl. Phycol.*, **28**, 599–607.

Kang, Y.H., Shin, J.A., Kim, M.S., and Chung, I.K. (2008). A preliminary study of the bioremediation potential of Codium fragile applied to seaweed integrated multi-trophic aquaculture (IMTA) during the summer. *J. Appl. Phycol.* **20**, 183–190. https://doi.org/10.1007/s10811-007-9204-5.

Kanno, K., Fujita, Y., Honda, S., Takahashi, S., and Kato, S. (2014). Urethane foam of sulfated polysaccharide ulvan derived from green-tide-forming chlorophyta: synthesis and application in the removal of heavy metal ions from aqueous solutions. *Polym. J.*, **46**, 813–818. https://doi.org/10.1038/pj.2014.70.

Karbalaei-Heidari, H.R., Shahbazi, M., and Absalan, G. (2013). Characterization of a novel organic solvent tolerant protease from a moderately halophilic bacterium and its behavior in ionic liquids. *Appl. Biochem. Biotechnol.*, **170**, 573–586. https://doi.org/10.1007/s12010-013-0215-1.

Karray, R., Hamza, M., and Sayadi, S. (2015). Evaluation of ultrasonic, acid, thermo-alkaline and enzymatic pre-treatments on anaerobic digestion of Ulva rigida for biogas production. *Bioresour. Technol.*, **187**, 205–213.

Karsten, U., Franklin, L.A., Tüning, K., and Wiencke, C. (1998). Natural ultra-violet radiation and photosynthetically active radiation induce formation of mycosporine-like amino acids in the marine macroalga *Chondrus crispus* (Rhodophyta). *Planta.* https://doi.org/10.1007/s004250050319.

Kazir, M., Abuhassira, Y., Robin, A., Nahor, O., Luo, J., Israel, A., Golberg, A., and Livney, Y.D. (2019). Extraction of proteins from two marine macroalgae, Ulva sp., and Gracilaria sp., for food application, and evaluating digestibility, amino acid composition and antioxidant properties of the protein concentrates. *Food Hydrocoll.*, **87**, 194–203. https://doi.org/10.1016/j.foodhyd.2018.07.047.

Kellogg, W.W. and Schware, R. (2019). Climate change and society: Consequences of increasing atmospheric carbon dioxide. *Climate Change and Society: Consequences of Increasing Atmospheric Carbon Dioxide.* Taylor and Francis. https://doi.org/10.4324/9780429048739.

Kelly, M. and Dworjanyn, S. (2008). The potential of marine biomass for anaerobic biogas production a feasibility study with recommendations for further research. The Crown Estate on Behalf of the Marine Estate, Scotland.

Kemper, N. (2008). Veterinary antibiotics in the aquatic and terrestrial environment. *Ecol. Indic.* https://doi.org/10.1016/j.ecolind.2007.06.002.

Kendel, M., Wielgosz-collin, G., Bertrand, S., Roussakis, C., Bourgougnon, N., and Bedoux, G. (2015). Lipid composition, fatty acids and sterols in the seaweeds *Ulva armoricana*, and *Solieria chordalis* from Brittany (France): An analysis from nutritional, chemotaxonomic, and antiproliferative activity perspectives. *Mar. Drugs.*, 5606–5628. https://doi.org/10.3390/md13095606.

Kerton, F.M., Liu, Y., Omari, K.W., and Hawboldt, K. (2013). Green chemistry and the ocean-based biorefinery. *Green Chem.*, **15**, 860–871. https://doi.org/10.1039/C3GC36994C.

Khambhaty, Y., Mody, K., Gandhi, M.R., Thampy, S., Maiti, P., Brahmbhatt, H., Eswaran, K., and Ghosh, P.K. (2012). *Kappaphycus alvarezii* as a source of bioethanol. *Bioresour. Technol.*, **103**, 180–185. https://doi.org/10.1016/j.biortech.2011.10.015.

Khambhaty, Y., Upadhyay, D., Kriplani, Y., Joshi, N., Mody, K., and Gandhi, M.R. (2013). Bioethanol from macroalgal biomass: Utilization of marine yeast for production of the same. *BioEnergy Res.*, **6**, 188–195. https://doi.org/10.1007/s12155-012-9249-4.

Kim, G.-Y., Seo, Y.H., Kim, I., and Han, J.-I. (2019). Co-production of biodiesel and alginate from *Laminaria japonica*. *Sci. Total Environ.*, **673**, 750–755. https://doi.org/10.1016/j.scitotenv.2019.04.049.

Kim, G.S., Shin, M.K., Kim, Y.J., Oh, K.K., Kim, J.S., Ryu, H.J., and Kim, K.H. (2014). Method of producing biofuel using sea algae. Patent. US8795994B2 https://patents.google.com/patent/US8795994/fr.

Kim, N.J., Li, H., Jung, K., Chang, H.N., and Lee, P.C. (2011). Ethanol production from marine algal hydrolysates using *Escherichia coli* KO11. *Bioresour. Technol.*, **102**, 7466–7469. https://doi.org/10.1016/j.biortech.2011.04.071.

Kissinger, G., Herold, M., and De Sy, V. (2012). A synthesis report for REDD+ policymakers drivers of deforestation and forest DegraDation drivers of deforestation and forest degradation: A synthesis report for REDD+ Policymakers [1], Kissinger, G Herold, M De Sy, Veronique.

Kivaisi, A.K. (2001). The potential for constructed wetlands for wastewater treatment and reuse in developing countries: A review. *Ecol. Eng.*, **16**, 545–560. https://doi.org/10.1016/S0925-8574(00)00113-0.

Kleitou, P., Kletou, D., and David, J. (2018). Is Europe ready for integrated multi-trophic aquaculture? A survey on the perspectives of European farmers and scientists with IMTA experience. *Aquaculture*, **490**, 136–148. https://doi.org/10.1016/J.AQUACULTURE.2018.02.035.

Koehl, M.A.R. (1999). Ecological biomechanics of benthic organisms: Life history, mechanical design and temporal patterns of mechanical stress. *J. Exp. Biol.*, **202**, 3469–3476.

Konda, N., Singh, S., Simmons B., Klein-Marcuschamer D. (2015). An investigation on the economic feasibility of macroalgae as a potential feedstock for biorefineries. *BioEnergy Research*, **8**, 1046–1056.

Korzen, L., Abelson, A., and Israel, A. (2015a). Growth, protein and carbohydrate contents in *Ulva rigida* and Gracilaria bursa-pastoris integrated with an offshore fish farm. *J. Appl. Phycol.* https://doi.org/10.1007/s10811-015-0691-5.

Korzen, L., Pulidindi, I.N., Israel, A., Abelson, A., and Gedanken, A. (2015b). Marine integrated culture of carbohydrate rich *Ulva rigida* for enhanced production of bioethanol. *RSC Adv.*, **5**, 59251–59256. https://doi.org/10.1039/C5RA09037G.

Korzen, L., Pulidindi, I.N., Israel, A., Abelson, A., and Gedanken, A. (2015c). Single step production of bioethanol from the seaweed *Ulva rigida* using sonication. *RSC Adv.*, **5**, 16223–16229.

Kostas, E.T., White, D.A., and Cook, D.J. (2017). Development of a bio-refinery process for the production of speciality chemical, biofuel and bioactive compounds from *Laminaria digitata*. *Algal Res.*, **28**, 211–219. https://doi.org/10.1016/j.algal.2017.10.022.

Kraan, S. (2013). Pigments and minor compounds in algae, in: *Functional Ingredients from Algae for Foods and Nutraceuticals*. https://doi.org/10.1533/9780857098689.1.205.

Kraan, S. and Guiry, M.D. (2001). Phase II: Strain hybridisation field experiments and genetic fingerprinting of the edible brown seaweed *Alaria esculenta*.

Kratky, L. and Jirout, T. (2011). Biomass size reduction machines for enhancing biogas production. *Chem. Eng. Techn.*, **34**, 391–399. https://doi.org/10.1002/ceat.201000357.

Krishnan, M. and Narayanakumar, R. (2013). Social and economicdimensions of carrageenanseaweed farming.

Kruse, A. and Faquir, M. (2007). Hydrothermal biomass gasification — Effects of salts, backmixing and their interaction. *Chem. Eng. Technol.*, **30**, 749–754. https://doi.org/10.1002/ceat.200600409.

Kruse, A., Forchheim, D., Gloede, M., Ottinger, F., and Zimmermann, J. (2010). Brines in supercritical biomass gasification: 1. Salt extraction by salts and the influence on glucose conversion. *J. Supercrit. Fluids,* **53**, 64–71. https://doi.org/10.1016/J.SUPFLU.2010.01.001.

Kuda, T., Tsunekawa, M., Goto, H., and Araki, Y. (2005). Antioxidant properties of four edible algae harvested in the Noto Peninsula, Japan, *Journal of Food*

Composition and Analysis, **18**(7), 625–633. https://doi.org/10.1016/j. jfca.2004.06.015.

Kumar, M., Kumari, P., Gupta, V., Reddy, C.R.K., and Jha, B. (2010). Biochemical responses of red alga *Gracilaria corticata* (*Gracilariales*, Rhodophyta) to salinity induced oxidative stress. *J. Exp. Mar. Bio. Ecol.* https://doi. org/10.1016/j.jembe.2010.06.001.

Kumar, S., Gupta, R., Kumar, G., Sahoo, D., and Kuhad, R.C. (2013). Bioethanol production from Gracilaria verrucosa, a red alga, in a biorefinery approach. *Bioresour. Technol.*,**135**,150–156.https://doi.org/10.1016/j.biortech.2012.10.120.

Kumari, P., Reddy, C.R.K., and Jha, B. (2011). Comparative evaluation and selection of a method for lipid and fatty acid extraction from macroalgae. *Anal. Biochem.*, **415**, 134–144. https://doi.org/10.1016/j.ab.2011.04.010.

Lamai, C., Kruatrachue, M., Pokethitiyook, P., Suchart Upatham, E., and Soonthornsarathool, V. (2005). Toxicity and accumulation of lead and cadmium in the filamentous green alga *Cladophora fracta* (O.F. Muller ex Vahl) Kutzing: A Laboratory Study, ScienceAsia.

Largo, D.B., Fukami, K., Nishijima, T., and Ohno, M. (1995). Laboratory-induced development of the ice-ice disease of the farmed red algae *Kappaphycus alvarezii* and Eucheuma denticulatum (*Solieriaceae, Gigartinales, Rhodophyta*). *J. Appl. Phycol.*, 7, 539–543.

Larsen, T.H., Williams, N.M., and Kremen, C. (2005). Extinction order and altered community structure rapidly disrupt ecosystem functioning. *Ecol. Lett.* https://doi.org/10.1111/j.1461-0248.2005.00749.x.

Lee, Y.R. and Row, K.H. (2016). Comparison of ionic liquids and deep eutectic solvents as additives for the ultrasonic extraction of astaxanthin from marine plants. *J. Ind. Eng. Chem.*, **39**, 87–92. https://doi.org/https://doi.org/10.1016/j. jiec.2016.05.014.

Legendre, L. (1990). The significance of microalgal blooms for fisheries and for the export of particulate organic carbon in oceans. *J. Plankton Res.*

Lehahn, Y., Ingle, K.N., and Golberg, A. (2016a). Global potential of offshore and shallow waters macroalgal biorefineries to provide for food, chemicals and energy: Feasibility and sustainability. *Algal Res.*, **17**, 150–160. https:// doi.org/10.1016/j.algal.2016.03.031.

Lehahn, Y., Ingle, K.N., and Golberg, A. (2016b). Global potential of offshore and shallow waters macroalgal biorefineries to provide for food, chemicals and energy: Feasibility and sustainability. *Algal Res.*, **17**, 150–160. https:// doi.org/10.1016/j.algal.2016.03.031.

Lei, Z., Li, C., and Chen, B. (2003). Extractive distillation: A review. *Sep. Purif. Rev.*, **32**, 121–213.

Lei, Z., Wang, H., Zhou, R., and Duan, Z. (2002). Influence of salt added to solvent on extractive distillation. *Chem. Eng. J.*, **87**, 149–156. https://doi.org/ http://dx.doi.org/10.1016/S1385-8947(01)00211-X.

Lenstra, W., van Hal, J., and Reith, J. (2011). Ocean seaweed biomass. For large scale biofuel production, in: *Ocean Seaweed Biomass*, Bremerhaven, Germany.

Leonelli, C. and Mason, T.J. (2010). Microwave and ultrasonic processing: Now a realistic option for industry. *Chem. Eng. Proc. Proc. Intensif.*, **49**, 885–900. https://doi.org/10.1016/J.CEP.2010.05.006.

Lepage, G. and Roy, C.C. (1984). Improved recovery of fatty acid through direct transesterification without prior extraction or purification. *J. Lipi. Res.*, **25**, 1391–1396.

Lepers, E., Lambin, E.F., Janetos, A.C., DeFries, R., Achard, F., Ramankutty, N., and Scholes, R.J. (2005). A synthesis of information on rapid land-cover change for the period 1981–2000. *Bioscience*. https://doi.org/10.1641/0006-3568(2005)055[0115:asoior]2.0.co;2.

Li, H., Qu, Y., Yang, Y., Chang, S., and Xu, J. (2016a). Microwave irradiation — A green and efficient way to pretreat biomass. *Bioresour. Technol.*, **199**, 34–41. https://doi.org/https://doi.org/10.1016/j.biortech.2015.08.099.

Li, H., Qu, Y., Yang, Y., Chang, S., and Xu, J. (2016b). Microwave irradiation — A green and efficient way to pretreat biomass. *Bioresour. Technol.*, **199**, 34–41. https://doi.org/https://doi.org/10.1016/j.biortech.2015.08.099.

Li, H.M., Zhang, C.S., Han, X.R., and Shi, X.Y. (2015). Changes in concentrations of oxygen, dissolved nitrogen, phosphate, and silicate in the southern Yellow Sea, 1980–2012: Sources and seaward gradients. *Estuar. Coast. Shelf Sci*. https://doi.org/10.1016/j.ecss.2014.12.013.

Lim, S.-F., Zheng, Y.-M., Zou, S.-W., and Chen, J.P. (2008). Characterization of copper adsorption onto an alginate encapsulated magnetic sorbent by a combined FT-IR, XPS, and mathematical modeling study. *Environ. Sci. Technol*. https://doi.org/10.1021/es7021889.

Lincoln, R.A., Strupinski, K., and Walker, J.M. (1991). Bioactive compounds from algae. *Life Chem. Repor.*, **8**, 97–183.

Lindequist, U. (2016). Marine-derived pharmaceuticals — challenges and opportunities **24**, 561–571.

Lindsey Zemke-White, W., and Ohno, M. (1999). World seaweed utilisation: An end-of-century summary. *J. Appl. Phyco.*, **11**, 369–376.

Liquidity Crisis (2016). *Econ. Artic. Appear. Brief. Sect.* print Ed. under *Headl. Liq. Cris.* 5 November.

Lirasan, T. and Twide, P. (2013). Fourteenth international seaweed symposium: Proceedings of the fourteenth international seaweed symposium held in brest, *Google Books*. Springer, pp. 353–355.

Liu, D., Keesing, J.K., Dong, Z., Zhen, Y., Di, B., Shi, Y., Fearns, P., and Shi, P. (2010). Recurrence of the world's largest green-tide in 2009 in Yellow Sea, China: Porphyra yezoensis aquaculture rafts confirmed as nursery for macroalgal blooms. *Mar. Pollut. Bull.*, **60**, 1423–1432. https://doi.org/10.1016/j.marpolbul.2010.05.015.

Liu, D., Keesing, J.K., Xing, Q., and Shi, P. (2009). World's largest macroalgal bloom caused by expansion of seaweed aquaculture in China. *Mar. Pollut. Bull.*, **58**, 888–895. https://doi.org/10.1016/j.marpolbul.2009.01.013.

Liu, K. (2019). Effects of sample size, dry ashing temperature and duration on determination of ash content in algae and other biomass. *Algal Res.*, **40**, 101486. https://doi.org/10.1016/j.algal.2019.101486.

Lo, W.-T., Purcell, J.E., Hung, J.-J., Su, H.-M., and Hsu, P.-K. (2008). Enhancement of jellyfish (*Aurelia aurita*) populations by extensive aquaculture rafts in a coastal lagoon in Taiwan. *ICES J. Mar. Sci.*, **65**, 453–461. https://doi.org/10.1093/icesjms/fsm185.

Lobban, C.S. and Harrison, P.J. (1994). Seaweed ecology and physiology. *Seaweed Ecol. Physiol.* https://doi.org/10.2307/2261617.

Lombrana, J. (2009). Fundamentals and tendencies in freez-drying of foods, in: Ratti, C. (Ed.), *Advances in Food Dehydration*. CRC Press., pp. 209–235. Boca Raton. FL.

Loureiro, R., Gachon, C.M.M., and Rebours, C. (2015). Seaweed cultivation : Potential and challenges of crop domestication at an unprecedented pace. *New Phytol.*, **206**, 489–492. https://doi.org/10.1111/nph.13278.

Lourenço, S.O., Barbarino, E., De-paula, J.C., Otávio, L., Pereira, S., Marquez, U.M.L., Marinha, D.D.B., Fluminense, U.F., and Postal, C. (2002). Amino acid composition, protein content and calculation of nitrogen-to-protein conversion factors for 19 tropical seaweeds. *Phycol. Res.*, **50**, 233–241.

Løvås, S.M. and Tørum, A. (2001). Effect of the kelp *Laminaria hyperborea* upon sand dune erosion and water particle velocities. *Coast. Eng.*, **44**, 37–63. https://doi.org/10.1016/S0378-3839(01)00021-7.

Lowry, O H., Rosebrough N J, Farr A L, Randall R J. (1951) Protein measurement with the folin phenol reagent. *J Biol Chem.*, **193**(1), 265–275.

Luo, L., Liu, F., Xu, Y., and Yuan, J. (2011). Hydrodynamics and mass transfer characteristics in an internal loop airlift reactor with different spargers. *Chem. Eng. J.*, **175**, 494–504. https://doi.org/10.1016/j.cej.2011.09.078.

Lyons, H. (2009). A review of the potential of marine algae as a source of biofuel in Ireland. FAO Report. http://www.fao.org/uploads/media/0902_SEI_-_A_Review_of_the_Potential_of_Marine_Algae.pdf.

Maceiras, R., Rodrí guez, M., Cancela, A., Urréjola, S., and Sánchez, A. (2011). Macroalgae: Raw material for biodiesel production. *Appl. Energ.*, **88**, 3318–3323. https://doi.org/10.1016/j.apenergy.2010.11.027.

Magnusson, M., Carl, C., Mata, L., de Nys, R., and Paul, N.A. (2016). Seaweed salt from Ulva: A novel first step in a cascading biorefinery model. *Algal Res.*, **16**, 308–316. https://doi.org/10.1016/j.algal.2016.03.018.

Mahmoud, A. (2010). Electrical field: A historical review of its application and contributions in wastewater sludge dewatering Electrical field : A historical review of its application and contributions in wastewater sludge dewatering. *Water Res.*, **44**, 2381–2407. https://doi.org/10.1016/j.watres.2010.01. 033.

Maiorella, B.L., Blanch, H.W., and Wilke, C.R. (1984). Feed component inhibition in ethanolic fermentation by Saccharomyces cerevisiae. *Biotechnol. Bioeng.*, **26**, 1155–1166. https://doi.org/10.1002/bit.260261004.

Malihan, L.B., Nisola, G.M., Mittal, N., Seo, J.G., and Chung, W.-J. (2014). Blended ionic liquid systems for macroalgae pretreatment. *Renew. Energ.*, **66**, 596–604. https://doi.org/10.1016/j.renene.2014.01.003.

Malinowski, J.J. and Daugulis, A.J. (1994). Salt effects in extraction of ethanol, 1-butanol and acetone from aqueous solutions. *AIChE J.*, **40**, 1459–1465.

Mandal, P., Pujol, C., Carlucci, M., Chattopadhyay, K., Damonte, E., and Ray, B. (2008). Anti-herpetic activity of a sulfated xylomannan from *Scinaia hatei*. *Phytochemistry*, **69**, 2193–2199.

Maneein, S., Milledge, J.J., Nielsen, B. V., and Harvey, P.J. (2018). A Review of Seaweed Pre-Treatment Methods for Enhanced Biofuel Production by Anaerobic Digestion or Fermentation. Fermentation. https://doi.org/10.3390/fermentation4040100.

Mann, K.H. (1991). Seaweeds: Their environment, biogeography, and ecophysiology. *Limnol. Oceanogr.* https://doi.org/10.4319/lo.1991.36.5.1066.

Manns, D., Andersen, S.K., Saake, B., and Meyer, A.S. (2016). Brown seaweed processing: enzymatic saccharification of Laminaria digitata requires no pretreatment. *J. Appl. Phycol.*, **28**, 1287–1294.

Margesin, R. and Schinner, F. (2001). Potential of halotolerant and halophilic microorganisms for biotechnology. *Extremophiles*, **5**, 73–83. https://doi.org/10.1007/s007920100184.

Marinho-Soriano, E., Moreira, W.S.C., and Carneiro, M.A. (2006). Some Aspects of the Growth of Gracilaria birdiae (Gracilariales, Rhodophyta) in an Estuary in Northeast Brazil. *Aquac. Int.*, **14**, 327–336. https://doi.org/10.1007/s10499-005-9032-z.

Marinho, G.S., Alvarado-Morales, M., and Angelidaki, I. (2016). Valorization of macroalga Saccharina latissima as novel feedstock for fermentation-based succinic acid production in a biorefinery approach and economic aspects. *Algal Res.*, **16**, 102–109. https://doi.org/10.1016/j.algal.2016.02.023.

Marinho, G.S., Holdt, S.L., and Angelidaki, I. (2015). Seasonal variations in the amino acid profile and protein nutritional value of Saccharina latissima cultivated in a commercial IMTA system. *J. Appl. Phycol.*, **27**, 1991–2000. https://doi.org/10.1007/s10811-015-0546-0.

Marroig, R.G. and Reis, R.P. (2015). Biofouling in Brazilian commercial cultivation of Kappaphycus alvarezii (Doty) Doty ex P. C. Silva. *J. Appl. Phycol.* https://doi.org/10.1007/s10811-015-0713-3.

Martín, L.A., Zaixso, A.L.B. De, and Leonardi, P.I. (2013). Epiphytism in a subtidal natural bed of Gracilaria gracilis of southwestern Atlantic coast

(Chubut, Argentina) 2003, 1319–1329. https://doi.org/10.1007/s10811-012-9961-7.

Martínez-Aragón, J.F., Hernández, I., Pérez-Lloréns, J.L., Vázquez, R., and Vergara, J.J. (2002). Biofiltering efficiency in removal of dissolved nutrients by three species of estuarine macroalgae cultivated with sea bass (Dicentrarchus labrax) waste waters 1. Phosphate. *J. Appl. Phycol.*, **14**, 365–374. https://doi.org/10.1023/A:1022134701273.

Martins, M., Vieira, F.A., Correia, I., Ferreira, R.A.S., Abreu, H., Coutinho, J.A.P., and Ventura, S.P.M. (2016). Recovery of phycobiliproteins from the red macroalga: Gracilaria sp. using ionic liquid aqueous solutions. *Green Chem.*, **18**, 4287–4296. https://doi.org/10.1039/c6gc01059h.

Maschek, J.A. and Baker, B.J. (2008). The chemistry of algal secondary metabolism, in: *Algal Chemi. Ecology.* https://doi.org/10.1007/978-3-540-74181-7_1.

Mason, D.M. and Gandhi, K. (1980). Formulas for calculating the heating value of coal and coal char: Development, tests and uses.

Mason, T.J., Chemat, F., and Vinatoru, M. (2011). The extraction of natural products using ultrasound or microwaves. *Curr. Org. Chem.*, **15**, 237–247.

Masri, M.A., Jurkowski, W., Shaigani, P., Haack, M., Mehlmer, N., and Brück, T. (2018). A waste-free, microbial oil centered cyclic bio-refinery approach based on flexible macroalgae biomass. *Appl. Ener.*, **224**, 1–12. https://doi.org/10.1016/J.APENERGY.2018.04.089.

Mata, L., Magnusson, M., Paul, N.A., and de Nys, R. (2016). The intensive land-based production of the green seaweeds *Derbesia tenuissima* and *Ulva ohnoi*: Biomass and bioproducts. *J. Appl. Phycol.*, **28**, 365–375. https://doi.org/10.1007/s10811-015-0561-1.

Mata, Y.N., Torres, E., Blázquez, M.L., Ballester, A., González, F., and Muñoz, J.A. (2009). Gold(III) biosorption and bioreduction with the brown alga *Fucus vesiculosus. J. Hazard. Mater.*, **166**, 612–618. https://doi.org/10.1016/j.jhazmat.2008.11.064.

Matsuhiro, B., Conte, A., Damonte, E., Kolender, A., Matulewicz, M., Mejías, E., Pujol, C., and EA., Z. (2005). Structural analysis and antiviral activity of a sulfated galactan from the red seaweed *Schizymenia binderi. Carboh. Res.*, **340**(15).

McGlade, J., Werner, B., Young, M., Matlock, M., Jefferies, D., Sonnemann, G., Aldaya, M., Pfister, S., Berger, M., Farell, C., Hyde, K., Wackernagel, M., Hoekstra, A., Mathews, R., Liu, J., Ercin, E., Weber, J.L., Alfieri, A., Martinez-Lagunes, R., and Edens, N. UNEP. (2012). Measuring Water Use in a Green Economy, A Report of the Working Group on Water Efficiency to the International Resource Panel. United Nations Environment Programme 9789280732207 https://research.utwente.nl/en/publications/measuring-water-use-in-a-green-economy-a-report-of-the-working-gr.

McGlathery, K., Sundbäck, K., and Anderson, I. (2007). Eutrophication in shallow coastal bays and lagoons: the role of plants in the coastal filter. *Mar. Ecol. Prog. Ser.*, **348**, 1–18. https://doi.org/10.3354/meps07132.

McHugh, D.J. (2003). A Guide to the Seaweed Industry. FAO Fisheries Technical Paper.

McHugh, D.J. (2006). The Seaweed Industry in the Pacific Islands. ACIAR Work. Paper.

Mcleod, E., Anthony, K.R., Andersson, A., Beeden, R., Golbuu, Y., Kleypas, J., Kroeker, K., Manzello, D., Salm, R. V, Schuttenberg, H., and Smith, J.E. (2013). Preparing to manage coral reefs for ocean acidification: Lessons from coral bleaching. *Front. Ecol. Environ.*, **11**, 20–27. https://doi.org/10. 1890/110240.

Meinita, M.D.N., Kang, J.Y., Jeong, G.T., Koo, H.M., Park, S.M., and Hong, Y.K. (2012). Bioethanol production from the acid hydrolysate of the carrageenophyte *Kappaphycus alvarezii* (cottonii). *J. Appl. Phycol.*, **24**, 857–862. https://doi.org/10.1007/s10811-011-9705-0.

Mekonnen, M.M. and Hoekstra, A.Y. (2016). Four billion people facing severe water scarcity. *Am. Assoc. Adv. Sci.* **2**, e1500323.

Mendez, F.J. and Losada, I.J. (2004). An empirical model to estimate the propagation of random breaking and nonbreaking waves over vegetation fields. *Coast. Eng.*, **51**, 103–118. https://doi.org/10.1016/J.COASTALENG.2003.11.003.

Merrill, A.L. and Watt., B.K. (173AD). Energy value of foods: Basis and derivation. ARS United States Department of the Agriculture.

Messyasz, B., Michalak, I., Łęska, B., Schroeder, G., Górka, B., Korzeniowska, K., Lipok, J., Wieczorek, P., Rój, E., Wilk, R., Dobrzyńska-Inger, A., Górecki, H., and Chojnacka, K. (2017). Valuable natural products from marine and freshwater macroalgae obtained from supercritical fluid extracts. *J. Appl. Phycol.*, 1–13. https://doi.org/10.1007/s10811-017-1257-5.

Messyasz, B., Michalak, I., and Schroeder, G. (2018). Valuable natural products from marine and freshwater macroalgae obtained from supercritical fluid extracts, 591–603. https://doi.org/10.1007/s10811-017-1257-5.

Mhatre, A.A., Gore, S., Mhatre, A.A., Trivedi, N., Sharma, M., Pandit, R., Anil, A., and Lali, A. (2018). *AC SC. Renew. Energy.* https://doi.org/10.1016/j. renene.2018.08.012.

Mhatre, A., Gore, S., Mhatre, A., Trivedi, N., Sharma, M., Pandit, R., Anil, A., and Lali, A. (2019). Effect of multiple product extractions on bio-methane potential of marine macrophytic green alga *Ulva lactuca. Renew. Ener.*, **132**, 742–751.

Michalak, I., Tuhy, Ł., and Chojnacka, K. (2015). Seaweed extract by microwave assisted extraction as plant growth biostimulant. *Open Chem.* **13**, 1183–1195.

Michler-Cieluch, T, Krause, G. and Buck, B.B. Marine Aquaculture within Offshore Wind Farms: Social Aspects of Multiple-Use Planning GAIA — Ecological Perspectives on Science and Society, **18**(2): 158–162.

Michler-Cieluch, T., Krause, G., and Buck, B.H. (2009). Reflections on integrating operation and maintenance activities of offshore wind farms and mariculture. *Oce. Coast. Manag.*, **52**, 57–68. https://doi.org/10.1016/j.ocecoaman. 2008.09.008.

Micklin, P. (2007). The aral sea disaster. *Annu. Rev. Ear. Planet. Sci.* https://doi. org/10.1146/annurev.earth.35.031306.140120.

Micklin, P.P. (1988). Desiccation of the aral aea: A water management disaster in the Soviet Union. *Science*, **241** (4870), 1170–1176. https://doi.org/10.1126/ science.241.4870.1170.

Milich, M. and Drimer, N. (2019). Design and analysis of an innovative concept for submerging open-sea aquaculture system. *IEEE J. Ocean. Eng.*, **44**, 707–718. https://doi.org/10.1109/JOE.2018.2826358.

Milledge, J., Smith, B., Dyer, P., and Harvey, P. (2014). Macroalgae-derived biofuel: A review of methods of energy extraction from seaweed biomass. *Energies*, 7, 7194–7222. https://doi.org/10.3390/en7117194.

Milledge, J.J. and Harvey, P.J. (2016). Potential process 'hurdles' in the use of macroalgae as feedstock for biofuel production in the British Isles. *J. Chem. Technol. Biotechnol.*, **91**, 2221–2234. https://doi.org/10.1002/jctb.5003.

Milledge, J.J., Nielsen, B. V., and Bailey, D. (2016). High-value products from macroalgae: the potential uses of the invasive brown seaweed, *Sargassum muticum. Rev. Environ. Sci. Biotechnol.*, **15**, 67–88. https://doi.org/10.1007/ s11157-015-9381-7.

Mitsch, W.J. and Jørgensen, S.E. (2003). Ecological engineering: A field whose time has come. *Ecol. Eng.*, **20**, 363–377. https://doi.org/10.1016/J. ECOLENG.2003.05.001.

Mittal, R., Sharma, R., and Raghavarao, K. (2019). Aqueous two-phase extraction of R-Phycoerythrin from marine macro-algae, *Gelidium pusillum. Bioresour. Technol.*, **280**, 277–286.

Möller, B., Hong, L., Lonsing, R., and Hvelplund, F. (2012). Evaluation of offshore wind resources by scale of development. *Energy*, **48**, 314–322. https:// doi.org/10.1016/J.ENERGY.2012.01.029.

Monagail, M.M., Cornish, L., Morrison, L., Araújo, R., and Critchley, A.T. (2017). Sustainable harvesting of wild seaweed resources. *Eur. J. Phycol.*, **52**, 371–390. https://doi.org/10.1080/09670262.2017.1365273.

Monteiro, C.M., Castro, P.M.L., and Malcata, F.X. (2009). Biosorption of zinc ions from aqueous solution by the microalga *Scenedesmus obliquus. Env. Chem Lett.*, **9**, 169–176. https://doi.org/10.1007/s10311-009-0258-2.

Monteiro, C.M., Castro, P.M.L., and Malcata, F.X. (2012). Metal uptake by microalgae: Underlying mechanisms and practical applications. *Biotechnol. Prog.*, **28**, 299–311. https://doi.org/10.1002/btpr.1504.

Montingelli, M. (2015). *Development and Application of a Mechanical Pretreatment to Increase the Biogas Produced from Irish Macroalgal Biomass*. Dublin City University.

Moore, C.M., Mills, M.M., Arrigo, K.R., Berman-Frank, I., Bopp, L., Boyd, P.W., Galbraith, E.D., Geider, R.J., Guieu, C., Jaccard, S.L., Jickells, T.D., La Roche, J., Lenton, T.M., Mahowald, N.M., Marañón, E., Marinov, I., Moore, J.K., Nakatsuka, T., Oschlies, A., Saito, M.A., Thingstad, T.F., Tsuda, A., and Ulloa, O. (2013). Processes and patterns of oceanic nutrient limitation. *Nat. Geosci.*, **6**, 701–710. https://doi.org/10.1038/ngeo1765.

Morand, P. (2005). Macroalgal population and sustainability. *J. Coast. Res.*, **21**(5), 1009–1020. https://doi.org/10.2112/04-700a.1.

Moreno-Mateos, D., Barbier, E.B., Jones, P.C., Jones, H.P., Aronson, J., López-López, J.A., McCrackin, M.L., Meli, P., Montoya, D., and Rey Benayas, J.M. (2017). Anthropogenic ecosystem disturbance and the recovery debt. *Nat. Commun.*, **20**(8), 14163. https://doi.org/10.1038/ncomms14163.

Morrissey, J., Kraan, S., Guiry, M.D., and Ryan, M. (2001). *A Guide to Commercially Important Seaweeds on the Irish Coast*. Bord Iascaigh Mhara/ Irish Sea Fisheries Board, Co. Dublin, pp. 1–64.

Morse, G., Brett, S., Guy, J., and Lester, J. (1998). Review: Phosphorus removal and recovery technologies. *Sci. Total. Environ.*, **212**, 69–81. https://doi.org/10.1016/S0048-9697(97)00332-X.

Moshfegh, M., Shahverdi, A.R., Zarrini, G., and Faramarzi, M.A. (2013). Biochemical characterization of an extracellular polyextremophilic α-amylase from the halophilic archaeon *Halorubrum xinjiangense*. *Extremophiles*, **17**, 677–687. https://doi.org/10.1007/s00792-013-0551-7.

Mouritsen, O.G., Dawczynski, C., Duelund, L., Jahreis, G., Vetter, W., and Schröder, M. (2013). On the human consumption of the red seaweed dulse (Palmaria palmata (L.) Weber & Mohr). *J. Appl. Phycol.*, **25**, 1777–1791. https://doi.org/10.1007/s10811-013-0014-7.

Msuya, F.E. and Neori, A. (2008). Effect of water aeration and nutrient load level on biomass yield, N uptake and protein content of the seaweed *Ulva lactuca* cultured in seawater tanks. *J. Appl. Phycol.*, **20**, 1021–1031. https://doi.org/10.1007/s10811-007-9300-6.

Mulazzani, L. and Malorgio, G. (2017). Blue growth and ecosystem services. *Mar. Policy.*, **85**, 17–24. https://doi.org/10.1016/j.marpol.2017.08.006.

Murphy, C., Hotchkiss, S., McKeown and Worthington, J. (2014). The potential of seaweed as a source of drugs for use in cancer chemotherapy. *Appl. Phycol.*, **26**, 2211–2264.

Murray, B., Pendleton, L., Jenkins, W., and Sifleet, S. (2011). *Green Payments for Blue Carbon: Economic Incentives for Protecting Threatened Coastal Habitats*. Nicholas Institute Environmental Policy Solutions.

Mussatto, S.I., Dragone, G., Guimarães, P.M.R., Silva, J.P.A., Carneiro, L.M., Roberto, I.C., Vicente, A., Domingues, L., and Teixeira, J.A. (2010).

Technological trends, global market, and challenges of bio-ethanol production. *Biotechnol. Adv.*, **28**, 817–830. https://doi.org/10.1016/j.biotechadv. 2010.07.001.

Mwangi, I.W. and Ngila, J.C. (2012). Removal of heavy metals from contaminated water using ethylenediamine-modified green seaweed (Caulerpa serrulata). *Phys. Chem. Earth.*, **50–52**, 111–120. https://doi.org/10.1016/j. pce.2012.08.015.

Nardelli, A.E., Chiozzini, V.G., Braga, E.S., and Chow, F. (2019). Integrated multi-trophic farming system between the green seaweed *Ulva lactuca*, mussel, and fish: A production and bioremediation solution. *J. Appl. Phycol.*, **31**, 847–856. https://doi.org/10.1007/s10811-018-1581-4.

Nazar, A., Anikuttan, P., and Kumar Samaland Sankar, A. (2019). Integrated Multi-Trophic Aquaculture (IMTA). A technology for uplifting rural economy and bio-mitigation for environmental sustainability Aquaculture Spectrum, **2**(5). 9–18.

Neethling, J.B. and Kennedy, H. (2018). potential nutrient reduction by treatment optimization, side-stream treatment, treatment upgrades, and other means. BACWA. Nutrient reduction study. https://bacwa.org/wp-content/uploads/2018/06/BACWA_Final_Nutrient_Reduction_Report.pdf.

Neori, A., Chopin, T., Troell, M., Buschmann, A.H., Kraemer, G.P., Halling, C., Shpigel, M., and Yarish, C. (2004). Integrated aquaculture: Rationale, evolution and state of the art emphasizing seaweed biofiltration in modern mariculture. *Aquaculture*. https://doi.org/10.1016/j.aquaculture.2003.11.015.

Neushul, M., Benson, J., Harger, B.W.W., and Charters, A.C. (1992). Macroalgal farming in the sea: water motion and nitrate uptake. *J. Appl. Phycol.*, **4**, 255–265. https://doi.org/10.1007/BF02161211.

Neushul, P. (1989). Seaweed for War: California's World War I Kelp Industry Linked references are available on JSTOR for this article: Seaweed for war: California's World War I Kelp Industry. *Technol. Cult.*, **30**, 561–583.

Neveux, N., Bolton, J.J., Bruhn, A., Roberts, D.A., and Ras, M. (2018). The bioremediation potential of seaweeds: Recycling nitrogen, phosphorus, and other waste products. *Blue Biotechnology*, Wiley Online Books. https://doi.org/doi:10.1002/9783527801718.ch7.

Neveux, N., Yuen, A.K.L., Jazrawi, C., He, Y., Magnusson, M., Haynes, B.S., Masters, A.F., Montoya, A., Paul, N.A., Maschmeyer, T., and de Nys, R. (2014). Pre- and post-harvest treatment of macroalgae to improve the quality of feedstock for hydrothermal liquefaction. *Algal Res.*, **6**, 22–31. https://doi.org/10.1016/J.ALGAL.2014.08.008.

Neveux, N, Yuen, A.K.L., Jazrawi, C., Magnusson, M., Haynes, B.S., Masters, A.F., Montoya, A., Paul, N.A., Maschmeyer, T., and de Nys, R. (2014). Biocrude yield and productivity from the hydrothermal liquefaction of marine and freshwater green macroalgae. *Bioresour. Technol.*, **155**, 334–341. https://doi.org/https://doi.org/10.1016/j.biortech.2013.12.083.

Newbold, T., Hudson, L.N., Hill, S.L.L., Contu, S., Lysenko, I., Senior, R.A., Börger, L., Bennett, D.J., Choimes, A., Collen, B., Day, J., De Palma, A., Díaz, S., Echeverria-Londoño, S., Edgar, M.J., Feldman, A., Garon, M., Harrison, M.L.K., Alhusseini, T., Ingram, D.J., Itescu, Y., Kattge, J., Kemp, V., Kirkpatrick, L., Kleyer, M., Correia, D.L.P., Martin, C.D., Meiri, S., Novosolov, M., Pan, Y., Phillips, H.R.P., Purves, D.W., Robinson, A., Simpson, J., Tuck, S.L., Weiher, E., White, H.J., Ewers, R.M., MacE, G.M., Scharlemann, J.P.W., and Purvis, A. (2015). Global effects of land use on local terrestrial biodiversity. *Nature.* https://doi.org/10.1038/nature14324.

Ng, P.O.A. (2002). Evaluation of antioxidative activity of extracts from a brown seaweed, *Sargassum siliquastrum*, 3862–3866. https://doi.org/10.1021/jf020096b.

Nielsen, M.M., Bruhn, A., Rasmussen, M.B., Olesen, B., Larsen, M.M., and Møller, H.B. (2012). Cultivation of *Ulva lactuca* with manure for simultaneous bioremediation and biomass production. *J. Appl. Phycol.*, **24**, 449–458. https://doi.org/10.1007/s10811-011-9767-z.

Niesenbaum, R.A. (1988). The ecology of sporulation by the macroalga *Ulva lactuca* L. (chlorophyceae). *Aquat. Bot.*, **32**, 155–166. https://doi.org/10.1016/0304-3770(88)90095-2.

Nikolaisen, L., Daugbjerg Jensen, P., Svane Bech, K., Dahl, J., Busk, J., Brødsgaard, T., Ramussen, M.B., Bruhn, A., Bjerre, A.-B., Bangsøe Nielsen, H., Rost Albert, K., Ambus, P., Kadar, Z., Heiske, S., Sander, B., and Ravn Schmidt, E. (2011a). Energy Production from Marine Biomass (*Ulva lactuca*). PSO Proj. No. 2008-1-0050 282–288. https://doi.org/10.1111/j.1399-302X.1996.tb00182.x.

Nikolaisen, L., Daugbjerg Jensen, P., Svane Bech, K., Dahl, J., Busk, J., Brødsgaard, T., Rasmussen, M.B., Bruhn, A., Bjerre, A.-B., and Bangsø Nielsen, H. (2011b). *Energy Production from Marine Biomass (Ulva lactuca)*. Danish Technological Institute.

Nikolaisen, L., Jensen, P.D., Bech, K.S., Dahl, J., Bus, J., Brødsgaard, T., Rasmussen, M.B., Bruhn, A., Bjerre, A.-B., Nielsen, H.B., Albert, K.R., Ambus, P., Kadar, Z., Heiske, S., Sander, B., and Schmidt, E.R. (2011c). *Energy Production from Marine Biomass (Ulva lactuca)*, PSO Project No. 2008-1-0050.

Nikolaisen, L., Jensen, P.D., Bech, K.S., Dahl, J., Busk, J., Brodsgaard, T., Bo, R.M., Bruhn, A., Bjerre, A.B., Nielsen, H.B., Albert, K.R., Ambus, P., Kadar, Z., Heiske, S., Sander, B., and Schmidt, E.R. (2008). *Energy Production from Marine Biomass (Ulva lactuca)*. PSO Project No. 2008-1-0050 1–72.

Nixon, S.W. (2009). Eutrophication and the macroscope. *Hydrobiologia.*, **629**, 5–19. https://doi.org/10.1007/s10750-009-9759-z.

Nixon, S.W. (1995). Coastal Marine Eutrophication: A definition, Social causes, and future concerns. *OPHELIA*, **41**, 199–219.

Noreen, A., Mahmood, K., Zuber, M., and Ali, M. (2016). A critical review of algal biomass : A versatile platform of bio-based polyesters from renewable resources. *Int. J. Biol. Macromol.*, **86**, 937–949. https://doi.org/10.1016/j.ijbiomac.2016.01.067.

Notoya, M. (2010). *Production of Biofuel by Macroalgae with Preservation of Marine Resources and Environment.* Springer, Dordrecht, pp. 217–228. https://doi.org/10.1007/978-90-481-8569-6_13.

Nourbakhsh, M., Sağ, Y., Özer, D., Aksu, Z., Kutsal, T., and Çağlar, A. (1994). A comparative study of various biosorbents for removal of chromium(VI) ions from industrial waste waters. *Proc. Biochem.*, **29**, 1–5. https://doi.org/10.1016/0032-9592(94)80052-9.

Nygård, H.S. and Olsen, E. (2012). Review of thermal processing of biomass and waste in molten salts for production of renewable fuels and chemicals. *Int. J. Low-Carbon Technol.* **7**, 318–324. https://doi.org/10.1093/ijlct/ctr045.

Oca, J., Cremades, J., Jiménez, P., Pintado, J., and Masaló, I. (2019). Culture of the seaweed *Ulva ohnoi* integrated in a Solea senegalensis recirculating system: Influence of light and biomass stocking density on macroalgae productivity. *J. Appl. Phycol.*, **31**, 2461–2467. https://doi.org/10.1007/s10811-019-01767-z.

Odum, H.T. (1962). Man in the ecosystem. Conn. Storrs Agric. *Exp. Stat. Bull.*, **652**, 75–75.

Odum, H.T., Siler, W.L., Beyers, R.J., and Armstrong, N. (1963). Experiments with engineering of marine ecosystems. Publ. Inst. Mar. Sci. Univ. Texas, **9**:374–403.

Olanrewaju, S.O., Magee, A., Kader, A.S.A., and Tee, K.F. (2017). Simulation of offshore aquaculture system for macro algae (seaweed) oceanic farming. Ships and Offshore Structures **12**(4): 1–10. https://doi.org/10.1080/17445302.2016.11 86861.

Oldeman, L.R. (1994). The global extent of land degradation. Land Resilience and Sustainable Land Use: ISRIC Bi-Annual Report 1991–1992.

Olofsson, K., Bertilsson, M., and Lidén, G. (2008). A short review on SSF — An interesting process option for ethanol production from lignocellulosic feedstocks. *Biotechnol. Biofu.*, **1**, 1–14. https://doi.org/10.1186/1754-6834-1-7.

Oren, A. (2010). Industrial and environmental applications of halophilic microorganisms. *Environ. Technol.*, **31**, 825–834. https://doi.org/10.1080/09593330903370026.

Oswald, A.R. and Scott Laurence, L.V.H. (1979). The utilization of cold, nutrient-rich deep ocean water for energy and mariculture. *Ocean Manag.*, **5**, 199–210.

Øverland, M., Mydland, L.T., and Skrede, A. (2019). Marine macroalgae as sources of protein and bioactive compounds in feed for monogastric animals. *Sci. Food Agric.*, **99**, 13–24. https://doi.org/10.1002/jsfa.9143.

Pacheco-Torgal, F., Ivanov, V., Karak, N., and Jonkers, H. (Eds.) (2016). *Biopolymers and Biotech Admixtures for Eco-Efficient Construction*

Materials, Number 63. ed. Woodhead Publishing Series in Civil and Structural Engineering.

Pacheco-Torgsl, F., Labeza, L., Labrincha, J., and De-Magalhaes, A. (2014). Eco-efficient construction and building materials. Life cycle assessment (LCA), eco-labelling and case studies. Woodhead Publishing Limited.

Pal, A., Kamthania, M.C., and Kumar, A. (2014). Bioactive Compounds and Properties of Seaweeds: A Review. https://doi.org/10.13140/2.1.1534.7845.

Palatnik, R. R. and Zilberman, D. (2017). Economics of Natural Resource Utilization — The Case of Macroalgae., in: Zilberman, A.P., and D. (Ed.), Modeling, Dynamics, Optimization and Bioeconomics II. Springer.

Palatnik, R.R., Eboli, F., Ghermandi, A., Kan, I., Rapaport-Rom, M., and Shechter, M. (2011). Integration of general and partial equilibrium agricultural land-use transformation for the analysis of climate change in the mediterranean. *Clim. Chang. Econ.* https://doi.org/10.1142/S2010007811000310.

Palatnik, R.R., Freer, M., Golberg, A., D.Zilberman and Levin, M. (2018). Economics of natural resource utilization: Case study of macroalgae biorefinery, in: Ecomod. Venice.

Palmqvist, E., Grage, H., Meinander, N.Q., and Hahn-Hägerdal, B. (1999). Main and interaction effects of acetic acid, Furfural and *p*-Hydroxybenzoic Acid on growth and ethanol productivity of yeast. *Biotechnol. Bioeng.*, **63**, 46–55. https://doi.org/10.1002/(SICI)1097-0290(19990405)63.

Pan, Y., Fan, W., Zhang, D., Chen, J., Huang, H., Liu, S., Jiang, Z., Di, Y., Tong, M., and Chen, Y. (2015). Research progress in artificial upwelling and its potential environmental effects. *Sci. China Ear. Sci.*, **59**, 236–248. https://doi.org/10.1007/s11430-015-5195-2.

Paredes, D., Kuschk, P., Mbwette, T.S.A., Stange, F., Müller, R.A., and Köser, H. (2007). New aspects of microbial nitrogen transformations in the context of wastewater treatment — A review. *Eng. Life Sci.*, 7, 13–25. https://doi.org/10.1002/elsc.200620170.

Parisi, C. and Ronzon, T. (2016). A global view of bio-based industries: benchmarking and monitoring their economic importance and future developments. *Publ. Off. Eur. Union*, DOI 10, 153649.

Park, J.-H., Yoon, J.-J., Park, H.-D., Lim, D.J., and Kim, S.-H. (2012). Anaerobic digestibility of algal bioethanol residue. *Bioresour. Technol.*, **113**, 78–82. https://doi.org/http://dx.doi.org/10.1016/j.biortech.2011.12.123.

Park, J.H., Hong, J.Y., Jang, H.C., Oh, S.G., Kim, S.H., Yoon, J.J., and Kim, Y.J. (2012). Use of Gelidium amansii as a promising resource for bioethanol: A practical approach for continuous dilute-acid hydrolysis and fermentation. *Bioresour. Technol.*, **108**, 83–88. https://doi.org/10.1016/j.biortech.2011.12.065.

Parris, K. (2011). Impact of agriculture on water pollution in OECD countries: Recent trends and future prospects. *Int. J. Wat. Resour. Dev.*, **27**, 33–52. https://doi.org/10.1080/07900627.2010.531898.

Patarra, R.F., Paiva, L., Neto, A.I., Lima, E., and Baptista, J. (2011). Nutritional value of selected macroalgae. *J. Appl. Phycol.*, **23**, 205–208. https://doi.org/10.1007/s10811-010-9556-0.

Patra, J.K., Kim, S.H., and Baek, K.-H. (2015). Antioxidant and free radical-scavenging potential of essential oil from enteromorpha linza L. Prepared by microwave-assisted hydrodistillation. *J. Food Biochem.*, **39**, 80–90. https://doi.org/10.1111/jfbc.12110.

Pattiya, A., Chaow-u-thai, A., and Rittidech, S. (2013). The influence of pretreatment techniques on ash content of Cassava residues. *Int. J. Gree. Ener.*, **10**, 544–552. https://doi.org/10.1080/15435075.2012.703629.

Pauli, G.A. (2009). *The Blue Economy: A Report to the Club of Rome*. Nairobi.

Pauli, G.A. (2010). *The Blue Economy: 10 Years, 100 Innovations, 100 Million Jobs*. Paradigm Publications.

Pavasant, P., Apiratikul, R., Sungkhum, V., Suthiparinyanont, P., Wattanachira, S., and Marhaba, T.F. (2006). Biosorption of Cu_{2+}, Cd_{2+}, Pb_{2+}, and Zn_{2+} using dried marine green macroalga *Caulerpa lentillifera*. *Bioresour. Technol.*, **97**, 2321–2329. https://doi.org/10.1016/J.BIORTECH.2005.10.032.

Peled, Y., Zemah Shamir, S., Shechter, M., Rahav, E., and Israel, A. (2018). A new perspective on valuating marine climate regulation: The Israeli Mediterranean as a case study. *Ecosyst. Serv.* https://doi.org/10.1016/j.ecoser.2017.12.001.

PEMSEA (2014). *Building a Blue Economy: Strategy, Opportunities and Partnerships in the Seas of East Asia The Road to a Sustainable East Asian Seas*, Tropical Coasts.

Peñuela, A., Robledo, D., Bourgougnon, N., Bedoux, G., Hernández-Núñez, E., and Freile-Pelegrín, Y. (2018). Environmentally friendly valorization of *Solieria filiformis* (Gigartinales, Rhodophyta) from IMTA using a biorefinery concept. *Mar. Drugs*, **16**, 487.

Pereira, H., Barreira, L., Figueiredo, F., Custódio, L., Vizetto-Duarte, C., Polo, C., Rešek, E., Aschwin, E., and Varela, J. (2012). Polyunsaturated fatty acids of marine macroalgae: Potential for nutritional and pharmaceutical applications. *Mar. Drugs*, **10**, 1920–1935. https://doi.org/10.3390/md10091920.

Pereira, L. and Neto, J.M. (2014). *Marine Algae: Biodiversity, Taxonomy, Environmental Assessment, and Biotechnology*. https://doi.org/10.1201/b17540.

Pereira, R., Yarish, C., and Critchley, A.T. (2013). Seaweed seaweed aquaculture seaweed aquaculture for human foods in land-based and IMTA systems, in: *Sustainable Food Production*. https://doi.org/10.1007/978-1-4614-5797-8_189.

Périno, S., Pierson, J.T., Ruiz, K., Cravotto, G., and Chemat, F. (2016). Laboratory to pilot scale: Microwave extraction for polyphenols lettuce. *Food Chem.*, **204**, 108–114. https://doi.org/10.1016/J.FOODCHEM.2016.02.088.

Peteiro, C. and Freire, Ó. (2012). Outplanting time and methodologies related to mariculture of the edible kelp *Undaria pinnatifida* in the Atlantic coast of Spain. *J. Appl. Phycol.*, **24**, 1361–1372. https://doi.org/10.1007/s10811-012-9788-2.

Peteiro, C., Sánchez, N., Dueñas-Liaño, C., and Martínez, B. (2014). Open-sea cultivation by transplanting young fronds of the kelp *Saccharina latissima*. *J. Appl. Phycol.*, **26**, 519–528. https://doi.org/10.1007/s10811-013-0096-2.

Pezoa-Conte, R., Leyton, A., Anugwom, I., von Schoultz, S., Paranko, J., Mäki-Arvela, P., Willför, S., Muszyński, M., Nowicki, J., Lienqueo, M.E., and Mikkola, J.-P. (2015). Deconstruction of the green alga *Ulva rigida* in ionic liquids: Closing the mass balance. *Algal Res.*, **12**, 262–273. https://doi.org/10.1016/J.ALGAL.2015.09.011.

Pfromm, P.H. (2008). The minimum water consumption of ethanol production via biomass fermentation. *Open Chem. Eng. J.*, **2**, 1–5.

Phillippi, A., Tran, K., and Perna, A. (2014). Does intertidal canopy removal of Ascophyllum nodosum alter the community structure beneath? *J. Exp. Mar. Bio. Ecol.*, **461**, 53–60. https://doi.org/10.1016/J.JEMBE.2014.07.018.

Pimentel, D. (2003). Ethanol fuels: Energy balance, economics, & environmental impacts are negative. *Nat. Resour. Res.*, **12**, 127–134. https://doi.org/1520-7439/03/0600-0127/1.

Pimentel, D. (2012). *Global Economic and Environmental Aspects of Biofuels*. CRC Press.

Pimentel, M. and Pimentel, M.H. (2008). *Food, Energy, and Society*. CRC Press, Boca Raton.

Pingali, P.L. (2012). Green Revolution: Impacts, limits, and the path ahead. *Proc. Natl. Acad. Sci.*, **109**, 12302–12308. https://doi.org/10.1073/pnas.0912953109.

Plank, J. (2004). Applications of biopolymers and other biotechnological products in building materials. *Appl. Microbiol. Biotechnol.*, **66**, 1–9. https://doi.org/10.1007/s00253-004-1714-3.

Ploechinger H. Construction material made of algae, method for cultivating algae, and algae cultivation plant. Patent US8685707B2.

Plouguerne, E., Hellio, C., Deslandes, E., Benoît, V., and Stiger-Pouvreau, V. (2008). Anti-microfouling activities in extracts of two invasive algae: *Grateloupia turuturu* and *Sargassum muticum*. *Bot. Mar.*, **51**, 202–208. https://doi.org/10.1515/BOT.2008.026.

Polikovsky, M., Fernand, F., Sack, M., Frey, W., Müller, G., and Golberg, A. (2016). Towards marine biorefineries: Selective proteins extractions from marine macroalgae *Ulva* with pulsed electric fields. *Innov. Food Sci. Emerg. Technol.*, **37**, 194–200. https://doi.org/10.1016/j.ifset.2016.03.013.

Polikovsky, M. and Golberg, A. (2019). Biorefinery of unique polysaccharides from *Laminaria* sp., *Kappaphycus* sp., and *Ulva* sp. Structure, enzymatic hydrolysis, and bioenergy from released monosaccharides, in: Enzymatic

Technologies for Marine Polysaccharides. https://doi.org/10.1201/9780429058653-9.

Ponce, N., Pujol, C., Damonte, E., Flores, M., and Stortz, C. (2003). Fucoidans from the brown seaweed Adenocystis utricularis: Extraction methods, antiviral activity and structural studies. *Carbohydr. Res.*, **338**, 153–165.

Popp, J., Harangi-Rákos, M., Gabnai, Z., Balogh, P., Antal, G., and Bai, A. (2016). Biofuels and their co-products as livestock feed: Global economic and environmental implications. *Molecules*, **21**, 285. https://doi.org/10.3390/molecules21030285.

Posen, I.D., Griffin, W.M., Matthews, H.S., and Azevedo, I.L. (2015). Changing the renewable fuel standard to a renewable material standard: Bioethylene case study. *Environ. Sci. Technol.*, **49**, 93–102. https://doi.org/10.1021/es503521r.

Posten, C. (2009). Design principles of photo-bioreactors for cultivation of microalgae. *Eng. Life Sci.*, **9**, 165–177. https://doi.org/10.1002/elsc.200900003.

Postma, P.R., Cerezo-Chinarro, O., Akkerman, R.J., Olivieri, G., Wijffels, R.H., Brandenburg, W.A., and Eppink, M.H.M. (2017). Biorefinery of the macroalgae *Ulva lactuca*: Extraction of proteins and carbohydrates by mild disintegration. *J. Appl. Phycol.*, 1–13. https://doi.org/10.1007/s10811-017-1319-8.

Potts, T., Du, J., Paul, M., May, P., Beitle, R. and Hestekin, J. (2012). The production of butanol from Jamaica bay macro algae, in: *Environmental Progress and Sustainable Energy*. pp. 29–36. https://doi.org/10.1002/ep.10606.

Prabhu, M., Chemodanov, A., Gottlieb, R., Kazir, M., Nahor, O., Gozin, M., Israel, A., Livney, Y.D., and Golberg, A. (2019). Starch from the sea: The green macroalga *Ulva* sp. as a potential source for sustainable starch production from the sea in marine biorefineries. *Algal Res.*, **37**, 215–227. https://doi.org/10.1016/j.algal.2018.11.007.

Prabhu, M., Levkov, K., Levin, O., Vitkin, E., Israel, A., Chemodanov, A., and Golberg, A. (2020a). Energy efficient dewatering of far offshore grown green macroalgae *Ulva* sp. biomass with pulsed electric fields and mechanical press. *Bioresour. Technol.*, **295**, 122229.

Prabhu, M.S., Israel, A., Palatnik, R.R., Zilberman, D., and Golberg, A. (2020b). Integrated biorefinery process for sustainable fractionation of *Ulva ohnoi* (Chlorophyta): Process optimization and revenue analysis. *J. Appl. Phycol.* https://doi.org/10.1007/s10811-020-02044-0.

Price, W.A., Tomlinson, K.W., and Hunt, J.N. (1969). The effect of artificial seaweed in promoting the build-up of beaches, in: *Coastal Engineering 1968*. American Society of Civil Engineers, New York, NY, pp. 570–578. https://doi.org/10.1061/9780872620131.036.

Qiao, L., Yang, X., Xie, R., Du, C., Chi, Y., Zhang, J., and Wang, P. (2020). Efficient production of ulvan lyase from *Ulva prolifera* by *Catenovulum* sp. LP based on stage-controlled fermentation strategy. *Algal Res.*, **46**, 101812. https://doi.org/10.1016/J.ALGAL.2020.101812.

Raghavan, V.G.S. (2015). Combined fields dewatering of seaweed (*Nereocystis luetkeana*). *Am. Soc. Agric. Eng.*, **37**(3), 899–906. https://doi.org/10.13031/2013.28157.

Raikova, S., Knowles, T.D.J., Allen, M.J., and Chuck, C.J. (2019). Co-liquefaction of macroalgae with common marine plastic pollutants. *ACS Sustain. Chem. Eng.*, **7**, 6769–6781. https://doi.org/10.1021/acssuschemeng.8b06031.

Raikova, S., Le, C.D., Beacham, T.A., Jenkins, R.W., Allen, M.J., and Chuck, C.J. (2017). Towards a marine biorefinery through the hydrothermal liquefaction of macroalgae native to the United Kingdom. *Bioma. and Bioener.*, **107**, 244–253. https://doi.org/10.1016/J.BIOMBIOE.2017.10.010.

Rajagopal, D., Sexton, S., Hochman, G., and Zilberman, D. (2009). Recent developments in renewable technologies: R&D investment in advanced biofuels. *Annu. Rev. Resour. Econ.*, **1**, 621–644. https://doi.org/10.1146/annurev.resource.050708.144259.

Ramachandra, T.V. and Hebbale, D. (2020). Bioethanol from macroalgae: Prospects and challenges. *Renew. Sustain. Ener. Rev.* https://doi.org/10.1016/j.rser.2019.109479.

Ramankutty, N., Evan, A.T., Monfreda, C., and Foley, J.A. (2008). Farming the planet: 1. Geographic distribution of global agricultural lands in the year 2000. *Global Biogeochem. Cycles.* https://doi.org/10.1029/2007GB002952.

Ramankutty, N., Mehrabi, Z., Waha, K., Jarvis, L., Kremen, C., Herrero, M., and Rieseberg, L.H. (2018). Trends in global agricultural land use: Implications for environmental health and food security. *Annu. Rev. Plant Biol.* https://doi.org/10.1146/annurev-arplant-042817-040256.

Ramos, E., Juanes, J.A., Galván, C., Neto, J.M., Melo, R., Pedersen, A., Scanlan, C., Wilkes, R., van den Bergh, E., Blomqvist, M., Karup, H.P., Heiber, W., Reitsma, J.M., Ximenes, M.C., Silió, A., Méndez, F. and González, B. (2012). Coastal waters classification based on physical attributes along the NE Atlantic region. An approach for rocky macroalgae potential distribution. *Estuar. Coast. Shelf Sci.* https://doi.org/10.1016/j.ecss.2011.11.041.

Rasyid, A. (2017). Evaluation of nutritional composition of the dried seaweed *Ulva lactuca* from pameungpeuk waters, Indonesia. *Trop. life Sci. Res.*, **28**, 119.

Raven, J.A. and Falkowski, P.G. (1999). Oceanic sinks for atmospheric CO2. *Plant, Cell Environ.* https://doi.org/10.1046/j.1365-3040.1999.00419.x.

Ray, B. and Lahaye, M. (1995). Cell-wall polysaccharides from the marine green alga *Ulva "rigida"* (*Ulvales*, Chlorophyta). Chemical structure of ulvan. *Carbohydr. Res.*, **274**, 313–318.

Ray, D.K., Mueller, N.D., West, P.C., and Foley, J.A. (2013). Yield trends are insufficient to double global crop production by 2050. *PLoS One.* https://doi.org/10.1371/journal.pone.0066428.

Reardon, T. and Zilberman, D. (2018). Climate smart food supply chains in developing countries in an era of rapid dual change in agrifood systems and the climate BT, in: *Climate Smart Agriculture: Building Resilience to Climate Change*, Lipper, L., McCarthy, N., Zilberman, D., Asfaw, S., Branca, G. (Eds.), Springer International Publishing, Cham, pp. 335–351. https://doi.org/10.1007/978-3-319-61194-5_15.

Reith, J.H., Curvers, A.P.W.M., Kamermans, P., Brandenburg, W., and Zeeman, G. (2005). Bio-offshore grootschalige teelt van zeewieren in combinatie met offshore windparken in de Noordzee, 1–137. Report ECN-C--05-008 https://edepot.wur.nl/120118.

Reith, J.H., Deurwaarder, E.P., Hemmes, K., Biomassa, E., Curvers, A.P.W.M., and Windenergie, E. (2005). BIO-OFFSHORE Grootschalige teelt van zeewieren in combinatie met offshore windparken in de Noordzee.

ReportLinker. (2019). *Global Algae Products Market By Source, By Application, By Region, Competition, Forecast & Opportunities, 2024.*

Ricardo, R., Neori, A., Valderrama, D., C.R.K., R., Cronin, H., and Forster, J. (2015). Farming of seaweeds, in: *Seaweed Sustainability*. Elsevier, pp. 27–57.

Richard, T.L. (2010). Challenges in scaling up biofuels infrastructure. *Science*, **329**, 793 LP – 796. https://doi.org/10.1126/science.1189139.

Rizzo, C., Genovese, G., Morabito, M., Faggio, C., Pagano, M., Spanò, A., Zammuto, V., Minicante, S.A., Manghisi, A., Cigala, R.M., Crea, F., Marino, F., and Gugliandolo, C. (2017). Potential antibacterial activity of marine macroalgae against pathogens relevant for aquaculture and human health. *J. Pure Appl. Microbiol.*, **11**(4), 1695–1706. https://doi.org/10.22207/JPAM.11.4.07.

Robert, P.C. (2002). Precision agriculture: A challenge for crop nutrition management, in: *Progress in Plant Nutrition: Plenary Lectures of the XIV International Plant Nutrition Colloquium*. Springer Netherlands, Dordrecht, pp. 143–149. https://doi.org/10.1007/978-94-017-2789-1_11.

Robic, A., Sassi, J.F., and Lahaye, M. (2008). Impact of stabilization treatments of the green seaweed *Ulva rotundata* (Chlorophyta) on the extraction yield, the physico-chemical and rheological properties of ulvan. *Carbohydr. Polym.*, **74**, 344–352. https://doi.org/10.1016/j.carbpol.2008.02.020.

Robic, M.A., Bertrand, D., Sassi, J.-F. Lerat, Y., and Lahaye, M. (2008). Determination of the chemical composition of ulvan, a cell wall polysaccharide from *Ulva* spp. (*Ulvales*, Chlorophyta) by FT-IR and chemometrics. *J. Appl. Phycol.*, **21**. https://doi.org/10.1007/s10811-008-9390-9.

Robin, A., Chavel, P., Chemodanov, A., Israel, A., and Golberg, A. (2017). Diversity of monosaccharides in marine macroalgae from the Eastern Mediterranean Sea. *Algal Res.*, **28**, 118–127. https://doi.org/10.1016/j.algal.2017.10.005.

Robin, A., Chavel, P., Chemodanov, A., Israel, A., and Golberg, A. (2018). Corrigendum to "Diversity of monosaccharides in marine macroalgae from the Eastern Mediterranean Sea". *Algal Res.*, **28**, 118–127. https://doi.org/10.1016/j.algal.2018.04.008.

Robin, A. and Golberg, A. (2016). Pulsed electric fields and electroporation technologies in marine Macroalgae Biorefineries. In: Miklavcic D. (eds) *Handbook of Electroporation*. pp. 1–6, Springer, Cham. https://doi.org/10.1007/978-3-319-26779-1_218-1.

Robin, A., Kazir, M., Sack, M., Israel, A., Frey, W., Mueller, G., Livney, Y.D., and Golberg, A. (2018a). Functional protein concentrates extracted from the green marine macroalga *Ulva* sp., by high voltage pulsed electric fields and mechanical press. *ACS Sustain. Chem. Eng.*, **6**, 13696–13705. https://doi.org/10.1021/acssuschemeng.8b01089.

Robin, A., Sack, M., Israel, A., Frey, W., Müller, G., and Golberg, A. (2018b). Deashing macroalgae biomass by pulsed electric field treatment. *Bioresour. Technol.*, **255**, 131–139. https://doi.org/10.1016/j.biortech.2018.01.089.

Robins, P.M. (1985). Aquatic species program review proceedings of the March 1985 principal investigators meeting. Golden, Colorado.

Rocca, S., Agostini, A., Giuntoli, J., and Marelli, L. (2015). Biofuels from algae: Technology options, energy balance and GHG emissions. https://doi.org/10.2790/125847.

Rodell, M., Velicogna, I., and Famiglietti, J.S. (2009). Satellite-based estimates of groundwater depletion in India. *Nature*. https://doi.org/10.1038/nature08238.

Rodrigues, D., Sousa, S., Silva, A., Amorim, M., Pereira, L., Rocha-Santos, T.A.P., Gomes, A.M.P., Duarte, A.C., and Freitas, A.C. (2015). Impact of enzyme- and ultrasound-assisted extraction methods on biological properties of red, brown, and green seaweeds from the Central West Coast of Portugal. *J. Agric. Food Chem.*, **63**, 3177–3188. https://doi.org/10.1021/jf504220e.

Roesijadi, A.G., Copping, A., and Huesemann, M. (2008). Techno-Economic feasibility analysis of offshore seaweed farming for bioenergy and biobased products. ARPA-E. Independent Research and Development Report. IR Number: PNWD-3931. Battelle Pacific Northwest Division.

Roesijadi, G., Copping, A.E., Huesemann, M.H., Forster, J., and Benemann, J.R. (2008). Techno-economic feasibility analysis of offshore seaweed farming for bioenergy and biobased products. PNNL IR Number: PNWD-3931.

Roesijadi, G., Jones, S.B.B., Snowden-Swan, L.J., and Zhu, Y. (2010). Macroalgae as a biomass feedstock: A preliminary analysis. Dep. Energy under Contract DE-AC05-76RL01830 by Pacific Northwest *Natl. Lab.* 1–50. https://doi.org/10.2172/1006310.

Rombaut, N., Tixier, A.-S., Bily, A., and Chemat, F. (2014). Green extraction processes of natural products as tools for biorefinery. *Biofu. Bioprod. Biorefin.*, **8**, 530–544. https://doi.org/10.1002/bbb.1486.

Romero-González, M.E., Williams, C.J., Gardiner, P.H., Gurman, S.J. and Habesh, S. (2003). Spectroscopic studies of the biosorption of Gold(III) by dealginated seaweed waste. *Environ. Sci. Technol.*, **37**, 4163–4169. https://doi.org/10.1021/es020176w.

Rose, M. and Palkovits, R. (2012). Isosorbide as a renewable platform chemical for versatile applications — Quo Vadis? *ChemSusChem.*, **5**, 167–176. https://doi.org/10.1002/cssc.201100580.

Rosenberg, C. and Ramus, J. (1982). Ecological growth strategies in the seaweeds *Gracilaria foliifera* (*Rhodophyceae*) and *Ulva* sp. (*Chlorophyceae*): Soluble nitrogen and reserve carbohydrates. *Mar. Biol.*, **66**, 251–259. https://doi.org/10.1007/BF00397030.

Routray, W. and Orsat, V. (2012). Microwave-assisted extraction of flavonoids: A Review. *Food Bioproc. Technol.*, **5**, 409–424. https://doi.org/10.1007/s11947-011-0573-z.

Ruttan, V.W. and Alexandratos, N. (1996). *World Agriculture: Towards 2010.* An FAO Study. *Popul. Dev. Rev.* https://doi.org/10.2307/2137724.

Saâ Nchez-Rodrõ Â Guez, I., Huerta-Diaz, M.A., Choumiline, E., Holguõ Â N-Quinä, O., and Zertuche-Gonzaâ Lez, J.A. (2001). Elemental concentrations in different species of seaweeds from Loreto Bay, Baja California Sur, Mexico: implications for the geochemical control of metals in algal tissue. Environmental Pollution **114**(2), 145–160.

Sadhukhan, J., Gadkari, S., Martinez-Hernandez, E., Ng, K.S., Shemfe, B., Garcia, E.T., Lynch, J. (2019). Novel macroalgae (seaweed) biorefinery systems for integrated chemical, protein, salt, nutrient and mineral extractions and environmental protection by green synthesis and life cycle sustainability assessments. *Green Chem.*, **21**, 2635.

Saha, B.C., Iten, L.B., Cotta, M.A., and Wu, Y.V. (2003). Dilute acid pretreatment, enzymatic saccharification and fermentation of wheat straw to ethanol § **40**, 3693–3700. https://doi.org/10.1016/j.procbio.2005.04.006.

Sales, M. and Ballesteros, E. (2012). Seasonal dynamics and annual production of *Cystoseira crinita* (*Fucales*: Ochrophyta)-dominated assemblages from the northwestern Mediterranean. *Sci. Mar.* https://doi.org/10.3989/scimar.03465.16d.

Sampath-Wiley, P., Neefus, C.D., and Jahnke, L.S. (2008). Seasonal effects of sun exposure and emersion on intertidal seaweed physiology: Fluctuations in antioxidant contents, photosynthetic pigments and photosynthetic efficiency in the red alga Porphyra umbilicalis Kützing (Rhodophyta, Bangiales). *J. Exp. Mar. Bio. Ecol.* https://doi.org/10.1016/j.jembe.2008.05.001.

Sanchez, O.J. and Cardona, C.A. (2008). Trends in biotechnological production of fuel ethanol from different feedstocks. *Bioresour. Technol.*, **99**, 5270–5295.

Sand-Jensen, K. and Borum, J. (1991). Interactions among phytoplankton, periphyton, and macrophytes in temperate freshwaters and estuaries. *Aquat. Bot.*, **41**, 137–175. https://doi.org/10.1016/0304-3770(91)90042-4.

Sanderson, J.C., Dring, M.J., Davidson, K., and Kelly, M.S. (2012). Culture, yield and bioremediation potential of *Palmaria palmata* (Linnaeus) Weber & Mohr and *Saccharina latissima* (Linnaeus) C.E. Lane, C. Mayes, Druehl & G.W. Saunders adjacent to fish farm cages in northwest Scotland. *Aquaculture*, **354–355**, 128–135. https://doi.org/10.1016/j.aquaculture.2012.03.019.

Santelices, B. (1999). A conceptual framework for marine agronomy. *Hydrobiologia*, **398**, 15–23.

Santelices, B. and Marquet, P.A. (1998). Seaweeds, latitudinal diversity patterns, and Rapoport's Rule. *Divers. Distrib.* https://doi.org/10.1046/j.1472-4642.1998.00005.x.

Sappati, P.K., Nayak, B., VanWalsum, G.P., and Mulrey, O.T. (2019). Combined effects of seasonal variation and drying methods on the physicochemical properties and antioxidant activity of sugar kelp (*Saccharina latissima*). *J. Appl. Phycol.*, **31**, 1311–1332. https://doi.org/10.1007/s10811-018-1596-x.

Saravana, P.S., Cho, Y.-N., Woo, H.-C., and Chun, B.-S. (2018). Green and efficient extraction of polysaccharides from brown seaweed by adding deep eutectic solvent in subcritical water hydrolysis. *J. Clean. Prod.*, **198**, 1474–1484.

Sargassum, A., Plouguerné, E., Ioannou, E., and Georgantea, P. (2010). Anti-microfouling activity of lipidic metabolites from the invasive brown anti-microfouling activity of lipidic metabolites from the invasive brown alga *sargassum muticum* (Yendo) Fensholt. *Mar. Biotechnol.*, 52–61. https://doi.org/10.1007/s10126-009-9199-9.

Scheuer, P., Haman, M., and Gravlos, D. (2000). Cytotoxic and antiviral compund. US Patent Number 6011010.

Schmitt, T.M., Hay, M.E., and Lindquist, N. (1995). Constraints on chemically mediated coevolution: Multiple functions for seaweed secondary metabolites. *Ecology.* https://doi.org/10.2307/1940635.

Schultz-Jensen, N., Thygesen, A., Leipold, F., Thomsen, S.T., Roslander, C., Lilholt, H., and Bjerre, A.B. (2013). Pretreatment of the macroalgae Chaetomorpha linum for the production of bioethanol — Comparison of five pretreatment technologies. *Bioresour. Technol.*, **140**, 36–42. https://doi.org/10.1016/j.biortech.2013.04.060.

Sciubba, E. (2012). A Thermodynamically Correct Treatment of Externalities with an Exergy-Based Numeraire. *Sustainability*, **4**, 933–957. https://doi.org/10.3390/su4050933.

Sciubba, E. (2003). Extended exergy accounting applied to energy recovery from waste: The concept of total recycling. *Energy*, **28**, 1315–1334. https://doi.org/10.1016/S0360-5442(03)00111-7.

Seghetta, M., Hou, X., Bastianoni, S., Bjerre, A.-B., and Thomsen, M. (2016a). Life cycle assessment of macroalgal biorefinery for the production of ethanol, proteins and fertilizers — A step towards a regenerative bioeconomy. *J. Clean. Prod.*, **137**, 1158–1169. https://doi.org/10.1016/j.jclepro.2016.07.195.

Seghetta, M., Marchi, M., Thomsen, M., Bjerre, A.-B., and Bastianoni, S. (2016b). Modelling biogenic carbon flow in a macroalgal biorefinery system. *Algal Res.*, **18**, 144–155. https://doi.org/10.1016/J.ALGAL.2016.05.030.

Seip, K.L. (1980). A computational model for growth a n d harvesting of the marine alga ascophyllum nodosum. *Ecol. Model.*, **8**, 189–199. https://doi.org/10.1016/0304-3800(80)90037-X.

Selvavinayagam, K.T. and Dharmar, K. (2017). Selection of potential method for cultivation and seed stock maintenance of Kappaphycus alvarezii during the northeast monsoon in southeast coast of India. *J. Appl. Phycol.* https://doi.org/10.1007/s10811-016-0967-4.

Sewell, J., Mayer, I., Langdon, S., Smyth, J., and Jodrell, D. (2005). The mechanism of action of Kahalalide F: variable cell permeability in human hepatoma cell lines. *Eur. J. Cancer*, **41**, 1637–1644.

Shacklock, P. and Croft, G.B. (1981). Effect of grazers on Chonder crispus in culture. *Aquaculture*, **22**, 331–342.

Shahar, B., Shpigel, M., Barkan, R., Masasa, M., Neori, A., Chernov, H., Salomon, E., Kiflawi, M., and Guttman, L. (2020). Changes in metabolism, growth and nutrient uptake of *Ulva fasciata* (*Chlorophyta*) in response to nitrogen source. *Algal Res.*, **46**. https://doi.org/10.1016/j.algal.2019.101781.

Shahid, A., Khan, A.Z., Liu, T., Malik, S., Afzal, I., and Mehmood, M.A. (2017). *Algae Based Polymers, Blends, and Composites*. Chapter 7: Production and processing of algal biomass.

Sharp, G.J., Ugarte, R., and Semple, R. (2006). The Ecological Impact of Marine Plant Harvesting in the Canadian Maritimes, Implications for Coastal Zone Management. *ScienceAsia*, **32**, 77–86. https://doi.org/10.2306/scienceasia1513-1874.2006.32(s1).077.

Shastri, Y. and Ting, K.C. (2013). Biomass feedstock production and provision: Overview, current status, and challenges, in: Shastri, Y., Hansen, A., Rodríguez, L., Ting, K.C. (Eds.), *Engineering and Science of Biomass Feedstock Production and Provision*. Springer New York, New York, NY, pp. 1–15. https://doi.org/10.1007/978-1-4899-8014-4_1.

Sheng, L.C. (2018). Challenges and Opportunities of IMTA in Hawaii and Beyond.

Shpigel, M., Shauli, L., Odintsov, V., Ben-Ezra, D., Neori, A., and Guttman, L. (2018). The sea urchin, Paracentrotus lividus, in an Integrated Multi-Trophic Aquaculture (IMTA) system with fish (*Sparus aurata*) and seaweed (*Ulva lactuca*): Nitrogen partitioning and proportional configurations. *Aquaculture*, **490**, 260–269. https://doi.org/10.1016/J.AQUACULTURE.2018.02.051.

Sievers, D.M. and Brune, D.E., 1978. Carbon/nitrogen ratio and anaerobic digestion of swine waste. *Trans. ASAE* **21**, 537–541.

Simpson, A.P. and Edwards, C.F. (2011). An exergy-based framework for evaluating environmental impact. *Energy*, **36**, 1442–1459. https://doi.org/10.1016/j.energy.2011.01.025.

Smil, V. (2008). *Energy in Nature and Society: The Energetics of Complex Systems.* MIT Press, Cambridge, MA.

Smit, A.J. (2004). Medicinal and pharmaceutical uses of seaweed natural products. *Appl. Phycol.*, 245–262. https://doi.org/10.1023/B:JAPH.0000047783.36600.ef.

Smith, C.T., Lattimore, B., Berndes, G., Bentsen, N.S., Dimitriou, I. (Hans) Langeveld, J.W.A., and Thiffault, E. (2017). Opportunities to encourage mobilization of sustainable bioenergy supply chains. *WIREs Energy Environ.*, **6**, e237. https://doi.org/10.1002/wene.237.

Smith, R.F., Apple, J.L., and Bottrell, D.G. (1976). The origins of integrated pest management concepts for agricultural crops, in: *Integrated Pest Management.* pp. 1–16.

Smith, R.G. and Bidwell, R.G.S. (1989). Mechanism of photosynthetic carbon dioxide uptake by the red macroalga, *Chondrus crispus. Plant Physiol.* https://doi.org/10.1104/pp.89.1.93.

Soberon, J. and Peterson, A.T. (2005). Interpretation of models of fundamental ecological niches and species' distributional areas. *Biodivers. Informatics.* https://doi.org/10.17161/bi.v2i0.4.

Solis, M.J.L., Draeger, S., and Cruz, E.T.E. dela (2010). Marine-derived fungi from *Kappaphycus alvarezii* and *K. striatum* as potential causative agents of ice-ice disease in farmed seaweeds. *Bot. Mar.*, 587–594. https://doi.org/10.1515/bot.2010.071.

Sondak, C.F.A., Ang, P.O., Beardall, J., Bellgrove, A., Boo, S.M., Gerung, G.S., Hepburn, C.D., Hong, D.D., Hu, Z., Kawai, H., Largo, D., Lee, J.A., Lim, P.E., Mayakun, J., Nelson, W.A., Oak, J.H., Phang, S.M., Sahoo, D., Peerapornpis, Y., Yang, Y., and Chung, I.K. (2017). Carbon dioxide mitigation potential of seaweed aquaculture beds (SABs), in: *J. Appl. Phycol.* https://doi.org/10.1007/s10811-016-1022-1.

Song, M., Duc Pham, H., Seon, J., and Chul Woo, H. (2015). Marine brown algae: A conundrum answer for sustainable biofuels production. *Renew. Sustain. Ener. Rev.*, **50**, 782–792. https://doi.org/10.1016/J.RSER.2015.05.021.

South, G.R. (1993). Edible seaweeds — An important source of food and income to indigenous Fijians. *NAGA, Iclarm Q.*, **16**(2–3), 4–6.

Spieler, R. (2003). Seaweed compound's anti-HIV efficacy will be tested in southern Africa. *Lancet.*, **359**, 1675.

Stagnol, D., Michel, R., and Davoult, D. (2016). Population dynamics of the brown alga Himanthalia elongata under harvesting pressure. *Estuar. Coast. Shelf Sci.*, **174**, 65–70. https://doi.org/10.1016/J.ECSS.2016.03.014.

Steen, H. (2004). Effects of reduced salinity on reproduction and germling development in *sargassum muticum* (Phaeophyceae, Fucales). *Eur. J. Phycol.*, **39**, 293–299. https://doi.org/10.1080/09670260410001712581.

Stefanidis, S.D., Heracleous, E., Patiaka, D.T., Kalogiannis, K.G., Michailof, C.M., and Lappas, A.A. (2015). Optimization of bio-oil yields by demineralization of low quality biomass. *Biomass and Bioenergy*, **83**, 105–115. https://doi.org/10.1016/j.biombioe.2015.09.004.

Stefels, J. (2000). Physiological aspects of the production and conversion of DMSP in marine algae and higher plants. *J. Sea Res.*, **43**, 183–197. https://doi.org/https://doi.org/10.1016/S1385-1101(00)00030-7.

Stengel, D.B. and Connan, S. (2015). Natural products from marine algae: Methods and protocols. *Natural Products From Marine Algae: Methods and Protocols*. https://doi.org/10.1007/978-1-4939-2684-8.

Stengel, D.B., Walker, J.M., Kenny, O., and Brunton, N.P. (2015). *Natural Products From Marine Algae*. https://doi.org/10.1007/978-1-4939-2684-8.

Stevens, C.J., Dise, N.B., Mountford, J.O., and Gowing, D.J. (2004). Impact of nitrogen deposition on the species richness of grasslands. *Science*, **303**, 1876–1879. https://doi.org/10.1126/science.1094678.

Stewart, H.L. and Carpenter, R.C. (2003). The effects of morphology and water flow on photosynthesis of marine macroalgae. *Ecology*, **84**, 2999–3012. https://doi.org/10.1890/02-0092.

Stichnothe, H., Meier, D., and de Bari, I. (2016). Biorefineries: Industry status and economics. *Dev. Glob. Bioecon.*, 41–67. https://doi.org/10.1016/B978-0-12-805165-8.00003-3.

Stonik, V. and Fedorov, S. (2014). Marine low molecular weight natural products as potential cancer preventive compounds. *Mar. Drugs*, **12**(2), 636–671.

Suarez Ruiz, C.A., Baca, S.Z., van den Broek, L.A.M., van den Berg, C., Wijffels, R.H., and Eppink, M.H.M. (2020). Selective fractionation of free glucose and starch from microalgae using aqueous two-phase systems. *Algal Res.*, **46**, 101801. https://doi.org/10.1016/J.ALGAL.2020.101801.

Suárez, Y., González, L., Cuadrado, A., Berciano, M., Lafarga, M., and Muñoz, A. (2003). Kahalalide F, a new marine-derived compound, induces oncosis in human prostate and breast cancer cell. *Mol. Cancer Ther.*, **2**, 863–872.

Suganya, T., Nagendra Gandhi, N., and Renganathan, S. (2013). Production of algal biodiesel from marine macroalgae Enteromorpha compressa by two step process: Optimization and kinetic study. *Bioresour. Technol.*, **128**, 392–400. https://doi.org/10.1016/j.biortech.2012.10.068.

Sugawara, I., Itoh, W., Kimura, S., Mori, S., and Shimada, K. (1989). Further characterization of sulfated homopolysaceharides as anti-HIV agents. *Experientia*, **45**, 996–998.

Sun, Y., Li, Z., and Xiu, Z. (2014). Method for salting-out extraction of acetone and butanol from a fermentation broth. Patent. US20130190536A1.

Susilorini, R.M.I.R., Hardjasaputra, H., Sri, T., Galih, H., Reksa, W.S., Ginanjar, H., and Joko, S. (2014). The advantage of natural polymer modified mortar with seaweed: Green construction material innovation for sustainable concrete. *Proce. Eng.*, **95**, 419–425. https://doi.org/10.1016/j.proeng.2014.12.201.

Suutari, M., Leskinen, E., Fagerstedt, K., Kuparinen, J., Kuuppo, P., and Blomster, J. (2015). Macroalgae in biofuel production. *Phycol. Res.*, **63**, 1–18. https://doi.org/10.1111/pre.12078.

Svirezhev, Y.M. and Steinborn, W.H. (2001). Exergy of solar radiation: Information approach. *Ecol. Modell.*, **145**, 101–110. https://doi.org/10.1016/S0304-3800(01)00409-4.

Synytsya, A., Copíková, J., Kim, W.J., and Park, Y.I. (2015). Cell wall polysaccharides of marine algae, in: *Springer Handbook of Marine Biotechnology*. https://doi.org/10.1007/978-3-642-53971-8_22.

Szargut, J. (1971). Anwendung der Exergie zur angenaherten wirtschaftlichen Optimierung. *Brennst. Wärme, Kraft*, **23**, 516–519.

Szetela, E.J., Krascella, N.L., Blecher, W.A., and Christopher, G.L. (1976). Evaluation of a marine energy farm concept. *Am. Chem. Soc., Div. Fuel Chem., Prepr.* (United States), **19**(4).

Tabassum, M.R., Wall, D.M., and Murphy, J.D. (2016). Biogas production generated through continuous digestion of natural and cultivated seaweeds with dairy slurry. *Bioresour. Technol.*, **219**, 228–238. https://doi.org/http://dx.doi.org/10.1016/j.biortech.2016.07.127.

Tang, C. and Tomlin, B. (2008). The power of flexibility for mitigating supply chain risks. *Int. J. Prod. Econ.*, **116**, 12–27. https://doi.org/10.1016/J.IJPE.2008.07.008.

Tang, C.S. (2006). Perspectives in supply chain risk management. *Int. J. Prod. Econ.*, **103**, 451–488. https://doi.org/10.1016/J.IJPE.2005.12.006.

Taziki, M., Ahmadzadeh, H., and Murry, M.A. (2016). Growth of *Chlorella vulgaris* in High Concentrations of Nitrate and Nitrite for Wastwater Treatment. *Curr. Biotechnol.* https://doi.org/10.2174/2211550104666150930204835.

Tekarslan-Sahin, S.H., Alkim, C., and Sezgin, T. (2018). Physiological and transcriptomic analysis of a salt-resistant Saccharomyces cerevisiae mutant obtained by evolutionary engineering. *Bosn. J. basic Med. Sci.*, **18**, 55–65. https://doi.org/10.17305/bjbms.2017.2250.

Tenorio, A.T., Kyriakopoulou, K.E., Suarez-Garcia, E., van den Berg, C., and van der Goot, A.J. (2018). Understanding differences in protein fractionation from conventional crops, and herbaceous and aquatic biomass — Consequences for industrial use. *Trends Food Sci. Technol.*, **71**, 235–245. https://doi.org/10.1016/j.tifs.2017.11.010.

The State of Food Security and Nutrition in the World (2019). https://doi.org/10.4060/ca5162en.

Tierney, M.S., Smyth, T.J., Hayes, M., Soler-Vila, A., Croft, A.K., and Brunton, N. (2013). Influence of pressurised liquid extraction and solid-liquid extraction methods on the phenolic content and antioxidant activities of Irish macroalgae. *Int. J. Food Sci. Technol.*, **48**, 860–869. https://doi.org/10.1111/ijfs.12038.

Tilman, D., Balzer, C., Hill, J., and Befort, B.L. (2011). Global food demand and the sustainable intensification of agriculture. *Proc. Natl. Acad. Sci. U.S.A.* https://doi.org/10.1073/pnas.1116437108.

Tilman, D. and Clark, M. (2014). Global diets link environmental sustainability and human health. *Nature.* https://doi.org/10.1038/nature13959.

Tinne, M., Preston, G., and Tiroba, G. (2006). Development of seaweed marketing and licensing arrangements, technical report 1, Project ST 98/009: Commercialisation of seaweed production in the Solomon Islands.

Tiwari, B.K. (2015). Ultrasound: A clean, green extraction technology. *Trends Anal. Chem.*, **71**, 100–109. https://doi.org/10.1016/J.TRAC.2015.04.013.

Tomlinson, I. (2013). Doubling food production to feed the 9 billion: A critical perspective on a key discourse of food security in the UK. *J. Rural Stud.* https://doi.org/10.1016/j.jrurstud.2011.09.001.

Toor, S.S., Rosendahl, L., and Rudolf, A. (2011). Hydrothermal liquefaction of biomass: A review of subcritical water technologies. *Energy*, **36**, 2328–2342. https://doi.org/10.1016/J.ENERGY.2011.03.013.

Torres, E., Mata, Y.N., Blá Zquez, M.L., Muñ Oz, J.A., Gonzá Lez, F., and Ballester, A. (2005). Gold and silver uptake and nanoprecipitation on calcium alginate beads. *Langmuir*, **21**, 7951–7958. https://doi.org/10.1021/la046852k.

Totti, C., Poulin, Æ.M., Romagnoli, Æ.T. (2009). Epiphytic diatom communities on intertidal seaweeds from Iceland. Polar Biology **32**, 1681–1691. https://doi.org/10.1007/ s00300-009-0668-4.

Trinchero, J., Ponce, N.M.A., Córdoba, O.L., Flores, M.L., Pampuro, S., Stortz, C.A., Salomón, H., Turk, G. (2009). Antiretroviral activity of fucoidans extracted from the brown seaweed adenocystis utricularis. *Phytother Res.* 2009 **23**(5), 707–712. doi: 10.1002/ptr.2723.

Trivedi, N., Baghel, R.S., Bothwell, J., Gupta, V., Reddy, C.R.K., Lali, A.M. and Jha, B. (2016). An integrated process for the extraction of fuel and chemicals from marine macroalgal biomass. *Sci. Rep.*, **6**, 30728. https://doi.org/10.1038/srep30728.

Trivedi, N., Gupta, V., Reddy, C.R.K. and Jha, B. (2013). Enzymatic hydrolysis and production of bioethanol from common macrophytic green alga Ulva fasciata Delile. *Bioresour. Technol.*, **150**, 106–112. https://doi.org/10.1016/J.BIORTECH.2013.09.103.

Trivedi, N., Reddy, C.R.K., Radulovich, R. and Jha, B. (2015). Solid state fermentation (SSF)-derived cellulase for saccharification of the green seaweed

Ulva for bioethanol production. *Algal Res.*, **9**, 48–54. https://doi.org/10.1016/j.algal.2015.02.025.

Troell, M., Halling, C., Neori, A., Chopin, T., Buschmann, A. H. and Kautsky, N. and Yarish, C. (2003). Integrated mariculture: Asking the right questions. *Aquaculture*, **226**, 69–90. https://doi.org/10.1016/S0044-8486(03)00469-1.

Troell, M., Joyce, A., Chopin, T., Neori, A., Buschmann, A.H. and Fang, J.G. (2009). Ecological engineering in aquaculture — Potential for integrated multi-trophic aquaculture (IMTA) in marine offshore systems. *Aquaculture*, **297**, 1–9. https://doi.org/10.1016/j.aquaculture.2009.09.010.

Tsubaki, S., Oono, K., Hiraoka, M., Onda, A. and Mitani, T. (2016). Microwave-assisted hydrothermal extraction of sulfated polysaccharides from Ulva spp., and Monostroma latissimum. *Food Chem.*, **210**, 311–316. https://doi.org/https://doi.org/10.1016/j.foodchem.2016.04.121.

Tsubaki, S., Oono, K., Hiraoka, M., Ueda, T., Onda, A., Yanagisawa, K. and Azuma, J. (2014). Hydrolysis of green-tide forming Ulva spp. by microwave irradiation with polyoxometalate clusters. *Green Chem.*, **16**, 2227–2233. https://doi.org/10.1039/C3GC42027B.

Tun, N.M., Mahanom, H., Nur, E.N. and S. Nur Intan, F. (2016). Optimization of spray drying process of Sargassum muticum color extract. *Dry. Technol.*, **34**, 1735–1744.

Turner, C. (2015). From supercritical carbon dioxide to gas expanded liquids in extraction and chromatography of lipids. *Lipi. Technol.*, **27**, 275–277. https://doi.org/10.1002/lite.201500060.

Tyner, W.E. (2008). The US Ethanol and Biofuels Boom: Its Origins, Current Status, and Future Prospects. *Bioscience*, **58**, 646–653. https://doi.org/10.1641/B580718.

Ubando, A.T., Felix, C.B. and Chen, W.-H. (2020). Biorefineries in circular bio-economy: A comprehensive review. *Bioresour. Technol.*, **299**, 122585. https://doi.org/10.1016/J.BIORTECH.2019.122585.

Uchida, M. and Murata, M. (2004). Isolation of a lactic acid bacterium and yeast consortium from a fermented material of Ulva spp. (Chlorophyta). *J. Appl. Microbiol.* https://doi.org/10.1111/j.1365-2672.2004.02425.x.

Ugarte, R. and Santelices, B. (1992). Experimental tank cultivation of gracilaria chihsis in central Chile. *Aquaculture*, **101**(1992) 7–16.

Uju, Wijayanta, A.T., Goto, M., and Kamiya, N. (2015). Great potency of sea-weed waste biomass from the carrageenan industry for bioethanol production by peracetic acid-ionic liquid pretreatment. *Biomass and Bioenergy*, **81**, 63–69. https://doi.org/10.1016/j.biombioe.2015.05.023.

United Nations (2019). World Population Prospects 2019, Department of Economic and Social Affairs.

Unsworth, R.K.F., Collier, C.J., Henderson, G.M., and McKenzie, L.J. (2012). Tropical seagrass meadows modify seawater carbon chemistry: implications for coral reefs impacted by ocean acidification. *Environ. Res. Lett.*, **7**, 024026. https://doi.org/10.1088/1748-9326/7/2/024026.

Uribe, E., Vega-Gálvez, A., García, V., Pastén, A., López, J., and Goñi, G. (2019). Effect of different drying methods on phytochemical content and amino acid and fatty acid profiles of the green seaweed, *Ulva* spp. *J. Appl. Phycol.*, **31**. https://doi.org/10.1007/s10811-018-1686-9.

Vairappan, C., Chung, C., Hurtado, A., Soya, F., Lhonneur, G., and Critchley, A. (2007). Distribution and symptoms of epiphyte infection in major carrageenophyte-producing farms. *J. Appl. Phycol.* https://doi.org/10.1007/978-1-4020-9619-8.

Vairappan, C.S., Razalie, R., Elias, U.M., and Ramachandram, T. (2014). Effects of improved post-harvest handling on the chemical constituents and quality of carrageenan in red alga, *Kappaphycus alvarezii* Doty. *J. Appl. Phycol.*, **26**, 909–916. https://doi.org/10.1007/s10811-013-0117-1.

Valderrama, D., Cai, J., Hishamunda, N., and Ridler, N. (2013). Social and economic dimantions of carrageenan seaweed farming. Fisheries and Aquaculture Technical Paper No. 580. Rome.

Valiela, I., Mcclelland, J., Hauxwell, J., Behr, P.J., Hersh, D., and Foreman, K. (1997). Macroalgal blooms in shallow estuaries: Controls and ecophysiological and ecosystem consequences. *Limnol. Oceanogr.*, **42**, 1105–1118. https://doi.org/10.4319/lo.1997.42.5_part_2.1105.

Van Alphen, B.J. and Stoorvogel, J.J. (2000). *A Methodology for Precision Nitrogen Fertilization in High-Input Farming Systems*. Precision Agriculture **2**, 319–332.

van den Burg, S., Stuiver, M., Veenstra, F., Bikker, P., López Contreras, A., Palstra, A., Broeze, J., Jansen, H., Jak, R., Gerritsen, A., Harmsen, P., Kals, J., Blanco, A., Brandenburg, W., van Krimpen, M., Pieter van Duijn, A., Mulder, W.J., and van, L.W.D. (2013). A Triple P review of the Feasibility of Sustainable Offshore Seaweed Production in the North Sea. LEI Wageningen UR. 105.

van den Burg, S.W.K., Dagevos, H., and Helmes, R.J.K. (2019). Towards sustainable European seaweed value chains: A triple P perspective. *ICES J. Mar. Sci.* https://doi.org/10.1093/icesjms/fsz183.

van der Wal, H., Sperber, B.L.H.M.H.M., Houweling-Tan, B., Bakker, R.R.C.C., Brandenburg, W., and López-Contreras, A.M. (2013). Production of acetone, butanol, and ethanol from biomass of the green seaweed *Ulva lactuca*. *Bioresour. Technol.*, **128**, 431–437. https://doi.org/10.1016/j.biortech.2012.10.094.

Van Grinsven, H.J.M., Ten Berge, H.F.M., Dalgaard, T., Fraters, B., Durand, P., Hart, A., Hofman, G., Jacobsen, B.H., Lalor, S.T.J., Lesschen, J.P., Osterburg, B., Richards, K.G., Techen, A.-K., Vertès, F., Webb, J. and Willems, W.J. (2012). Management, regulation and environmental impacts of nitrogen fertilization in northwestern Europe under the Nitrates

Directive; a benchmark study. *Biogeosciences*, **9**, 5143–5160. https://doi.org/10.5194/bg-9-5143-2012.

Van Hal, J.W., Huijgen, W.J.J., and López -Contreras, A.M. (2014). Opportunities and challenges for seaweed in the biobased economy. *Trends Biotechnol.* https://doi.org/10.1016/j.tibtech.2014.02.007.

van Maris, A.J.A., Abbott, D.A., Bellissimi, E., van den Brink, J., Kuyper, M., Luttik, M.A.H., Wisselink, H.W., Scheffers, W.A., van Dijken, J.P., and Pronk, J.T. (2006). Alcoholic fermentation of carbon sources in biomass hydrolysates by *Saccharomyces cerevisiae*: Current status, *Anton. Leeuw. Int. J. G.* pp. 391–418. https://doi.org/10.1007/s10482-006-9085-7.

Van Oosten, M.J., Pepe, O., De Pascale, S., Silletti, S., and Maggio, A. (2017). The role of biostimulants and bioeffectors as alleviators of abiotic stress in crop plants. *Chem. Biol. Technol. Agric.*, **4**, 1–12. https://doi.org/10.1186/s40538-017-0089-5.

Vásquez, J.A., Zuñiga, S., Tala, F., Piaget, N., Rodríguez, D.C., and Vega, J.M.A. (2014). Economic valuation of kelp forests in northern Chile: values of goods and services of the ecosystem. *J. Appl. Phycol.*, **26**, 1081–1088. https://doi.org/10.1007/s10811-013-0173-6.

Veresoglou, S.D., Barto, E.K., Menexes, G., and Rillig, M.C. (2013). Fertilization affects severity of disease caused by fungal plant pathogens. *Plant Pathol.* https://doi.org/10.1111/ppa.12014.

Villalba, R.G., Bastida, J.A.G., María, T.G.C., Francisco, A.T.B., Espín, J.C., and Larrosa, M. (2012). Alternative method for gas chromatography-mass spectrometry analysis of short-chain fatty acids in faecal samples. *J. Sep. Sci.*, **35**, 1906–1913. https://doi.org/10.1002/jssc.201101121.

Vitkin, E., Gillis, A., Polikovsky, M., Bender, B., Golberg, A., and Yakhini, Z. (2020). Distributed flux balance analysis simulations of serial biomass fermentation by two organisms. *PLoS One*, **15**, e0227363. https://doi.org/10.1371/journal.pone.0227363.

Vitkin, E., Golberg, A., and Yakhini, Z. (2015). BioLEGO — a web-based application for biorefinery design and evaluation of serial biomass fermentation. *Technology*, **3**, 89–98. https://doi.org/10.1142/S2339547815400038.

Vitousek, P.M., Aber, J.D., Howarth, R.W., Likens, G.E., Matson, P.A., Schindler, D.W., Schlesinger, W.H., and Tilman, D.G. (1997). Human alteration of the global nitrogen cycle: Sources and consequences. *Ecol. Appl.* https://doi.org/10.1890/1051-0761(1997)007[0737:HAOTGN]2.0.CO;2.

Vo Dinh, T., Saravana, P.S., Woo, H.C., and Chun, B.S. (2018). Ionic liquid-assisted subcritical water enhances the extraction of phenolics from brown seaweed and its antioxidant activity. *Sep. Purif. Technol.*, **196**, 287–299. https://doi.org/10.1016/J.SEPPUR.2017.06.009.

Wagener, K. (1981). Mariculture on land: A system for biofuel farming in coastal deserts. *Biomass*, **1**, 145–158. https://doi.org/10.1016/0144-4565(81)90022-6.

Wahidi, M. El., Amraoui, B.El., Amraoui, M.El., and Bamhaoud, T. (2014). Screening of anti microbial activity of macroalgae extracts from the Moroccan Atlantic coast. *Ann. Pharm. Fr.*, **73**, 190–196. https://doi.org/10. 1016/j.pharma.2014.12.005.

Walker, C., Cole, A., and Sheehan, M. (2016). Modelling of thin layer drying of macroalgae, in: *Congress Incorporating Chemeca 2015*.

Wang, B., Tong, G.Z., Qu, Y. Le., and Li, L. (2011). Microwave-assisted extraction and in vitro antioxidant evaluation of polysaccharides from *Enteromorpha prolifera*, in: *Chemical, Mechanical and Materials Engineering, Applied Mechanics and Materials*. Trans Tech Publications, pp. 204–209. https://doi. org/10.4028/www.scientific.net/AMM.79.204.

Wang, D., Xu, Y., Hu, J., and Zhao, G. (2004). Fermentation kinetics of different sugars by apple wine yeast *Saccharomyces cerevisiae. J. Inst. Brew.*, **110**, 340–346. https://doi.org/10.1002/j.2050-0416.2004.tb00630.x.

Wang, G., Chang, L., Zhang, R., Wang, S., Wei, X., Rickert, E., Krost, P., Xiao, L., and Weinberger, F. (2019). Can targeted defense elicitation improve seaweed aquaculture? *J. Appl. Phycol.*, **31**, 1845–1854. https://doi. org/10.1007/s10811-018-1709-6.

Wang, J. and Chen, C. (2009). Biosorbents for heavy metals removal and their future. *Biotechnol. Adv.*, **27**, 195–226. https://doi.org/10.1016/J. BIOTECHADV.2008.11.002.

Wang, K., Zhang, J., Shanks, B.H., and Brown, R.C. (2015). The deleterious effect of inorganic salts on hydrocarbon yields from catalytic pyrolysis of lignocellulosic biomass and its mitigation. *Appl. Ener.*, **148**, 115–120. https://doi.org/10.1016/j.apenergy.2015.03.034.

Wang, L., Wang, X., Wu, H., and Liu, R. (2014). Overview on biological activities and molecular characteristics of sulfated polysaccharides from marine green algae in recent years. *Mar. Drugs.* https://doi.org/10.3390/ md12094984.

Wang, S. and Wai, C.M. (1996). Supercritical fluid extraction of bioaccumulated mercury from aquatic plants. *Environ. Sci. Technol.*, **30**, 3111–3114.

Wang, T., Jónsdóttir, R., and Ólafsdóttir, G. (2009). Total phenolic compounds, radical scavenging and metal chelation of extracts from Icelandic seaweeds. *Food Chem.*, **116**, 240–248. https://doi.org/https://doi.org/10.1016/j. foodchem.2009.02.041.

Wang, X., Liu, X., and Wang, G. (2011). Two-stage hydrolysis of invasive algal feedstock for ethanol fermentation. *J. Integr. Plant Biol.*, **53**, 246–252. https://doi.org/10.1111/j.1744-7909.2010.01024.x.

Wang, Z., Xiao, J., Fan, S., Li, Y., Liu, X., and Liu, D. (2015). Who made the world's largest green tide in china? — An integrated study on the initiation and early development of the green tide in yellow sea. *Limnol. Oceanogr.* https://doi.org/10.1002/lno.10083.

Wargacki, A.J., Leonard, E., Win, M.N., Regitsky, D.D., Santos, C.N.S., Kim, P.B., Cooper, S.R., Raisner, R.M., Herman, A., Sivitz, A.B., Lakshmanaswamy, A., Kashiyama, Y., Baker, D., and Yoshikuni, Y. (2012). An engineered microbial platform for direct biofuel production from brown macroalgae. *Science*, **335**, 308-313. https://doi.org/10.1126/science.1214547.

Wartenberg, R., Feng, L., Wu, J.J., Mak, Y.L., Chan, L.L., Telfer, T.C., and Lam, P.K.S. (2017). The impacts of suspended mariculture on coastal zones in China and the scope for integrated multi-trophic aquaculture. *Ecosyst. Heal. Sustain.* https://doi.org/10.1080/20964129.2017.1340268.

Wei, N., Quarterman, J., and Jin, Y. (2013). Marine macroalgae: An untapped resource for producing fuels and chemicals. *Trends Biotech.*, **31**. https://doi.org/10.1016/j.tibtech.2012.10.009.

Weldemhret, T.G., Bañares, A.B., Ramos, K.R.M., Lee, W.-K., Nisola, G.M., Valdehuesa, K.N.G., and Chung, W.-J. (2020). Current advances in ionic liquid-based pre-treatment and depolymerization of macroalgal biomass. *Renew. Energy*, **152**, 283–299.

Wen, H., Gao, H., Zhang, T., Wu, Z., Gong, P., Li, Z., Chen, H., Cai, D., Qin, P., and Tan, T. (2018). Hybrid pervaporation and salting-out for effective acetone-butanol-ethanol separation from fermentation broth. *Bioresour. Technol. Repor.*, **2**, 45–52. https://doi.org/10.1016/J.BITEB.2018.04.005.

West, P.C., Gibbs, H.K., Monfreda, C., Wagner, J., Barford, C.C., Carpenter, S.R., and Foley, J.A. (2010). Trading carbon for food: Global comparison of carbon stocks vs. crop yields on agricultural land. *Proc. Natl. Acad. Sci. U.S.A.* https://doi.org/10.1073/pnas.1011078107.

Whittick, A. (1983). Spatial and temporal distributions of dominant epiphytes on the stripes of Lamznarza hyperborea (gum.) Fosl. (Phaeophyta: Laminabiales) in S. E. Scotland. *J. Exp. Mar. Biol. Ecol.* DOI:10.1016/0022-0981(83)90002-3.

Wichard, T., Charrier, B., Mineur, F., Bothwell, J.H., Clerck, O. De., and Coates, J.C. (2015). The green seaweed *Ulva*: A model system to study morphogenesis. *Front. Plant. Sci. B.*, 72. https://doi.org/10.3389/fpls.2015.00072.

Wilkie, A.C. and Mulbry, W.W. (2002). Recovery of dairy manure nutrients by benthic freshwater algae. *Biores.*, **84**, 81–91.

William J.O. (2003). My sixty years in applied algology. *J. Appl. Phycol.*, **16**, 99–106.

Williams, R.G. and Follows, M.J. (2002). Physical transport of nutrients and the maintenance of biological production. *Oce. Biogeochem.* 19–51. https://doi.org/10.1007/978-3-642-55844-3_3.

Williams, S.L. (1986). The Physiological Ecology of Seaweeds. Christopher S. Lobban, Paul J. Harrison, Mary Jo Duncan. *Q. Rev. Biol.* https://doi.org/10.1086/414969.

Witman, J.D. and Roy, K. (2013). *Marine Macroecology*. https://doi.org/10.7208/chicago/9780226904146.001.0001.

Woertz, I., Feffer, A., Lundquist, T., and Nelson, Y. (2009). Algae grown on dairy and municipal wastewater for simultaneous nutrient removal and lipid production for biofuel feedstock. *J. Environ. Eng.*, **135**, 1115–1122. https://doi.org/10.1061/(ASCE)EE.1943-7870.0000129.

Wong, K. and Chikeung Cheung, P. (2001). Influence of drying treatment on three Sargassum species 2. Protein extractability, *in vitro* protein digestibility and amino acid profile of protein concentrates. *J. Appl. Phycol.*, **13**, 51–58. https://doi.org/10.1023/A:1008188830177.

Worm, B., Lotze, H.K., and Sommer, U. (2000). Coastal food web structure, carbon storage, and nitrogen retention regulated by consumer pressure and nutrient loading. *Limnol. Oceanogr.*, **45**, 339–349. https://doi.org/10.4319/lo.2000.45.2.0339.

World Bank (2015). Oceans, fisheries and coastal economies. Environment. https://www.worldbank.org/en/topic/oceans-fisheries-and-coastal-economies.

Wu, C.H., Chien, W.C., Chou, H.K., Yang, J., and Lin, H.T.V. (2014). Sulfuric acid hydrolysis and detoxification of red alga Pterocladiella capillacea for bioethanol fermentation with thermotolerant yeast Kluyveromyces marxianus. *J. Microbiol. Biotechnol.*, **24**, 1245–1253. https://doi.org/10.4014/jmb.1402.02038.

Wu, M., Mintz, M., Wang, M., and Arora, S. (2009). Water consumption in the production of ethanol and petroleum gasoline. *Environ. Manage.*, **44**, 981. https://doi.org/10.1007/s00267-009-9370-0.

Wu, S.-C. (2017). Antioxidant activity of sulfated seaweeds polysaccharides by novel assisted extraction, in: Xu, Z. (Ed.), Solubility of polysaccharides. InTech, Rijeka. https://doi.org/10.5772/intechopen.69633.

Xie, E.Y., Liu, D.C., Jia, C., Chen, X.L., and Yang, B. (2013). Artificial seed production and cultivation of the edible brown alga Sargassum naozhouense Tseng et Lu. *J. Appl. Phycol.*, **25**, 513–522. https://doi.org/10.1007/s10811-012-9885-2.

Xing, Q., An, D., Zheng, X., Wei, Z. and Wang, X. (2019). Monitoring seaweed aquaculture in the Yellow Sea with multiple sensors for managing the disaster of macroalgal blooms Remote Sensing of Environment Monitoring seaweed aquaculture in the Yellow Sea with multiple sensors for managing the disaster of macroal. *Remo. Sens. Environ.*, **231**. https://doi.org/10.1016/j.rse.2019.111279.

Xu, D., Gao, Z., Zhang, X., Qi, Z., Meng, C., Zhuang, Z., and Ye, N. (2011). Evaluation of the potential role of the macroalga Laminaria japonica for alleviating coastal eutrophication. *Bioresour. Technol.*, **102**, 9912–9918. https://doi.org/10.1016/J.BIORTECH.2011.08.035.

Xu, F.L., Dawson, R.W., Tao, S., Li, B.G., and Cao, J. (2002). System-level responses of lake ecosystems to chemical stresses using exergy and structural exergy as ecological indicators. *Chemosphere*, **46**, 173–185. https://doi. org/10.1016/S0045-6535(01)00127-8.

Yaich, H., Garna, H., Besbes, S., Paquot, M., Blecker, C., and Attia, H. (2011). Chemical composition and functional properties of *Ulva* lactuca seaweed collected in Tunisia. *Food Chem.*, **128**, 895–901.

Yamamoto, K., Endo, H., Yoshikawa, S., Ohki, K., and Kamiya, M. (2013). Various defense ability of four sargassacean algae against the red algal epiphyte Neosiphonia harveyi in Wakasa Bay, Japan. *Aquat. Bot.*, **105**, 11–17. https://doi.org/10.1016/j.aquabot.2012.10.008.

Yanagisawa, M., Nakamura, K., Ariga, O., and Nakasaki, K. (2011). Production of high concentrations of bioethanol from seaweeds that contain easily hydrolyzable polysaccharides. *Proce. Biochem.*, **46**, 2111–2116. https://doi. org/10.1016/j.procbio.2011.08.001.

Yang, L.-E., Lu, Q.-Q., and Brodie, J. (2017). *Rhodophyta* in China: History, culture and taxonomy. *Eur. J. Phycol.*, **52**, 251–263. https://doi.org/10. 1080/09670262.2017.1309689.

Yang, L.E., Lu, Q.Q., and Brodie, J. (2017). A review of the bladed Bangiales (Rhodophyta) in China: history, culture and taxonomy. *Eur. J. Phycol.* https://doi.org/10.1080/09670262.2017.1309689.

Yang, X., Wu, X., Hao, H., and He, Z. (2008). Mechanisms and assessment of water eutrophication. *J. Zhejia. Univ. Sci. B.*, **9**, 197–209. https://doi. org/10.1631/jzus.B0710626.

Yantovski, E. (2000). Exergonomics in education. *Energy*, **25**, 1021–1031. https://doi.org/10.1016/S0360-5442(00)00027-X.

Yantovsky, E.I. (1989). Non-equilibrium thermodynamics in thermal engineering. *Energy*, **14**, 393–396. https://doi.org/10.1016/0360-5442(89)90134-5.

Yates, J.L. and Peckol, P. (1993). Effects of nutrient availability and herbivory on polyphenolics in the seaweed fucus vesiculosus. *Ecology*. https://doi. org/10.2307/1939934.

Ye, J., Huang, W., Wang, D., Chen, F., Yin, J., Li, T., Zhang, H., and Chen, G.-Q. (2018). Pilot Scale-up of Poly(3-hydroxybutyrate-co-4-hydroxybutyrate) Production by Halomonas bluephagenesis via cell growth adapted optimization process. *Biotechnol. J.*, **13**, 1800074. https://doi.org/10.1002/biot.201800074.

Ye, N.H., Zhang, X. wen, Mao, Y. Ze, Liang, C. Wei, Xu, D., Zou, J., Zhuang, Z. M. and Wang, Q. Y. (2011). "Green tides" are overwhelming the coastline of our blue planet: Taking the world's largest example. *Ecol. Res.* https://doi. org/10.1007/s11284-011-0821-8.

Yokoyama, S. (2007). Energy production from marine biomass: Fuel power generation driven by methane produced from seaweed. *World Acad.*, **5**, 1131–1151.

Yu, Q. and Kaewsarn, P. (2001). Desorption of Cu_{2+} from a biosorbent derived from the marine Alga *Durvillaea potatorum*. *Sep. Sci. Technol.*, 1495–1507. https://doi.org/10.1081/SS-100103884.

Yu, Q., Kaewsarn, P., and Ma, W. (2008). Long-term desorption kinetics of Cu_{2+} from biosorbent in water and seawater long-term desorption kinetics of Cu^{2+} from biosorbent in water and seawater. *Toxicoh. Environ. Chem.*, **79**, 117–125. https://doi.org/10.1080/02772240109358981.

Yuan, Y. and Macquarrie, D.J. (2015). Microwave assisted step-by-step process for the production of fucoidan, alginate sodium, sugars and biochar from Ascophyllum nodosum through a biorefinery concept. *Bioresour. Technol.*, **198**, 819–827. https://doi.org/https://doi.org/10.1016/j.biortech.2015.09.090.

Zaky, A.S., Tucker, G.A., Daw, Z.Y., and Du, C. (2014). Marine yeast isolation and industrial application. *FEMS Yeast Res.*, **14**, 813–825. https://doi.org/10.1111/1567-1364.12158.

Zeraatkar, A.K., Ahmadzadeh, H., Talebi, A.F., Moheimani, N.R., and McHenry, M.P. (2016). Potential use of algae for heavy metal bioremediation: A critical review. *J. Environ. Manage.* https://doi.org/10.1016/j.jenvman.2016.06.059.

Zeroual, Y., Moutaouakkil, A., Zohra, D.F., Talbi, M., Ung Chung, P., Lee, K. and Blaghen, M. (2003). Biosorption of mercury from aqueous solution by *Ulva lactuca* biomass. *Bioresour. Technol.*, **90**, 349–351. https://doi.org/10.1016/S0960-8524(03)00122-6.

Zhang, J. (2018). *Seaweed Industry in China.* https://www.submariner-network.eu/images/grass/Seaweed_Industry_in_China.pdf.

Zhang, J., Gurkan, Z., and Jørgensen, S.E. (2010). Application of eco-exergy for assessment of ecosystem health and development of structurally dynamic models. *Ecol. Modell.* https://doi.org/10.1016/j.ecolmodel.2009.10.017.

Zhang, N., Wang, M., and Wang, N. (2002). Precision agriculture: A worldwide overview. *Comput. Electron. Agric.*, **36**, 113–132.

Zhang, T., Datta, S., Eichler, J., Ivanova, N., Axen, S.D., Kerfeld, C.A., Chen, F., Kyrpides, N., Hugenholtz, P., and Cheng, J.-F. (2011). Identification of a haloalkaliphilic and thermostable cellulase with improved ionic liquid tolerance. *Green Chem.*, **13**, 2083–2090.

Zhang, Y., He, P., Li, H., Li, G., Liu, J., Jiao, F., Zhang, J., Huo, Y., Shi, X., Su, R., Ye, N., Liu, D., Yu, R., Wang, Z., Zhou, M., and Jiao, N. (2019). *Ulva prolifera* green-tide outbreaks and their environmental impact in the Yellow Sea, China. *Natl. Sci. Rev.* https://doi.org/10.1093/nsr/nwz026.

Zhao, Y., Hao, Y., and Ramelow, G.J. (1994). Evaluation of treatment techniques for increasing the uptake of metal ions from solution by nonliving seaweed algal biomass. *Environ. Monit. Assess.* **33**, 61–70.

Zheng, Y., Jin, R., Zhang, X., Wang, Q., and Wu, J. (2019). The considerable environmental benefits of seaweed aquaculture in China. *Stoch. Environ. Res. Risk Assess.*, **33**, 1203–1221. https://doi.org/10.1007/s00477-019-01685-z.

Zhou, D., Zhang, L., Zhang, S., Fu, H., and Chen, J. (2010). Hydrothermal liquefaction of macroalgae *Enteromorpha prolifera* to Bio-oil. *Energy & Fuels*, **24**, 4054–4061. https://doi.org/10.1021/ef100151h.

Zilberman, D., Rajagopal, D., and Kaplan., S. (2017). Effect of biofuel on agricultural supply and land use., in: M. Khanna and D. Zilberman (Ed.), *Handbook of Bioenergy Economics and Policy. Natural Resource Management and Policy*. New York, NY.: Springer Publishing Company, pp. 163–182.

Zollmann, M., Robin, A., Prabhu, M., Polikovsky, M., Gillis, A., Greiserman, S., and Golberg, A. (2019a). Green technology in green macroalgal biorefineries. *Phycologia*. https://doi.org/10.1080/00318884.2019.1640516.

Zollmann, M., Traugott, H., Chemodanov, A., Liberzon, A., and Golberg, A. (2019b). Deep water nutrient supply for an offshore *Ulva* sp. cultivation project in the Eastern Mediterranean Sea: Experimental Simulation and Modeling. *Bioener. Res*. https://doi.org/10.1007/s12155-019-10036-3.

Zollmann, M., Traugott, H., Chemodanov, A., Liberzon, A., and Golberg, A. (2018). Exergy efficiency of solar energy conversion to biomass of green macroalgae *Ulva* (*Chlorophyta*) in the photobioreactor. *Ener. Convers. Manag.*, **167**, 125–133.

Index